Patrick Moore's Practical Astronomy Series

Other Titles in this Series

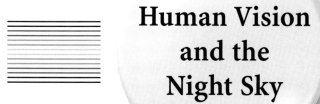

Human Vision and the Night Sky

Hot to Improve Your Observing Skills

Michael P. Borgia

 Springer

Library of Congress Control Number: 2005938491

ISBN-10: 0-387-30776-1   Printed on acid-free paper.
ISBN-13: 978-0387-30776-3

Printed in the United States of America.   (EB/BP)

9 8 7 6 5 4 3 2 1

springer.com

"Daddy, I wanna get a ladder and touch the Moon"

Robert Michael Borgia
Age 3

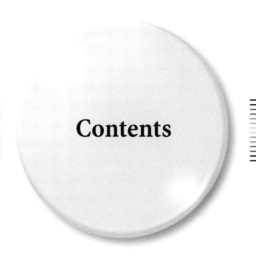

Contents

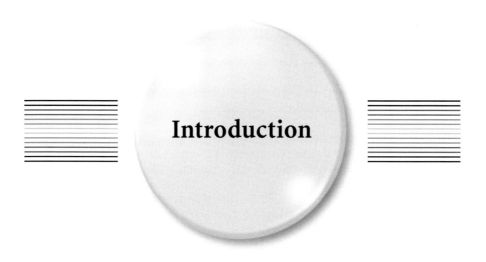

Introduction

For years, the images have blazed through your imagination. They are the magnificent full-color photographs returned by the Hubble Space Telescope and its sister Great Observatories[1] of the grand depths of the cosmos. From the "pillars of creation," considered to be Hubble's signature image, to the incomprehensible depths of the Hubble Deep Fields to the intricate details imaged in the surface and cloud tops of Mars or Jupiter, the power of the Hubble Telescope to turn on the public to science is unparalled in the history of modern culture. They also have spurred new telescope sales to unimagined highs. And after years of watching the heavens through the eyes of NASA, you've decided it's time to see it for yourself. You make the trip to the department store and pick up that shiny new "500×" telescope, set it up and soon you're in business.

Unfortunately, the high initial expectations usually give way to disappointment. Instead of seeing the magnificent swirling clouds of gas in the Orion Nebula, you see a pale green-gray cloud with a couple of nondescript stars lurking nearby. The swirling red, yellow and brown storms of Jupiter are nowhere to be seen; only varying shades of gray in the planet's cloud bands, assuming you can see bands at all! And Mars? After waiting all night for the red planet to rise up over the morning horizon, you are greeted by nothing more than a featureless reddish-orange dot. After a few weeks of this, the telescope suddenly is no longer making the nightly trip outside. Soon the scope only gets outside one night a week and not long after that, it becomes a place to hang laundry. It need not be that way for the sky you long to see is out there. You just need to learn *how* to see it. If you are that person,

---

[1] NASA's "Great Observatories" include the Hubble Space Telescope (launched 1990), the Compton Gamma Ray Observatory (launched 1991, de-orbited 2001), the Chandra X-Ray Observatory (launched 1999) and the Spitzer Space Telescope (launched late 2003).

then this book is for you. If you own a larger telescope and feel you have run out of challenges, then you've got the right book too. One thing that many people who do not study the sky don't understand is that astronomy is the one and only science where ordinary people with an ordinary education can make the discoveries that electrify the public and even alter the course of modern science. Amateurs discovered the great comets Hale–Bopp, Hyakutake, West and Ikeya–Seki. Amateurs today are even helping to discover new planets around distant suns. Indeed, with the attention of most professional astronomers focused on non-visible wavelengths, most major discoveries made in visible-light astronomy today are made by amateur astronomers just like us.

The opportunities for discovery, learning and wonder are absolutely endless, but it also takes an enormous amount of work. If you're willing to do it, then please read on. I wrote this book to share with you what I have had to learn through hard trial and (a lot of) error. I hope to share with you so as to limit your frustration, increase learning and most of all expand your joy in this amazing hobby and limitless science. We'll begin in the pages ahead by discussing the critical elements of the *integrated observing system*. This system has three critical components all of which must work correctly and in harmony for you to have success. These are the observer's eyes, his equipment, and lastly his brain. A perfect scope and flawless vision are useless without the knowledge of how to use it and of what it is you are looking for, what to expect when observing and why that particular object is of such interest. A well-trained mind and a perfect scope are of little use if the eyes are in poor health or are adversely affected by factors external to the eyes or external to the body. Perfect eyes and a well-trained mind will not perceive very much if the telescope cannot produce a sharp image because it is poorly maintained or its optics or mounting are of poor quality. Our first three chapters are about preparing and training the eyes, acquiring the right equipment for your particular needs including some frank advice about how to shop for that first serious telescope, then we will talk about training the mind, the need to gain knowledge and then putting it all together to make observing fun, enriching, and satisfying.

Once you have all the tools in place, we'll go out in the field for a test run and put our eyes, brain and telescope to work. We'll walk through a typical first night in the field by planning and executing an observing session where time can be an issue, both in terms of being ready for a precise moment and making use of time of limited quantity. The first night out can be the most wonderful night of your life as an amateur astronomer, or the night that turns you off the hobby completely. We'll talk about how to make it the former rather than the latter by teaching you to manage your time, your equipment, yourself and perhaps most importantly, your expectations.

Now that you've put it all together, in the next ten chapters, we will take the grand tour of the universe, starting close in with the Moon then making our way further and further out into space. In each chapter, we will do three things. First we'll talk in depth about each object as a physical entity. We'll then talk some about the history of that object from the point of view of the human experience, how did we come to know what we know and why is it important to us? Knowledge is what in turn makes us curious; it is as much a part of being human as breathing. A small primer to arouse curiosity makes us seek more knowledge. That in turn makes us more curious. The desire to gain knowledge is therefore self-perpetuating so long as we can continue to satisfy our curiosity. As long as we can satisfy that urge, then the hobby will remain satisfying and self-fulfilling. Finally we will help manage expectations. You will never see in the telescope what the amazing pictures returned by the Hubble or Keck telescopes can. For this reason, all the images produced in this book are my own. I am a *very* amateur astrophotographer and I'm still working after many years on mastering the art of image processing (unsharp mask, anyone?). The pictures are far from perfect in many cases because most of what you see in a telescope is far from perfect, not to mention the photographer. The motion of Earth's atmosphere distorts the planets and the nebulae and galaxies are washed out by light pollution. Astrophotography is an enormous challenge, as my own images prove over and over. The pictures more accurately represent what you might see in an actual telescope.

Finally we will challenge you. Each chapter ends with a series of projects that will show you how to do so much more than simply gaze through a scope. You will challenge and train your eyes, learn how to pick the right equipment for what you want to do, how to organize yourself and how to gain knowledge. You will track sunspots, locate the Apollo landing sites, study the geography of Mars, wonder at the remarkable resonance of Jupiter's moons, and discover why Mercury and Venus behave so differently from each other. We'll learn the techniques that amateurs just like you and me use to hunt the sky for comets or rouge asteroids. We'll go into deep space and discover how astronomers learned to measure grand distances in the universe, watch stars brighten and fade both predictably and unpredictably. We'll take the grand view and the up-close view of nebulae, galaxies and clusters and learn from where each type came and what makes each object important to us. Then we will discuss Messier's famous catalog and learn how to earn amateur astronomy's ultimate right of passage, finding every object on that list in one single night.

Ready? Then let's go stargazing!

CHAPTER ONE

The Integrated Observing System. Part I: Your Eyes

One of the most terrible misconceptions about astronomy that those who are first getting into the hobby have is that it's an easy source of instant gratification. Set the scope up, look through it and be amazed. Astronomy, like many hobbies just does not work out that way. Your ability to be successful and have enjoyment in astronomy is based entirely on your willingness to work at the art-form of observing and the quality of your equipment. If you've ever played golf (I have and I use the word loosely), you will understand this. It takes many years of practice, consistent effort and a willingness to study the game to make a good golfer. A good player also needs the proper equipment. He needs clubs that are the right length for both his body and arms. The clubs must be of the proper flexibility for your game's strengths and weaknesses. Stiff shafts deliver more accurate shots while flexible shafts deliver greater distance. You just do not walk on a golf course and expect to play "all-square" with Tiger Woods. A good golfer is a complete integrated system, the perfect marriage of clubs, player, practice and ability. Just the same, you should not expect to step up to a department store telescope and be able to instantly see all the grandeur that the heavens have to offer on the first night. You need to have the right equipment and you need to have the willingness to learn how to use it and care for it. Your eyes, your brain and your telescope are all part of an integrated observing system. Visual astronomy is an art-form as much as it is a science. Success is based on equal parts of quality equipment, carefully honed skills and good fortune. Astronomy, like any challenging hobby, is very hard work.

About twenty years ago, I had reached that critical mass point in amateur astronomy. My "500×" Tasco department store telescope had basically become a coat hangar in my bedroom. Its 50 mm (2 inch) objective lens adequately showed the disk of Jupiter, the rings of Saturn and the phases of the Moon and Venus. But under the moderately light polluted skies of northwestern New Jersey, even the sky's brightest deep sky wonders were virtually invisible to me. Anything fainter than the naked eye threshold was invisible. Though I was twenty years old and working my way through college, I saved carefully and purchased a Celestron Super C8 Plus in May 1986. I was able to turn it into the night sky in time to catch the retreating Comet 1P/Halley, just before it disappeared into the cosmic deep for another seventy-six years. This proved to be my first great disappointment. Though it appeared larger and brighter in the new telescope, it still appeared dull and featureless. I was puzzled at reading the descriptions of the comet written by the leading amateur astronomical observers of the time who were using equipment much the same as I was, many in the same general area of the country. They described it as dynamic with many differing features in both the coma and tail that I just could not begin to make out. Why was this? It was because I did not know how to see.

The Apollo-era geologist Farouk El Baz once said, "Anyone can look, but few really *see*." "Seeing" is the great skill that makes an astronomer successful at the eyepiece. Learning how to see requires a great deal of patience and practice. It also requires some understanding of how the most important piece of astronomical equipment you own works, your own eyes! The human eye is an amazing evolution in biological optics. It is one of the few sight organs belonging to any species that is capable of imaging both faint light and in color. Cats, for example, have extremely keen night vision, but are completely color-blind. What is sometimes difficult to understand about the human eye is that the eye cannot image color and faint light at the same time. First, lets take a closer look at the eye and the way it is built. We will then discuss several important observing considerations and techniques that affect the way the eye works and the way the brain perceives. These factors include dark adaptation, the use of averted vision, light pollution, the condition of both the physical organism (you) and the condition of the atmosphere.

The Eyes in the Dark

The operating principles of the eye are rather simple. A clear organic lens (the cornea) focuses light entering the eye. This lens refracts light onto the retina, a projection surface at the back of the eye. The amount of light that reaches the retina

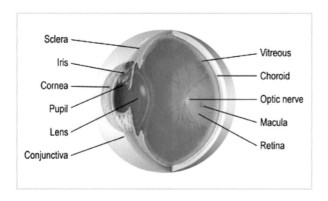

Figure 1.1. Anatomy of the human eye.

is controlled by an involuntary muscle array called an "iris." The iris opens and closes the pupil, the center opening of the eye. The pupil dilates between a diameter of anywhere between 1 and 8 millimeters. The wider the pupil can open, the more light can enter. An involuntary reflex centered in the rearmost portion of the brain controls the opening of the pupil. As ambient light levels decline, the brain opens the pupil wider to allow more light to enter. This process of *dark adaptation* requires time and care. If you step out of your home straight to the eyepiece of a telescope aimed at a faint galaxy near the limit of the telescope's reach, I promise that you will not see it, even if the object is directly in the center of your field of view. The most important step to take in preparing yourself to view the sky is to allow you a proper opportunity to adapt to the dark. Your pupil will dilate to near-maximum aperture (opening) within a few seconds. It will still require a considerable amount of time for the eye to complete electrochemical changes that will occur in the retina. Successfully completing this requires removing yourself from any bright light source for a minimum of thirty minutes before any serious observing can begin. Take extra care to avoid any exposure to light ranging from white to the higher energy end of the spectrum (toward blue). Avoiding all extraneous light is better yet. Over the first fifteen minutes, your eyes will gain nearly two magnitudes in sensitivity. By thirty minutes, your eyes are nearly fully dark-adapted. Although the eyes will grow slightly more sensitive over the next ninety minutes, the huge majority of your dark-adaptation is completed within that first half-hour. If you must use a light, always make sure that it is red. The eye is least sensitive to the lower energy, longer wavelengths of red light. This is not to suggest that any light is a good thing at all, but if you must use one, keep it red. You can purchase a quality red lens flashlight through many astronomy warehouses or pilot supply shops (pilots must also take care to dark adapt before flying at night). Any exposure to white light will spoil your dark-adaptation almost instantly. You will then have to start over again.

Once your night vision is fully adapted, you are ready to make maximum use of your eyes. Take a look at the changes in the night sky. Now that you can see stars that are only one-sixth as bright as when you started, the sky looks a lot more crowded than it did just thirty minutes ago. How many stars can you actually see? There are some 6,000 stars in the sky that are brighter than the naked-eye limit of magnitude 6.0. The most sharp-eyed observer might be able to see to magnitude 6.5 if there is no light pollution present. A good test of both the acuity of your dark vision and the transparency of your sky is to count the stars within the Great Square of Pegasus. This asterism is visible during late summer, fall and early winter in the northern hemisphere nearly directly overhead. Later in this chapter we'll give you the details of a simple naked-eye observing project to use this area of sky as a test-bed to check the darkness of your sky and sharpness of your night vision and preparation.

If you are disappointed with your results, try this little trick. Look slightly away from the observing target. Move the center of your gaze off to the east or west of the Great Square. Keep your attention centered on the contents of the Great Square. How many more stars leap into view now? The reason why is because of the construction of the retina. Your retina is composed of two different types of photoreceptive cells. The type of cell that dominates the center of the retina are called "cones." These cells are not very common among other species. They have the unique ability to perceive color. The signals transmitted by the cone cells are then transmitted via the optic nerve and assembled into a complete image in the brain. These cells do have one important weakness however. They require a substantial amount of light energy to stimulate the photochemical reaction within them. The faint light generated by stars and other deep-sky phenomena usually are not up to the job of firing the cones. As a result of this, turning to look at a faint object in the corner of your vision will often cause it to disappear altogether. To see faint objects, one must use a different type of photoreceptor.

The cells that make up the outer 80% of the retina are different both in shape and character from the cones. These cells, called "rods," are designed to detect very faint levels of light. They cannot however see in color. The rod cells see only in gray; the average viewer can detect approximately forty different shades. To see the faintest objects in the sky, it is critical to use the rods effectively. When striving to detect the faintest objects by looking straight at them, you are focusing that object on the cones of the retina. If there is not enough light to stimulate the cones, you will never detect the object. By directing your gaze slightly off to the side, you allow the light of the object to fall on the rods where you are far more likely to detect the object. This technique is called *averted vision* and is a crucial skill to be mastered. It is, unfortunately, counterintuitive. That is to say, it is a technique that goes against everything that seems natural for you to do. Averted vision requires a great deal of practice. An experienced observer, hunting down a faint galaxy, never looks at anything straight on.

Light Pollution

A composite photograph of Earth's surface at night taken from space shows the alarming spread of surface lighting at night. The east coast of the United States is clearly recognizable in any wide-angle picture of the planet at night. The problem is that most of this light goes to waste. Instead of shining down on the ground to illuminate our streets at night, they instead scatter huge amounts of light into the night sky. This is what astronomers call "light pollution." Light pollution is no good for anybody. It wastes uncountable billions of dollars per year in electricity, scatters unwanted light in the eyes of drivers and ruins the natural beauty of the night sky. In urban inner cities, so much light from unshielded high-pressure sodium streetlights is scattered into the sky that only the brightest stars and planets can be seen. I remember one particularly horrid night in New York where despite a clear sky and unrestricted visibility, I found the limiting visual magnitude to be just slightly better than +2.0. I could barely make out Polaris to the north, but Zubenelgenubi (the brightest star in Libra, magnitude +2.8) could not be seen at all. The sky that night in Flushing Meadow resembled the view one might have from inside a milk bottle. It did not help on that late August night that the U.S. Open tennis tournament was being played to my south while the bright lights of Shea Stadium flared not far to the north. The city skyline loomed brilliantly to the west and the lights of suburban Long Island soared the view to the east. That is the horror of light pollution at its very worst. Unless you're content with views of the planets and their brightest moons, a view from the city will surely spoil anyone's enthusiasm for astronomy. In urban areas, this background sky glow from city lights is brighter than the stars trying to shine through it. When this is the case, there's no way for that object to shine through.

In some suburban communities, light pollution is being pushed back. Some communities are replacing high-pressure sodium lights with low-pressure sodium. These lights shine at wavelengths that are not quite as damaging as are high-pressure lights. Many of these lights are also shielded so that the light they emit is directed at the ground where it is needed, not into eyes of drivers or into the evening sky. This can yield as much as two full magnitudes of improvement in the transparency of the sky. Many communities have passed laws mandating the use of lights that restrict sky glow, and recently the Massachusetts state legislature passed a law mandating such measures for the entire state.[2] The best solution however for dealing with light pollution is to get away from it altogether. In many areas, driving about an hour away from bright city lights will do wonders to clear the view. Stars down to near the naked-eye threshold creep into view and the sky turns a deep clear black except for one milky colored band that refuses to vanish. It may take a city-based observer exposed to his first dark sky a moment or two to realize what that glow is. When he does however, it becomes obvious why our galaxy is named the "Milky Way." If you are serious about astronomy, do whatever it takes to find a good dark-sky site. No matter how good your eyes, your technique and your equipment it will not do you a lick of good if the glory of the sky is hidden behind streetlight glare. If you do not belong to one, now is the time to find a good astronomy club. Nearly all clubs have a line on a good dark-sky site where their members go to observe on a regular basis, scheduled or otherwise. For your telescope, you may wish to consider a light-pollution rejection filter; but for your eyes, there is no better way to deal with light pollution than to be rid of it.

Hypoxia and the Physical Organism

As crucial as the condition of the sky is the physical condition of the observer. The observer may have the perfect combination of sky conditions and observing technique yet is unable to see the faintest details of objects or see at all if he is betrayed by his own physical condition. Two key contributors to the physical health and efficiency of the eyes are adequate supplies of vitamin A and the ability to respirate and metabolize oxygen. In any modern industrial society, vitamin A deficiencies are almost unheard of. Taking massive quantities of it will not give anyone super-human vision and can in fact be harmful. Still, a healthy diet is a must for good ocular health.

A greater factor that affects many people is the body's ability to take in (respirate), distribute and metabolize oxygen. The inability of the body to deliver oxygen

---

[2] In an unfortunate "mistake", the state Senate "lost" the bill and failed to physically deliver it to Governor Jean Smart for her signature. Since it went missing for more than ten days and the Legislature was no longer in session, the promising legislation was accidentally "pocket vetoed" under the provisions of the Massachusetts constitution. That's politics!

to the tissues is called "hypoxia." Hypoxia can result from many potential factors. The eyes are the most oxygen hungry organs in the body and their efficiency rapidly deteriorates when an adequate supply of oxygen is not available. To make matters worse, the first part of the eye to be affected by hypoxia is the rod cells of the retina, the detectors of low-level light. Several potential factors can be involved in determining how efficiently the body can distribute and metabolize oxygen. The first such variable is the amount of oxygen available in the atmosphere. Hypoxia caused by a lack of oxygen is called "hypoxic hypoxia." At sea level, the partial pressure of oxygen in the atmosphere is approximately 220 millibars. This amount of oxygen is plenty for the body to use but in order to get the oxygen into the bloodstream; it must be forced through the walls of the aveoli. This is done by ambient air pressure, which at sea level is approximately 1,013 millibars (29.92 inches Hg). As you climb higher in the atmosphere, the ratio of oxygen to other gasses remains the same.[3] But the pressure of the air falls off dramatically with increasing altitude. This is of interest to us because many astronomers seek to flee the effects of light pollution and weather by climbing the mountains. Today, many of the world's premier observatories are on extremely high mountaintops ranging 14,000 feet or higher above sea level. At this altitude, the partial pressure of oxygen in the atmosphere is only 130 millibars. This is only about 60% of what was available at sea level. From this, you might initially assume that you are getting only 60% of your sea level efficiency from your lungs. This assumption falls flat when you stop to consider that the total air pressure at 14,000 feet is only 600 millibars. That means not only is there less oxygen to breathe but the efficiency of the body in delivering that oxygen to the blood is greatly impaired. The eyes will be among the first of the body's users of oxygen to feel the effects as the rods begin to fail due to oxygen deprivation. The key to preventing this if you climb the mountains to see the stars is to breathe supplemental oxygen. But what if you're not going to Mauna Kea to observe? At what altitude should one consider this step? A lot lower than you might think. The Federal Aviation Administration's guidelines for pilots seems to be a good reference since the eyes are affected so early and night vision in particular is affected. Pilots are not required to begin using oxygen until they are higher than 12,500 feet after thirty continuous minutes but because of the detrimental effects of hypoxia on night vision the FAA strongly recommends that oxygen be used at night any time operating above 5,000 feet. If it works for pilots, it will work for you.

Even while observing at sea level, hypoxia symptoms can become an impediment to good night vision due to several potential factors. Anything that comes between an oxygen molecule and the cells of the body can cause hypoxia. Hypoxia can affect you even though plenty of oxygen is available. If you are in poor cardiovascular health, your night vision will suffer from effects very similar to those induced by high altitude. If the heart does not circulate blood adequately, the oxygen cannot reach the tissues with the speed the body demands resulting in what is called *stagnant hypoxia*. Though the cause is different the result is very much

---

[3] The atmosphere is about 78% nitrogen, 21% oxygen with the last 1% divided up between carbon dioxide, water vapor and other noble gasses. Water vapor can, during extremely humid conditions, make up more than 4% of the atmosphere.

the same. The heart does not pump the blood at an adequate rate to keep the tissues of the body oxygenated and night vision begins to suffer. A hypoxic reaction can also be induced by poor hematological health. *Anemic hypoxia* is caused by the inability of the red blood cells to absorb and transport oxygen even though oxygen itself may be plentiful. *Sickle cell anemia*, a condition that primarily occurs in African-American males, is a leading cause of this. The rest of us should be aware that many other forms of anemia exist that are not so racially discriminatory. Anemia can also be induced by poor diet. You can also induce hypoxic symptoms yourself by taking certain substances into your own body. Alcohol and many drugs can reduce the ability of the red blood cells to carry oxygen, a form of hypoxia called *histotoxic hypoxia*.

The health of the eye itself is critical as well. Many individuals suffer from some form of ocular pathology. The most common include myopia (near-sightedness), presbyopia (far-sightedness) and astigmatism. Myopia often occurs at a fairly young age (I began wearing glasses at age 9). It is caused by a misshaped cornea, which causes light from distant objects to come to focus prior to reaching the retina. The result is badly distorted images of distant objects. The problem is easily corrected by using corrective lenses such as glasses or contact lenses which induce a focusing error in the light path that is exactly the opposite of that created by the flawed cornea. The result is a properly focused image. Presbyopia commonly affects older individuals. As the eyes age, the cornea begins to become rigid and less flexible. This causes the eye to have difficulty focusing on objects that are relatively close. The cure is the same as for myopia: corrective lenses that eliminate the fault by introducing an equal and opposite error. Astigmatism is a somewhat more difficult and complex problem for ophthalmologists to deal with. Astigmatism is a flaw in the cornea that causes images to appear distorted, even though they are well focused. A properly focused star appears as a perfect pinpoint of light while the same star might appear to have a "diffraction spike" radiating from it when viewed through an astigmatic eye. Correcting the flaw using contact lenses is particularly difficult and most users with severe astigmatisms must use eyeglasses to correct them. This is particularly painful at the eyepiece of the telescope and even more so during astrophotography. The telescope and camera see and focus at "20/20"[4]. In order to properly focus them, you must also see 20/20. Glasses make seeing through a camera or telescope particularly awkward. Most astronomers who need eye correction prefer to use contact lenses.

Many people are taking action today to repair eyesight damaged by age or genetics. Surgical fixes for eye problems have been with us for many years. The first such procedure for correcting bad eyesight was called "radial keratotomy." This surgical procedure involved slicing the cornea, pizza style with a scalpel and refiguring it into the proper shape. Radial keratotomy did work but often resulted in heavy scarring and if the results were not perfect, it could not be reversed, changed or performed again. Another technique that enjoyed some popularity in the early 1990s was "orthokeratology." Orthokeratology involved the use of special hard

---

[4] The first number tells how close the viewer must be to an object to focus on it. The second number refers to the distance that a person with normal vision can focus on the object. The larger the second number is, the more near-sighted you are.

contact lenses that functioned in a way not unlike a dental retainer. These lenses over time reshape the cornea, gradually forcing it toward the proper shape. The user would regularly change to a new set of lenses that would continue to refine the correction until it is completed. The corrective process takes about six months. Once 20/20 vision was restored, the patient would continue to periodically wear contact lenses to maintain the shape of his corneas. The advantages of this procedure were obvious in that no surgery was necessary and no scarring ever took place. If it did not work, the effects were completely reversible. The disadvantages were that if the user stopped using his retainer lenses, the eyes would eventually return to their original flawed state.

The technique most favored today is called "laser keratotomy." In laser keratotomy a scalpel is used only to make an initial slot in the cornea. The slot allows the lens to relax while a precision laser figures the lens to the correct shape. Like in other forms of keratotomy, the effects are irreversible so if you are unhappy with the outcome, too bad. Unlike traditional surgery, there is no scarring and recovery time is quick. Usually the patient is out of the office within an hour and seeing normally within two or three days without optical aid. Some surgeons are now performing newer forms of the surgery without using any blades at all. There is one important drawback to laser keratotomy. Current Food and Drug Administration rules limit the scope of the surgery to the inner 6-millimeter radius from the pupil center. The problem is if you are one of those people who have a very wide opening pupil after dark adaptation. If the pupil can open wider than the surgically repaired area of the cornea, the result can be severely distorted night vision. One common effect is "haloing" of bright lights. Your vision is normal in the center but as light travels across the non-repaired outer cornea into the eye, the fringes of lights become distorted and create halos around the light source. As an amateur astronomer and a pilot, my night vision is crucial to both my livelihood and my hobby. Because of this, I have decided for myself to forgo any surgery or other vision correction until the medical procedure evolves further. Contact lenses do an adequate job for my near-sightedness, and evolving technology has finally created a contact lens that corrects my particular astigmatism.

The Restless Atmosphere

Despite the fact that its relatively paltry 94-inch primary mirror is dwarfed by monstrous telescopes on Earth that now are more than four times larger, the best images in astronomy still come from the Hubble Space Telescope. The hulking monsters on Mauna Kea measure from the 8-meter Subaru telescope to the 10-meter Keck twins. These incredibly powerful instruments are still only second best to Hubble.

The reason why is because these Earth-based monsters must deal with an impediment that Hubble does not. They must peer through Earth's relentlessly turbulent atmosphere. The layer of gases that surrounds the planet and makes life possible here is a major obstacle to a clear view of the stars. The atmosphere's gases distort and disrupt starlight as it passes through and in fact completely block out large segments of the spectrum, particularly in the ultraviolet range. The atmosphere also disrupts the view by continually moving, expanding and contracting in response to changing weather and heating patterns. The more turbulent the air becomes, the more difficult it will be to gain a quality view of the sky. The level of quality of the view is what astronomers call *seeing*. When there is good seeing, the air is still and calm. The stars do not appear to twinkle, but burn steady and unchanging. These are the nights that serious amateurs live for. When the stars do twinkle, it may inspire classic poetry and children's songs, but it is the bane of the astronomer's existence. All telescopes can do under these conditions is magnify the twinkle of the stars into a bigger brighter blur.

Astronomical seeing is almost entirely a function of atmospheric stability. In meteorology, stability refers to the tendency of a sample of air to rise when it is lifted. Air that wants to rise is considered to be unstable. Air can be induced to rise by heating it from below, by flowing up a mountain, or by a colder, heavier air mass burrowing underneath. A simple measure taken by meteorologists accurately determines the stability of the air. Adiabatic lapse rate is the measure of how fast the air temperature falls with increasing altitude. The faster the air cools, the more unstable it is. These values are then compared against those of the International Standard Atmosphere (ISA)[5]. If the temperature is cooling faster than ISA calls for with increasing altitude, the air is considered to be unstable. The rising air currents climb into the higher altitudes, cool off, then sink back. This churning motion in the atmosphere, combined with the changing density of moving air masses, distorts the light of the stars passing through it resulting in blurred planets and twinkling stars. If the lapse rate is less than that suggested by ISA, then there will be only minimum motion in the air. Better yet is if the air grows warmer with increasing altitude, causing warm light air to sit atop cold heavy air. These conditions lead to great stability in the atmosphere. Heavy on bottom and light on top means everything wants to stay put. These are the nights that astronomers live for. Get that scope out, but do it fast because if adequate moisture is available, this type of weather condition will eventually give rise to fog.

The sky gives valuable visual clues during the day that tip you as to what kind of conditions to expect after sunset. Rising air releases moisture as it cools to the dew point and makes clouds. If the clouds lie low and flat with no vertical development (stratiform), that means the air in which they are created is stable. When the sun sets and fair weather clouds dissipate, the seeing will be good. If the clouds are puffy and billow high into the sky (cumuliform), then the air that built them is rising and the air mass is unstable. This will resort in distorted seeing and reduced optical performance. What is deceptive is that the rapidly circulating air gives rise to a sky that will have extraordinary clarity at night, fooling one into thinking that it is going to be a great night for observing. In fact, the nights with the best seeing just might be those nights where there is just the tiniest bit of haze in the air. Lack of motion in the atmosphere allows the haze to hang around. Overall there is not much you can do about the condition of the atmosphere except to wait it out. If you keep looking at your target long enough, there will inevitably

---

[5] ISA is defined as a sea level temperature of 15°C and a pressure of 29.92″ Hg. Temperature declines by 2°C and pressure by 1.06″ Hg initially per 1,000′ of altitude gain. Temperature levels off and reverses trend at 36,000′. Pressure continues to decline by approximately 3.2% per 1,000′.

come that split second where all is calm and the image is perfect. In that magical instant, subtle details of the planets leap into view, faint stellar companions wink at you and spiral arms and nebular wisps suddenly appear. Then the air churns again and they all vanish again in the instability. It is easy to get frustrated after waiting for a long time to see those Jovian cloud bands, but the wait will be worth it. That magical instant where all is calm is what the serious amateur astronomer waits all night for.

Observing Projects I – Putting the Eyes to Work

Observing Project 1A – Pressing the Dawes Limit

Some of the most enjoyable sights in astronomy are those you can see without a telescope. There are many objects and phenomena in the sky that are both beautiful to see and test the ability of the observer to push his or her eyes to their limits. In this next section, we'll take a look at some of these and find out just how good you are.

An oncoming car has two headlights. If the car were to back away from you, those headlights would appear to move closer together with increasing distance. Eventually the lights would become so close together that it would become impossible for you to separate them from each other and they would merge into a single point of light. The minimum angular distance between two light sources at which you can still distinguish two lights is called the *Dawes limit*. The ancient American Indians used to test the visual acuity of their children by having them look at a star and see if they could tell if there was only one or maybe more.

A favorite target used for this test is the star at the bend in the handle of the Big Dipper. The star is named Mizar and at first glance appears to be a rather ordinary star, shining bluish-white at magnitude +2.2. If your eyes are good, you may detect that Mizar is not alone, but has a companion nearby. Shining about 12 arc minutes[6] to Mizar's northwest is a fainter star that is nearly lost in Mizar's glow. The star is called Alcor and shines at magnitude +4.0. Splitting Mizar and Alcor is a very tough challenge, especially if your sky is light polluted, which may make seeing Alcor difficult under any circumstances. Though these two stars appear to be a pair, in fact Alcor lies more than three light years farther in the background than Mizar does. Mizar is about 78 light years distant while Alcor is just over 81 light years distant. The two stars only appear to be close together, but in fact this system is not a double star at all. Each one is independent. Mizar in actuality is a double star by itself. Its partner is easily visible in any telescope shining about 14 arc seconds to Mizar's southwest at magnitude 3.9. If your eyes are sharp enough to split Mizar

---

[6] One arc minute equals 1/60th of one degree. A degree is divided into 60 minutes. Each minute in turn is divided into 60 seconds, just as it is on a clock.

and Alcor, you're ready to try and take on more difficult challenges. Twelve minutes is tough under any circumstances. Can you split a pair of stars right near the Dawes limit? If you can manage Mizar and Alcor, then move southward to Libra and try to split Zubenelgenubi. The brightest point of light in the constellation Libra glimmers at magnitude 2.7 and is also a pair of stars. Zubenelgenubi is much tougher to split than is the Mizar and Alcor pair. The stars are much closer together, only about 230 arc seconds apart, not far above the Dawes limit for the unaided eye. The secondary star is also much fainter than is Alcor, shining at only magnitude +5.1. The brighter component shines so much brighter that it can drown out the fainter star in its glare. Unless your sky is fairly dark, spotting the companion will be very difficult. Splitting close double stars near the limit of the resolution of your observing instrument requires time and patience. It does not matter whether you are using the human eye or the mighty Keck reflectors, when you are working at the limits of your equipment, you need to wait for just that right moment when the eye is relaxed and the air is still for just that one precious second.

Observing Project 1B – Deep Sky Visual Acuity (Stellar Objects)

Riding high in the winter sky is the beautiful open star cluster called the Pleiades, listed in the Messier catalog as M45. The cluster, also known as the "seven sisters," is so named for the seven fairly bright stars that can be discerned within the hazy patch of light. The cluster itself shines with an overall brightness of about magnitude +1.5. These are relatively newborn stars, just recently born from the stellar womb (though the gas that surrounds them is not part of the cloud from which the stars formed). The stars are surrounded by a bluish glow that consists of interstellar dust and gas interacting with the stars. The stars and gas glow together to form one of the deep sky's most beautiful jewels and by far the brightest object in Charles Messier's legendary catalog. It also makes yet another ideal testing ground for your night visual acuity. How many of the seven sisters can you actually see? The brightest member of the cluster is Alcyone, a young blue white star shining at

Figure 1.2. M45 at the upper right of the totally eclipsed Moon. Twelve members are visible. 35 mm SLR piggybacked on a Celestron Super C8 Plus.

magnitude +2.8. Atlas is the star farthest to the east shining at magnitude +3.6. Working to the west, Merope is magnitude +4.1, Maia shines at magnitude +3.8 and Electra is magnitude +3.6. These five are the only ones that I have been able to see from my northern New Jersey home. There are four other stars, Taygeta at magnitude +4.3, Pleione at magnitude +5.0, Celaento at magnitude +5.5 and Asterope at magnitude +5.8 that are visible to the unaided eye in dark conditions, but I have never been able to detect them without binoculars. Through binoculars, the Pleiades will number more than a dozen stars. A telescope will reveal yet even more within the hazy bluish white patch of light. How many can you find with the unaided eye? If you can find more than four members, then your eyesight is most keen and your sky fairly transparent. If you can bag seven, then you are seeing to magnitude 5.5 and your sky is extremely dark. If you can see Asterope, then you should consider having astronomy club meetings where you live, for your eyes and sky are good for seeing clear down to the naked-eye limit.

Observing Project 1C – The Light Pollution Census

A great test of the darkness of your sky that we alluded to earlier in this chapter is to do a census of the Great Square of Pegasus. How many stars can you see within the Square? If you can count only two, you are seeing to only magnitude +4.4. Picking up four stars takes you to magnitude +5.0. There are a total of eight stars brighter than magnitude +5.5 and if you can count as many as 16, you are seeing to magnitude +6.0. If you have extremely good night vision and a very dark sky, you might be able to see to magnitude +6.5 in which case you may count as many as 35 stars within the square. The Great Square makes a great test because it is most prominent in the sky during late autumn and winter. During this time of year, the atmosphere is normally very stable because the ground cools very rapidly at night and chills the air close to the ground. This has the effect of greatly diminishing the

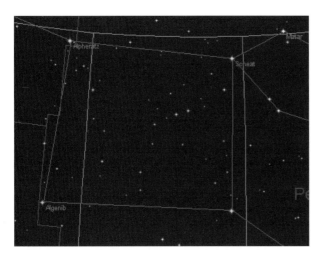

Figure 1.3. Great Square of Pegasus mag 6.5. Graphic created by author with *Redshift 4*.

temperature lapse rate and lends great stability to the lower atmosphere. With stable conditions prevailing on many nights, use the Great Square of Pegasus to judge the darkness of your autumn sky and the sharpness of your eyes. Count carefully, be discriminating and above all, be honest with yourself. Remember you are doing a serious scientific project and you must take great care not to let your desire to prove how clear your sky is interfere with an honest and objective result. How many stars can you see in the square?

Observing Project 1D – Deep Sky Visual Acuity (Non-Stellar Objects)

Stars are fairly easy to make out because all of their luminosity is concentrated into a nearly immeasurable point of light. Try to imagine how much more difficult it would be to see if the total light were to be spread out over several square degrees. Such is the case with deep sky objects such as nebulas, star clusters and galaxies. That is what we will test with this next drill. How many objects in Charles Messier's famous catalog can you find with the unaided eye? Messier was a comet hunter in eighteenth century France. To aid his fellow comet hunters and him, Messier set out to catalog all the objects in the sky, which could easily be mistaken for comets. By the time he was finished, he had logged and listed over one-hundred of the deep sky's grandest wonders. In this project, we will attempt to see how many of these we can find without optical aid. Most are well below the threshold required for naked-eye visibility, but some can be seen easily from any location. The open star cluster M45, which we explored earlier, is the brightest single object in the catalog at total magnitude +1.5. This light is scattered across more than 6 square degrees of sky, making the overall brightness seem much fainter than the number would suggest. Magnitude estimates of an object's brightness in astronomy are measures of total light. The more the total light is spread out, the lower the overall surface brightness becomes. To simulate this effect, pour some sugar out on a table and concentrate the grains as tightly as you can. Now see the difference in apparent brightness when the sugar grains are spread out over a wide area. There is the same amount of sugar crystals just as a 1.5 magnitude star produces the same number of photons as the Pleiades cluster. The light of the star is much more concentrated, therefore it appears much brighter because it is concentrated into one single point.

Check out the Messier catalog list printed at the back of this book in Appendix A. How many of the objects brighter than magnitude six can you see? To the south of M45 in the winter sky is the bright emission nebula ("bright" is both an adjective and a technical classification of this object) M42. The Great Orion Nebula hangs from the belt of mighty Orion the hunter. The glowing gas cloud shines at magnitude +2.5 and cannot be mistaken for any other object in the sky. This showpiece of the heavens is located nearly on the celestial equator and thus visible to astronomers all over the world. Over to the northwest, try for the Andromeda Galaxy (M31). Andromeda shines at only magnitude +4.6 and that light is spread over a wide area of sky. Andromeda measures about 1.8 degrees long by 29 minutes wide. This is about the area covered by three full moons. Under moderate light pollution conditions, you might not be able to see it at all without binoculars. Finding

Andromeda is deeply satisfying for many amateurs for it represents the farthest place you can possibly see with the unaided eye. Remember that averted vision!

Star clusters make inviting targets in the summer sky. Here are three globular clusters brighter than magnitude 6. In Hercules, try M13 at magnitude +5.8. This cluster will require a dark sky and a good eye, but it is distinguishable covering a span of about 17 arc minutes. Through a telescope, M13 is the grandest globular cluster in the northern sky, but is a great test for the unaided eye. To the south is the globular cluster M4, near Antares in Scorpio. Though it lists as slightly brighter than M13 at magnitude +5.6, its light is spread out over a wider area making it more difficult to see. In addition, M4 never gets very high above the horizon for northern hemisphere viewers, unlike M13, which sails nearly overhead during summer nights. Riding a little higher in the sky to the northwest of M4 in the constellation Serpens is the globular cluster M5. Also listed at magnitude +5.6, it is somewhat more compact than M4 making viewing a little easier. Still both of these clusters represent challenges even in a dark sky. In the southern hemisphere are two of the skies grandest jewels, the globular clusters 47 Tucanae and Omega Centauri. Both of these monster globulars are easily visible to the unaided eye and outshine any cluster in the north. Because they are only visible from the southern hemisphere, Messier could never add them to his catalog.

Open star clusters can be even more difficult because they do not have the bright cores that characterize globulars. Still there are some that you can hunt down without any optical aid. Next to M45, the sky's next brightest open cluster is called Praesepe in Cancer. Also known as the "Beehive Cluster" it is listed as M44 in the Messier catalog. The cluster lists at magnitude +3.5, but the light is scattered over more than a full degree of sky. A tighter grouping lies near the tail of Scorpius called M6. This open cluster near the Scorpion's tail is only 21 minutes across, about the size of many globulars and its total light measures magnitude +4.6. M6 has a nearby partner called M7. This open cluster lists more than a full magnitude brighter at +3.3 but it covers more than four times as much sky so the two clusters appear to have a very similar surface brightness.

The Messier catalog offers a treasure trove of deep-sky wonders. How many can you find on your own? If you can find ten, you've got a great sky and keen eyes.

Observing Project 1E – The Sixth Naked-Eye Planet: Uranus

In the late eighteenth century, William Herschel was scanning the skies from his observatory looking for comets when he chanced upon a sixth-magnitude star that his charts did not show. After observing the object for several nights it appeared to move across the sky in a slow but steady eastward motion. Herschel though he had discovered a new comet. But further observations yielded even more startling results. The object was moving in an orbit that was not very comet-like. The object's orbit appeared to be circular, like that of a planet and not the highly elongated path followed by a comet. The object also seemed to exhibit more of a planet-like appearance, without the fuzzy coma characteristic of comets and it had no tail to be seen at all. After several weeks of careful research, Herschel stunned the world

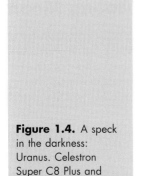

Figure 1.4. A speck in the darkness: Uranus. Celestron Super C8 Plus and Meade Pictor 216 XT at f/10. Photograph by author.

by announcing the discovery of his new planet. Astronomers around the world raced to their telescopes to see the new wonder of the solar system. Many modern names were floated for the new world, including "Herschel" for its discoverer and "Georgian Star" for Britain's King George III, but the name that stuck was "Uranus," the mythical father of Saturn and grandfather of Jupiter.

Uranus' discovery was historic, for the solar system had its first ever new member. The new world measured in at 29,000 miles in diameter, about the size of three and one-half Earth's. Uranus for most of its history was considered to be a gas giant much like Jupiter and Saturn, consisting mainly of gaseous hydrogen and helium. The Voyager 2 flyby in 1986 changed all that. It showed the planet to be substantially different from the first two Jovian worlds in that it was made up of large quantities of ices, thus earning the new moniker "ice giant." Neptune also turned out to be more like Uranus after Voyager 2's 1989 flyby of that planet.

Though Herschel needed a telescope and a lot of good luck to find Uranus, all you need is that dark sky, good eyes and a star chart to find the seventh planet from the sun. The planet generally shines at magnitude +5.8, placing it just above the threshold for naked-eye visibility. Finding it among the other 6,000 points of light in the sky magnitude six or brighter is an intimidating challenge. Popular astronomy magazines such as *Sky & Telescope* or *Astronomy* will print charts annually, showing the location and path of Uranus through the evening sky for the year. To find such a faint nondescript object, work to become intimately familiar with the sky in that area. That way, anything that appears unusual will immediately jump out at you. As you carefully scan that area of sky on a nightly basis, you will see one of the faintest objects in the area move slightly. The motion is normally towards the east, but as Earth overtakes Uranus in its orbit each year, our point of view causes the illusion called *retrograde motion*. The planet appears to stop in its path and then begin to track backwards towards the west for several months before resuming normal motion again. In the darkest of skies, the planet may also betray its true green color, but for most of us it will just appear to be its normal shade of gray.

Observing Project 1F – The Very Young Moon

Each month as the Moon swings around Earth in its orbit it passes between Earth and the Sun (most of the time slightly above or below it). As the Moon passes this position it reemerges into the evening sky as a crescent, which grows fatter each night, as the Moon appears to move farther away from the Sun in the sky. In this project we will attempt to determine how soon after new moon can you first detect the crescent.

There are several factors that improve or degrade your prospects for success in this project. The most important variable is the angle that the *ecliptic* makes against the sky. The ecliptic is the imaginary projection of the plane of Earth's orbit against the celestial sphere. All the planets and our Moon orbit in a path around the Sun that always remains on or close to the ecliptic. Since the Moon never strays far from this line, the angle the ecliptic makes with the horizon is vital to your observing prospects. An object can be very close to the Sun and yet be easily visible if the ecliptic makes a steep angle with the horizon. At other times, the object can be twice as far from the Sun and yet be nearly invisible if the ecliptic makes a shallow angle with the Sun. The best time to try and bag a very young moon is during the months of February, March and April. During these months, the ecliptic makes its steepest angle with respect to the horizon because Earth's north pole is tipped so that it is pointing behind the planet with respect to its orbital path. Solar system objects close to the Sun appear to stand almost directly above it at sunset. During August, September and October the reverse situation occurs when Earth's north pole leads in our orbit and the ecliptic appears to lie almost flat with respect to the western horizon at sunset. The reverse of this relationship occurs if you are trying to locate objects at sunrise. The most favorable months are August, September and October while the late winter and early spring months present poor viewing geometry at dawn. These relationships are also inverted if you are reading in the southern hemisphere.

Any good astronomy publication will tell you not just the day when new moon occurs but in fact the precise moment of new moon, usually in Universal Time. If the moon is new precisely at the time of your local sunset, you will have the opportunity to view a twenty-four hour old moon at sunset the next night. On nights where the viewing geometry is favorable, go out and find a spot with a clear view of the western horizon. Note the spot where the Sun sets and about thirty minutes later look about one fist-width (an average person's fist at arm's length subtends an angle of about 10 degrees) above the sunset point or slightly to the south. A twenty-four hour old moon should be fairly easy to spot, showing a hairline crescent hanging in an almost ghostly fashion just above the western horizon. Most people will not notice the moon until the next night. The extremely thin crescent, just 7% illuminated and only 11 degrees from the Sun, can be very difficult if any haze exists and makes for a good observing challenge.

Once you've succeeded at this, try an even younger moon. At twenty hours, the moon is barely 6% illuminated. The youngest crescent that anyone has reportedly sighted with the unaided eye (that can be verified) was only fourteen hours old. At that age the Moon is only about 6 degrees east of the Sun and is only 4% illuminated. Such a feat requires a very clear sky, perfect eyesight and viewing technique and a good shot of good luck. Can you do it? Check your ephemeris for next month's new moon and see if you can give it a try.

Observing Project 1G – Elusive Mercury

We all learned in grade school that there are five classic planets visible to the unaided eye. Venus and Jupiter flare brilliantly in the night sky and are unmistakable. Saturn also shines brightly and prominently all night long when favorably placed. Mars is unique with its orange-red glow. All of these planets shine high in a dark sky when favorably placed and require no effort to find. Poor Mercury is left to suffer in anonymity, spinning around the Sun at breakneck speed completing a circuit every eighty-eight days. As it passes between Earth and Sun (inferior conjunction) it enters the morning sky for several weeks, then passes behind the Sun (superior conjunction) and enters the evening sky. The period of visibility when the planet is in the morning or evening sky is referred to as an *apparition*. The planet will remain in the morning or evening sky for seven to nine weeks at a time depending on its distance from the Sun. The planet however will only be visible for a fraction of the time, for two to three weeks with the unaided eye while it is farthest from the Sun and reasonably bright.

Observers wishing to sight Mercury must learn to deal with two important contradictions. First is the assumption that Mercury must be easiest to see when it ranges farthest from the Sun. Mercury orbits the Sun in a path that has a greater eccentricity (deviation from circular) than any other planet besides Pluto. This causes the planet at some apparitions to roam much farther from the Sun than at others. As a result when Mercury reaches its greatest elongation (farthest angular distance from the Sun for a particular apparition) at the same time it is closest to the Sun (perihelion), it never strays more than 18 degrees from the Sun. If Mercury reaches greatest elongation at the time of its aphelion (farthest from the Sun) it can appear as far as 28 degrees from the Sun. Northern hemisphere observers are at an unfortunate disadvantage. The problem is that when Mercury reaches greatest elongation at aphelion, it is always at a time when viewing geometry is at its most unfavorable. Greatest elongations when viewing geometry is most favorable occur only with Mercury near perihelion. As a result, when Mercury wanders farthest from the Sun, the ecliptic lies flat with respect to the horizon and the planet is buried in the glare of twilight. When Mercury sticks close to the Sun, the viewing angle is at its best. This is the best time to look for the elusive innermost planet even though it is much closer to the Sun than it would otherwise be.

The second contradiction is that it is best to look for Mercury between the time of greatest elongation and inferior conjunction (in an evening apparition, in a morning apparition inferior conjunction comes first). This is when the planet is closest to Earth and therefore should be at its brightest. In fact, nothing could possibly be farther from the truth. Mercury is a unique object in all the cosmos in that it is the only celestial body that grows *fainter* as it draws *closer*! No other body in the universe does this. No other body changes brightness through such a range as Mercury either. Mercury is actually at its brightest (magnitude −2.0, brighter than Sirius) when it is near superior conjunction, when it is farthest from Earth. As it draws nearer to Earth, it fades to magnitude zero by greatest elongation. Over the next two weeks as it nears its closest point to Earth, it completely fades out of sight, growing as faint as magnitude +4.6 just before inferior conjunction. We'll discuss this remarkable aspect of Mercury's behavior a little later on when we look at it through the telescope.

Finding the planet is a simple matter of knowing exactly where to look and then being there at the right time. Trying to find the planet too soon after sunset will cause frustration because the sky is still too bright. Waiting too long will cause Mercury to sink too low to the horizon. The best time to look is during the time period starting about seven to ten days before greatest elongation in the evening sky or during the same period after greatest elongation in the morning sky. Begin looking about thirty minutes after sunset. During times of favorable viewing geometry, look just above and a little to the south of the sunset or sunrise point. As the pink sunset sky fades to deep blue, pinkish Mercury will appear. During this time, the planet will be shining at brighter than magnitude zero and will stand out nicely. During an evening apparition the planet will quickly retreat to the horizon and is lost within the next half-hour. If you are observing in the morning sky, see how long you can keep Mercury in sight as dawn approaches. The planet is at magnitude zero at greatest elongation but about two weeks later, you should be able to see it as bright as magnitude −1.0. This should be bright enough to allow any observer to track the planet until sunrise.

When you succeed, you will join a rather exclusive club. Everyone on Earth has seen the four other bright planets, but not one person in a thousand has consciously looked at Mercury and realized what it is. Now with our eyes well trained and exercised, let's begin the work of assembling the equipment needed to complete the integrated observing system.

CHAPTER TWO

The Integrated Observing System. Part II: Your Equipment

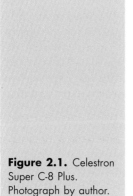

Figure 2.1. Celestron Super C-8 Plus. Photograph by author.

After spending my childhood and teen years observing under bright city or suburban lights with a small-aperture department store refractor telescope, I entered my college years with the dream of a larger instrument that would bring the deep sky into view for me. The two-inch Tasco refractor served its purpose well for showing the visible planets, Moon, and Sun. Seeing the deep sky beyond the solar system, the outer planets, faint comets, or the asteroid belt's largest denizens was hopelessly beyond my department store telescope's limited reach. As I began to search among the various manufacturers, the vast number of choices available rapidly overwhelmed me. Should I consider another refractor or a reflector? If I were to choose a reflector, what type should it be? What I decided to do was make a list of qualities that I needed to have in a telescope and then set about finding the one that would come closest to meeting my unique needs. Here are the items that made up my list:

(1) *Aperture*: It had to be big enough to permit viewing deep sky objects and stars to a reasonably faint magnitude. Getting to magnitude 14 (Pluto) would require at least eight inches.

(2) *Portable*: It had to be small enough to permit me to move it up or down stairs, transport easily by car and light enough to carry over moderate distances.

(3) *Optical Quality*: I was going to pay what amounted to a lot of money for a college student so it had to be a serious instrument that would provide the best, brightest, crispest images possible for my investment.

(4) *Versatile*: The telescope had to be capable of a full range of operations from wide-field viewing to providing high-power detail both as a visual and photographic instrument.

(5) *Durability*: This was going to be my telescope for many, many years. It would have to be solidly built and be able to withstand the rigors of decades of use.

(6) *Upgradeable*: As my needs and skills in amateur astronomy grew through the years the telescope had to be able to grow with me.

(7) *Serviceable*: In the unlikely event that there was ever a mechanical problem with the scope or a part needed to be replaced or if I just needed advice on how to work something, the manufacturer of the telescope had to be able to provide it.

When you create this list, make certain that you organize the items in the order in which they are important to you in case you need to or are willing to compromise. In my case, all seven of these items were absolutely non-negotiable. For example, without aperture there is absolutely no reason for me to be buying a telescope. I already owned one and wanted to replace it because the one I had was not big enough. It would have made no sense to replace one small scope with another. Portability was number two because I had only a limited amount of space in which to store a telescope and needed to be able to move it as necessary to be able to get it to the observing site. All my equipment had to fit into a college student's car (1981 K-Car, don't laugh at me). It then had to be hauled from the car to the place where the telescope would be set up, often a distance of a hundred feet from where I would park. A huge telescope weighing more than a hundred pounds just would not get the job done. The quality of the optics was another issue. My plan was to spend about $2,000 on my new scope. For that money, it had to have the best optics I could afford. This required that I put in a fair amount of time learning about the strengths and weaknesses of various optical systems that were available. I also needed a scope that was versatile. It had to be of the right focal length and size so that it could make photographic images of the deep sky with reasonable speed yet also provide crisp images at high power when called for visually. It also had to be durable, capable of withstanding not only the test of time, but also the rigors of repeated transportation, bumping and jostling without not only remaining intact, but without requiring constant readjustment each time I put it in the car. It had to be upgradeable, meaning capable of accepting add-on accessories such as drive motors, eyepieces and camera adaptors. As my skill grew and I wanted to probe deeper into the heavens, the telescope had to be able to grow with me and be capable of using advanced equipment beyond its own equipment package. Finally, I demanded a telescope that would be easily serviceable in the rare event that something went wrong with it. I set my sights on a scope with a good warranty and a company with a good chain of dealers where a telescope could be taken for repair or at least for shipment back to the factory for maintenance.

Remember that not everything on my list may appear on yours and you may in turn have needs that I did not. If you're mounting the scope on a permanent pier, then its not likely that portability is important to you, especially if that pier is to be surrounded by a dome. You may desire the capability for full computer control of your telescope or Global Positioning System capability to eliminate any need for manual navigation of the heavens. I have always found these tasks among the most pleasurable in astronomy and I take great pride in finding my way around the sky. If you would prefer just to hit the GO TO button, there's nothing wrong with that if it increases your pleasure in astronomy or at least lowers your frustration level. I would just prefer to find it myself. Anyway, when I bought my scope in 1986, there was no such technology available. You may have an absolute price cap to live with or other restrictions. Make out your own list and set your priorities accordingly.

There are some things you should always remember about buying a telescope. If you are about to lay out money for a serious telescope, be aware that it is an investment that will hold good value. Telescopes are not cars. Optical systems do not degrade in quality from being used so plan to take very good care of it because its going to be with you for many years. Mine is now nineteen years old, still looks like its fresh out of the box even though we've been all over the universe together. I expect we will travel the cosmos together for many more years to come. Though I'm planning to add a second telescope to my collection, it is going to be a smaller scope for quick low-power viewing of the skies on short notice. Secondly, the scope is only as good as the quality of its optics and mount. A telescope with poor optics will produce images that are distorted either in shape, color or both. A telescope also can have the world's best optics but if you put it on a shaky mount, an image magnified 100 times will only show you a vibration magnified 100 times.

By now, you probably realize that I've spent some time taking backhanded swipes at "500×" department store telescopes. They may serve their purpose reasonably well as entry-level instruments, but if you buy one realistically believing it will produce 500×, forget it. My third key thing to remember is that the primary function of a telescope is *not* to magnify. Its primary function is to collect large amounts of light and bring that light to a clear, sharp focus. Magnification is a secondary consideration and in fact is not performed by the telescope at all, but rather by the eyepiece. A telescope's ability to clearly present magnified images is directly related to how much light it can collect. If you try to magnify an object 500 times using a 2-inch telescope (that's 250× per inch of aperture), all you will wind up with is a blur magnified 500 times. A telescope of this size cannot magnify with that much power and bring the image to focus. Even medium-size telescopes of 8 inches aperture (that's 62.5× per inch) cannot withstand the use of that much power. In reality, a telescope with good optics should be able to focus an image at a *maximum* of 50× per inch of aperture. Any more than that results in a loss of image brightness, clarity and contrast that grows progressively worse as more power is used. If you try to push that two-inch department store refractor to more than 100×, you are going to be sorely disappointed. Never buy a telescope that is marketed on the basis of its magnifying power. If you look at ads for Celestron or Meade telescopes, you will never ever see their instruments advertised on the basis of magnifying power. With an eyepiece of the proper focal length, any telescope can be made to magnify an object 500 times. Aggravating the situation further is the possibility that the lenses of that department store telescope may not be made of the best material available. Top-quality refractors are made with optical quality glass and top-end scopes may use a lens element made of calcium fluorite. Cheap scopes may have lenses made of plate glass, Pyrex or even worse, plastic! This will create imperfect or even badly distorted images even at low powers.

Refractors vs. Reflectors

Now that you have made the decision to step up to a medium to large aperture telescope from your department store model, you need to consider the various types of optical designs and decide what best suits your needs. The telescope you own now is most likely a *refractor*. This design, also called a *dioptric* telescope, is based directly upon the original opera glass telescope designed by Galileo in 1610. That original telescope used a simple convex lens to gather and focus light and a concave lens at the opposite end of the tube to bring that light to a crude focus. The lens at the front of the scope is called the *objective*. In any telescope the objective is the lens or mirror element that gathers starlight and directs it to a focus point. With this simple design, Galileo discovered that Venus exhibited phases like the Moon and that Jupiter had satellites circling it. These discoveries led Galileo to realize that the geocentric model of the solar system was incorrect. Earth was not at the center of the solar system, but the Sun was! Galileo would pay dearly for his blasphemy. The church would torment him, ruin him, excommunicate him and finally forced him to recant. He was of course correct. The Sun was at the center of the solar system and all the planets circled it. It would be nearly 400 years before the church would come to admit it early in the reign of Pope John Paul II when Galileo was formally brought back into the Roman Catholic Church.

Science has refined the design of the refractor over the years allowing it to create sharper images with better focus. For an amateur shopping for a telescope, the chief advantage of a refractor is that its lenses are rigidly held in place in the telescope tube making the telescope virtually maintenance free. The lenses never need to be adjusted and are in fact often cemented in place. The refractor lenses generally are of long focal length, often producing f ratios[7] of f/12 to f/14. These long-focus telescopes provide sharp images with high magnification without using excessively short focal-length eyepieces. The refractor does have some drawbacks that prevent it from being used as a design for large-aperture telescopes. By the time the scope reaches about 4 inches (100 mm) in diameter, the objective lens starts to become too heavy and the tube too long for use in a design that can still be considered portable. Lenses can be designed that provide a shorter focal length, but that results

---

[7] *Focal length* is the distance from the objective lens to the point at which light rays come to focus. *F-ratio* is the focal length divided by the diameter of the objective.

in some loss of image sharpness and they don't get any lighter. Refractors also are subject to an error that is an inherent byproduct of the design. As light passes through the glass lenses, the colors are separated in the same way that a prism might and not transmitted evenly by the lens elements. This causes an error known as *chromatic aberration*. Objects viewed through a refractor will often have a fringe of false color surrounding them caused by the optical separation of the differing wavelengths of light through the telescope objective lens. To try to minimize the effects of this color distortion, the telescopes lenses may be coated with metallic-based compounds that also improve overall light transmission. For those who desire perfect color in their images, the closest you can get is a telescope with a *calcium fluorite* lens element. Calcium fluorite is not glass, but a mineral that must be ground and polished in a very time-consuming and expensive procedure. Fluorite lenses however are not subject to the chromatic aberrations that plague glass lens refractors. Refractors do have other limitations as well. The long focal lengths produce very narrow fields of view that cannot contain entire deep sky objects. Long focal lengths are also tough for astrophotography work. The longer the focal ratio of the telescope becomes, the longer exposure times are required for imaging a given object. Refractors can be very frustrating telescopes for taking pictures.

The limitations and chromatic aberrations of the refractor began to lead early astronomers to look for other solutions to the problems of building large-aperture telescopes. By 1681, the noted early physicist Sir Issac Newton invented a telescope using a large mirror as a light-collecting surface rather than a lens. This type of telescope is called a "reflector" or *catropic* telescope. Reflectors use a large concave spherical mirror (primary mirror or objective mirror) to gather light and bring it to focus on a small, optically flat mirror near the front of the telescope tube. This secondary mirror turns the light 90 degrees to an image-forming eyepiece at the side of the tube. This original type of reflector is called a Newtonian reflector in honor of its inventor. A later type of reflector, called a Cassegrain telescope, focuses light through a hole in the center of the primary mirror to an eyepiece in the back of the telescope. Because light does not actually pass through any of the telescopes surfaces, they do not actually need to be made out of pure glass. Telescope mirrors

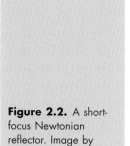

Figure 2.2. A short-focus Newtonian reflector. Image by author.

can be highly polished glass, Pyrex or even in some very early instruments, aluminum. Glass or Pyrex is generally used today because it is easiest to polish and figure to exactly the precise shape needed, then once the mirror is correctly figured, an aluminum overcoat is applied and polished until ready for use. The glass or Pyrex only needs to be thick enough to form and maintain the shape of the mirror. Reflector telescope mirrors are therefore much lighter per inch of aperture than are refractors. This removed many of the limitations that existed on telescope size. The world's largest refractor measures only 40 inches in diameter but reflectors now exist that are ten times that size.

Like refractors, reflectors also have an inherent flaw. Because a spherical mirror is used to reflect light to a flat one, some points of the secondary mirror are farther from the primary mirror's point of focus than others. Usually the objective is designed to focus light to the center of the secondary mirror. Portions of the image that fall on the outer parts of the secondary mirror tend to be distorted slightly. This error inherent in the reflector design is called *spherical aberration*. Stars at the center of the field focus to sharp points but stars near the edge may appear slightly streaked from the center of the field towards the edge. As telescope sales began to grow among the amateur public, two different types of fixes became available. One is to use a primary mirror with a different type of shape called a *parabolic* mirror. This mirror has a much deeper curve than a spherical mirror. A spherical mirror if continued in shape would eventually form a perfect sphere. A parabolic mirror has a much sharper curve to it and would close into a much more oblate form. This shape causes light to reach the secondary mirror in a more uniform manner creating sharper images. It can also be a very expensive mirror to produce because its curve is so complex. The second type of correcting mechanism that has evolved over the years is called a *corrector plate*. The corrector plate is a glass lens introduced at the opening of the telescope. The corrector introduces an error into the light path that is exactly opposite of that introduced by the primary mirror. Thus in spite of what your mother told you, in this case two wrongs do make a right. There are two types of correctors commonly in use. Smaller telescopes may use a *Maksutov* corrector. This is a thick lens element with a heavy concave shape. Maksutov correctors are highly efficient and provide super-sharp images, but like refractor lenses, they become impractical to use once larger than about four inches in diameter. *Schmidt* correctors are not quite as effective as Maksutovs but are thin and very light. They are also relatively inexpensive to produce. Schmidt correctors are preferred for use in any reflector telescope design larger than four inches. Reflectors do have some other disadvantages. The classic Newtonian or Cassegrain design has an optical tube that is open to the elements and therefore must be carefully cared for. Mirror surfaces must be kept meticulously clean on a regular basis. The secondary mirror is in the path of light to the primary mirror and is held in place by a spider support that also blocks some light from reaching the primary mirror. This so-called *secondary obstruction* created by the mirror and its support can block as much as 15% of the light entering the telescope. The mirrors must be carefully kept in line with each other through a process called *collimation*. This is particularly true of the secondary mirror, which must be in perfect alignment to properly redirect light to the eyepiece. An improperly collimated reflector can rapidly become a source of great frustration. Telescopes that employ Schmidt or Maksutov correctors have the advantage of being sealed at the front end, protecting the telescope mirrors. These telescopes are also less prone to (but not immune from) collimation problems. The secondary mirror housing of a Schmidt corrector telescope is fixed in the center of the corrector plate and is easily adjustable with an Allen-head wrench. The Maksutov is even simpler. The secondary "mirror" is actually an aluminized spot in the center of the corrector and never needs to (and cannot) be adjusted.

A telescope is only as good as the mounting it sits on. Mounts come in two basic types. The simpler type is called the *alt-azimuth* mounting. The mount rotates up–down and left–right allowing the telescope to be adjusted in both altitude and azimuth. To follow an object across the sky, one must follow the object in both axes. My 4.5-inch Bushnell sits in a very simple type of alt-azimuth mount, which is simply a bowl in which the bulbous base of the telescope sits and rotates. At the other end of the complexity scale, most GO TO scopes are alt-azimuth mounted with a computer issuing corrections as they track across the sky. *Equatorial* mounted scopes also move left–right and up–down, but the left–right axis can be pointed directly at the north pole, allowing the telescope to track an object with only a single motion. A drive motor that turns the scope at the same rate as Earth rotates in the opposite direction will enable that scope to keep an object centered in the field so long as the scope is both properly aligned and perfectly level. The most important thing though about a mounting is that is must always be perfectly rigid and not move. If the mount vibrates, no matter how well the scope performs at high power, all you will see is a highly magnified vibration if the mount is not stable.

Reflectors and refractors have other characteristics that cannot fairly be described as either strengths or weaknesses, but will play an important role in determining what type of telescope you will eventually settle on. Refractors tend to have fairly high f-ratios compared with reflectors. A typical refractor will have an f-ratio of anywhere from f/12 to as high as f/15. These telescopes will produce crisp images and high magnification with long focal-length eyepieces. But because the f-ratios are so large, they will be difficult to use for long-exposure astrophotography. Exposures for deep sky objects will be impractically long. Reflectors, particularly Newtonian types, have very short f-ratios that mean that high powers are not easily usable. Newtonian reflectors will have f-ratios around f/6. They will produce very wide field images that are bright. Reflectors are also very good for long-exposure astrophotography, often requiring less than half the time for an exposure that a refractor of the same size. Refractors will always have the eyepiece at the rear end of the optical tube. With the use of a right-angle prism, the refractor user will always be able to find a comfortable observing position. Newtonian reflectors have the eyepiece near the other end of the telescope on the side of the tube. This can cause an observer to torque his body into many unusual positions while trying to see through his telescope. In the largest Dobsonian designs, a ladder may become necessary when viewing objects near the zenith.

The Decision

After many months of careful consideration, the telescope I finally settled on was the 1986 model Celestron Super C8 Plus. This telescope is a Cassegrain design reflector with a Schmidt corrector. This combination Schmidt–Cassegrain design is the most popular in use today among amateurs purchasing medium to large size telescopes. Let us revisit my original list of important qualities and see how the Celestron meets my needs.

(1) *Aperture*: At 8 inches in diameter, the Celestron could reach objects as faint as magnitude +13.8, the brightness of Pluto. The telescope came with optional coatings to improve light transmission to boost the limit to magnitude +14.0

(2) *Portability*: The Schmidt–Cassegrain design uses a convex secondary mirror and an f/10 objective. This design allows an 80-inch focal-length telescope to be accommodated by a tube that is only 17 inches long. The entire telescope assembly, the fork-arm mount and clock drive base weighs only twenty-three pounds.

(3) *Optical Quality*: At that time, there was no mass producer of consumer telescopes that had a better reputation for quality than Celestron. Remember that as I write this, this was nineteen years ago. Though Celestron certainly has not slipped, its competition has certainly gotten better.

(4) *Versatility*: At f/10 and 8 inches of aperture, the Schmidt–Cassegrain has enough focal length to provide crisp images visually, but it is fast enough to yield reasonably short photographic exposure times while reaching a theoretical photographic limit of magnitude +16.

(5) *Durability*: The Schmidt–Cassegrain optical design is nearly as rigid and maintenance free as any refractor, though some collimation is occasionally needed. The fork mount and drive base are made of cast iron and will last forever.

(6) *Upgradeable*: The Celestron product line comes with a wide range of visual and photographic accessories for use in any application. The scope comes with a star diagonal mirror and two eyepieces (a 26-mm Plossl and a 7-mm Orthoscopic) providing 77× and 286×. This got me off to a good start and I have been able to add many accessories since purchasing the scope, including some that could not have been imagined in 1986.

(7) *Serviceable*: The telescope is covered by Celestron's limited warranty. If anything breaks, it gets fixed for free (assuming that I did not take a hammer to it).

Other telescope purchasers as I said may have other criteria that are important to them while others may not be interested in some of the things that I want. One observer may want aperture in a telescope and nothing else. He may have no interest in astrophotography or any other complex applications or may already have a smaller telescope that can do that. This observer may want a moderate cost second scope solely for the purpose of pulling out the faintest objects out of the sky he possibly can for the lowest cost. This observer might be interested in a type of Newtonian reflector called a "Dobsonian." These scopes come mounted on the simplest type of mount, often on wheels for ease of movement. They are built for one thing only, size. Dobsonian buyers can get scopes as large as 16 inches or more for the same price as a Celestron Schmidt–Cassegrain half the size. There are also users who want a small telescope that can be whipped out with only a few minutes notice. Such telescopes as Bushnell's Voyager and Edmund Scientifics Astroscan cost around $200 and provide a 5-inch Schmidt–Newtonian reflector that sits in your lap or on a special cradle. Whatever your needs are, there is a telescope to suit your purpose. Make sure you put in the research and thought needed to ensure you get the scope you need, not the one the salesman says you need. List what you need for a telescope to do and get one that does it. Don't get a Dobsonian if a computer controlled GO TO telescope is what you really need. The only result of that will be that you will own a telescope that can detect sixteenth magnitude galaxies, but is completely useless to you if you don't know how to navigate. The modern GO TO scope is wonderful if you are lost, particularly if they have a GPS receiver. All you have to do is turn it on and the GPS will automatically update your position and time so that the telescope can now point with arc second accuracy to any point in the sky without any other input from the user other that being told what the target is. Both Celestron and Meade sell their largest scopes with this technology, but you will pay over $5,000 for it. If you know your way around the sky, a 20-inch Dobsonian will do a better job optically and cost you only about 40% of the price. To use the analogy of the golfer again as we did in the first pages of this book, if you don't have the right equipment to do the job, all you will experience is frustration. A golfer who uses clubs that are even 2 inches too long for him will be completely unable to perform with them. The clubs must be perfect your game and your body. So must your optical equipment for the job you will ask it to do.

Accessories

My Celestron Super C8 Plus came equipped with several accessories out of the box (it was actually shipped in three boxes). These included an 8×50 polar axis finder scope with illuminated reticule, a star-diagonal prism, a 26-mm Plossl eyepiece and a 7-mm Orthoscopic eyepiece. This initial collection of equipment was adequate to allow me to begin exploring the heavens with my new telescope. As the next few months went my, I began to discover the need to expand my equipment inventory. With a focal length of 2,000 mm, the 26-mm eyepiece yielded 77× while the 7-mm produced 286×. The low-power eyepiece gave good wide field views across an area of about 45 arc minutes. The higher power eyepiece was so powerful that even with the Celestron's superior optics, getting a sharp steady view was a very rare event. I needed eyepieces that would allow me to use more intermediate magnifications for nights when the seeing was not as good as I would otherwise desire. After that came the desire to record on film, the sights that my eyes beheld. That led to my expanding into camera mounts and eventually into amateur CCD imaging. Lets now take a look at things that you will be eventually adding to your inventory.

(1) Eyepieces
(2) Barlow lenses and focal reducers
(3) Solar filter
(4) 35-mm SLR camera
(5) CCD camera
(6) Camera and CCD mounts and accessories
(7) Color filters
(8) Spectroscope
(9) Carrying case

(10) Laptop computer

(11) Portable DC power supply

(12) Paper or software star charts

This is a list of equipment needed to bring your telescope up to a solid level of versatility both for planetary observing and deep sky work as well as visual astronomy and photography. Let's take a look now at what each different type of accessory does for us.

Eyepieces

The eyepiece is the lens assembly at the focal point of the telescope that forms and magnifies the image. Most telescopes, even department store types, come with at least one for initial use. Sadly today many of the major manufacturers deliver their telescopes with just that one eyepiece. Today's Celestar 8 and Nexstar 8 telescopes, the current editions of the legendary Celestron C8, come with only a single 25-mm Plossl eyepiece (though Celestron does offer a nice observing kit that adds five more eyepieces and a filter set for under $100). When you purchase eyepieces, remember to get the proper size for your telescope. Eyepieces are sold in three industry-standard sizes (0.965 inch, 1.25 inch, and 2 inches) encompassing several different designs. The major things that distinguish eyepieces from one another are *eye relief* and *field of view*. Eye relief is the maximum distance from the telescope that you can see the entire field of view. The farther away you can keep your eyes, the more comfortable you will be. Field of view is the angular measure of how far you can see from the left edge to the right edge of the field.

If you own an economy or department store type telescope it likely came with an eyepiece of the Huygens or Ramsden optical design. These are two-element eyepieces designed to be low cost and produce satisfactory images to the undemanding eye. Eye relief is minimally adequate and the field of view is usually less than

Figure 2.3. Plossl eyepieces. Photograph by author.

30 degrees. The perception that I always had with these eyepieces is that I was looking through a pinhole. If my head wandered slightly, my eye would wander out of the field and if my eye were to tear at the eyepiece, I would lose sight of the target. These eyepieces are almost universally sized at 0.965 inch and while they serve department store telescopes adequately, they are not suited for use in high-performance instruments.

Telescopes intended for high-quality astronomical work are equipped with 1.25-inch eyepieces. The cast iron cell that seals the end of a Celestron or Meade telescope has an opening in it with a universal thread to allow attachment of 2-inch accessories. This opening is usually fitted with an optional adapter called a *visual back* that stops that opening down to 1.25 inches and has a thumbscrew to secure accessories with a non-threaded drawtube. This arrangement allows the telescope to accept 1.25-inch or two-inch accessories. The preferred size is 1.25 inch since there are a much wider variety of accessories at more modest cost and they are much lighter. Using 2-inch accessories usually requires that the telescope be carefully counterbalanced. For these telescopes, the lowest cost eyepiece one should consider is the Kellner design. These eyepieces generally cost less than $50 per unit and provide a field of view of about 40 degrees. Kellners provide bright images and good eye relief as long as you do not use too much power. At higher powers, eye relief becomes uncomfortably short. The Kellner eyepiece uses three lenses in its design. Light first strikes a convex element that then focuses light on a second convex lens. That lens is directly mated to a concave third element.

If you need to add more magnification flexibility to your telescope at a modest cost, orthoscopic eyepieces might be the way to go. Orthoscopic eyepieces utilize four elements in their design. Light first contacts a three-stage lens stack of convex, concave and convex lenses. These focus light onto a concave lens at the viewing end. Eye relief is better than in cheaper designs and, combined with their field of view, the Orthoscopic eyepiece was for many years considered the best all-around telescope eyepiece. It has over the years lost its top-of-the-line stature to more modern designs. Orthoscopics may also suffer from field curvature near the edges of the field of view. This may cause star images to appear slightly streaked rather than perfect pinpoints near the edge of the field of view. I added two Orthoscopic eyepieces to my equipment box after buying the Celestron, an 18-mm (111×) and a 12.5-mm (160×). These additions allowed me more magnification in conditions where my 7-mm would be useless.

For those with a little more money to spend after the big purchase, consider adding more Plossl eyepieces to your collection. This design has replaced the Orthoscopic as the most popular among discriminating observers. The Plossl also utilizes four optical elements. The first and second elements are mated and are concave and convex. This stack sends light to a second stack that is convex and concave bringing the image to focus. The design has the advantage of providing an image that is uniformly sharp from center to edge. Each one can cost somewhere between $60 and $100 depending upon size and manufacturer. Plossl designs feature excellent eye relief since the exit end of the eyepiece is considerable wider than less costly designs. The field of view is about 50 degrees on a typical model. Unlike with the Kellner design, the Plossl provides excellent views at both high and low powers. Since the Plossl design is usually the featured eyepiece in most manufacturers product lines, you may have choices of equipment with special overcoatings to cut glare and improve light transmission. Eyepieces such as Celestron's Ultima line provide improved image quality and sharpness at a considerably increased cost. Celestron now offers Plossls as part of a discount equipment kit. This allowed me to add 32-mm, 20-mm, 15-mm, 9-mm, 6-mm and 4-mm eyepieces to my collection (along with a Barlow and a filter set) for minimal cost of less than $100. You can now buy six of them for what a single Plossl cost twenty years ago.

Two other highly popular high-end eyepieces are the Erfle and Nagler designs. These are super-wide field eyepieces that more resemble looking through a window than a telescope. This is accomplished using what is usually a six-element array providing a field of view as great as 70 degrees. With such extreme width in the field of view, these eyepieces are generally best used for low-power views. Some loss of image sharpness may be experienced when using these eyepieces with very short focal lengths.

For those desiring the most field of view and maximum eye relief for maximum comfort, consider adding 2-inch accessories for your telescope. A special visual back allows 2-inch accessories to be added easily. Newtonian users might have to install a new focuser. Some elements might be directly attached to the rear cell of the telescope. Two-inch accessories are much larger and therefore more expensive to purchase. A single eyepiece can run to well over $200. They are also much heavier than 1.25-inch equipment and require counterbalancing. This in turn will cost more money.

Barlow Lenses and Focal Reducers

Adding a Barlow lens and a focal reducer to your equipment box will effectively triple the number of eyepieces at your disposal. These two accessories are remarkably simple and fairly low cost. They have the effect of also changing the character of your telescope. A focal reducer effectively shortens the f-ratio of the telescope

Figure 2.4. Barlow lens. Photograph by author.

creating a lower magnification and wide field of view. The Barlow lens does exactly the opposite. It will double or even triple the magnification provided by a typical eyepiece.

The Barlow lens utilizes a single element concave lens that has the effect of doubling the magnification of the eyepiece–telescope combination. Adding one Barlow effectively doubles the number of eyepieces at your disposal. Barlows are usually inexpensive to add to a telescope collection making it a great addition for any scope. They are however a bit on the long side and that means there will be a dramatic change in focus from where the focuser of the telescope was originally set. Barlows also are available in a triple magnification design. Some manufacturers also offer a short Barlow lens that will not affect focus position as severely as more traditional designs.

Another variant on the Barlow theme is the zoom lens that introduces a movable negative element into the light path. By rotating a knob, the magnification of the telescope can be varied over a wide range allowing the user to work at different magnifications without changing eyepieces. Once you start handling expensive oculars with numb hands in freezing cold, you will quickly come to appreciate the value of a zoom eyepiece. In nearly all cases the Barlow lens or zoom lens slips into the drawtube of your focuser or visual back.

The focal reducer is a positive lens that does the opposite of the typical Barlow. Instead of doubling the magnification, the focal reducer shortens the focal length of the telescope and reduces the magnification and creates a wider field of view. In addition to the wide field of view, low f-ratios mean brighter photography and shorter exposure times, both issues of great importance to astrophotographers. The telecompressor offered by Celestron reduces the f-ratio of my Schmidt–Cassegrain from f/10 to f/6.3. That means that my 26-mm Plossl's magnification is reduced from 77× to 48× while the field of view increases from less than three-quarters of a degree to better than 1.2 degrees. The telecompressor has disadvantages as well. Its large lens size makes it unsuitable for a draw tube insertion. Mine attaches directly to the Celestron's rear cell, then the visual back screws onto the back of the telecompressor. The telecompressor also requires a lot of back focus. Some observers may run out of focus travel before the telecompressor–eyepiece stack can come to a sharp focus. Observers who wish to turn their telescopes into real wide field instruments may opt for ultra-low f-ratio reducers such as the model designed by Optec. This reducer takes an f/10 telescope down to f/3.3. That would, using the eyepiece discussed previously with my Celestron, reduce magnification from 77× to 26× and take that three-quarter degree field of view up to over 2.1 degrees. The Optec reducer does introduce severe field curvature and Optec does not advise using their product for visual observing. It is strictly a photographic accessory.

Users of Barlows and telecompressor also must remember one other thing. Any time you introduce a piece of glass into the light path, not all the light that enters that lens will come out the other side. Introducing any additional lens into the light path will reduce the brightness of the image at final focus. Given a choice of using a 20-mm ocular or a 40-mm with a Barlow to produce 100×, I would prefer to use the 20 mm. Telecompressor users should also be aware that if they are doing astrophotography or CCD imaging that using a shorter f-ratio also means shortening the range of critical focus, the precise range in which the focuser must be placed to produce maximum image sharpness on film or the CCD chip. This range, measured in microns, can be cut in half by a telecompressor. These two accessories can add great flexibility to your equipment box but just as you would not use a screwdriver on a nail, make sure you use the right tool for the right job. A Barlow costs around about $40. Telecompressors are more expensive, usually around $140.

Solar Filter

Viewing and photographing the Sun can be a very rewarding endeavor, but it is one that requires a great deal of caution. The surface brightness of the Sun at Earth's distance is equal to an eye-blowing 1.5 million candlepower per square inch. That much light will very quickly destroy the retina of the eye if you even glimpse at it for any length of time. Imagine that light amplified many times by the size of a telescope objective. To safely view the Sun, you must have a properly designed solar filter. There are two ways in which this type of filter can be added to the telescope. You can use either an aperture filter, which fits over the front end to the telescope, eliminating more than 99% of the Sun's surface brightness and total light, or an eyepiece filter. This is a filter that screws into the open end of the ocular barrel. The only type of filter that you should consider using is the aperture filter. The reason why is because this filter is not required to absorb sunlight that has been amplified by the telescope. An eyepiece filter must absorb the sunlight amplified many times by the telescope objective. This can cause the filter to become extremely hot and under that kind of heat the glass element may crack or shatter while you're looking through it leading to disaster.

Aperture filters come in different types. The simplest type is a full-aperture filter that filters light passing through the full width of the objective. Most filters utilize a neutral density glass that renders the Sun in its natural yellow color. Because the Sun is the brightest object in the sky, a large aperture is not necessary to produce

Figure 2.5. Mylar solar filter. Photograph by author.

a bright, crisp image. Celestron offers a 3-inch off-axis filter that fits over the 8-inch objective of the C8. This reduces the total amount of light reaching the eye-piece providing a more comfortable view. A disadvantage of this is that it will limit the amount of magnification you can use. Another type of filter that many people use now is a Mylar filter. This type of filter uses two sheets of Mylar each of which is aluminized on a single side. The two Mylar sheets are laid over each other with the aluminum on the inside, protecting it from scratches. These types of filters produce sharp images of the Sun at a fraction of the cost of their glass counter-parts. If you're considering a Mylar filter, don't be turned off by your first look at one. The aluminized Mylar looks like lightly crumpled aluminum foil leaving one to wonder how this could ever transmit a sharp image. Believe me, it does. Mylar filters do have one disadvantage. The aluminized Mylar strongly absorbs the red end of the spectrum rendering images of the Sun that are nearly powder blue in color. When I work with solar images, I always convert them to grayscale anyway so this does not bother me all that much. The Sun has been burning with the same color for all of human history and that value is very precisely known. The exact color that you see in the scope has no scientific importance at all, but if the aes-thetic value of a yellow Sun is important to you then you should avoid a Mylar filter. Solar filters run in price from about $80 and up in today's market.

35-mm SLR Camera

Capturing pictures of what you see in your telescope is one of the most rewarding and at the same time challenging tasks you can undertake in astronomy. In order to accomplish this you need a 35-mm SLR camera with a removable lens. A ring then is attached to the camera that will allow the camera to be mated to an adapter that directly threads onto the rear cell of the telescope in place of the visual back. The telescope then becomes the camera lens. This type of camera is the only type of camera that allows you to view directly through the light path rather than

Figure 2.6. 35-mm camera and lenses. Photograph by author.

through a parallel finder. This will allow you to directly sight and focus the target. Prior to reaching the film, the light is deflected by two 45-degree mirrors to the viewfinder. When the shutter is tripped, the first mirror is retracted out of the way either electrically or mechanically. This allows the light to reach the film when the shutter is opened. When the exposure is completed, the shutter closes and the mirror extends to its normal position.

Another important benefit of this type of camera is that it allows you to select various exposure time lengths. A typical setup will allow automatic exposure times from as fast as 1/500 of a second to one full second. Of even greater importance is the bulb (B) setting. This allows the shutter to be manually held open for as long as the user desires. This is critical since most deep sky photographs are taken at the full focal length of the telescope (in my case, that's f/10). Taking an exposure of a ninth-magnitude nebula might take as long as an hour.

When buying a camera for astrophotography, buy as much quality as you can afford. If you buy a good camera, you will only have to do it once. Manual cameras are relatively inexpensive and have the advantage of not needing batteries that can die on you in extreme cold. Electric cameras operate much more smoothly and precisely. The quality of the lens of your camera is important since you will not always use it to image through your telescope. You may wish to mount the camera on top of the telescope "piggyback" style. This allows a clock-driven telescope to serve as mount for beautiful wide-field panoramic shots. Also consider a camera that offers an array of telephoto lenses. This allows modest magnifications to be used while retaining a wide field view. The only other type of accessory that is an absolute must is a cable release for the shutter. This allows the shutter to be opened without you touching the camera and thus creating vibrations. This can either be a manual type of release with a spring or an air-loaded type. The air-release is smoother, but also more expensive. If you choose to settle for a spring cable release for your camera, make sure you use the camera's delay timer if you are not using the "B" setting. This will allow any vibrations created by the spring to damp out before the shutter trips. New cameras can cost $500 and up with just a single lens. When camera shopping, a trip to an Internet auction site is a great idea to get good-quality equipment at a cheap price. I recently bought a used Minolta and three lenses for under $100 in mint condition.

CCD Camera

The charge-coupled device (CCD) camera has been a part of astronomy for decades but until recently has been an exclusive province of professionals. The cameras were large, power-hungry and incredibly expensive. During the mid-1990s, advances in electronics allowed CCD chips to be produced commercially on a large scale. This advance has made possible the digital imaging revolution that so many of us take advantage of today. Your camcorder, the camera in your cell phone and your digital camera all use CCD technology. The CCD chip is a tiny silicon wafer covered with an array of light-sensitive detectors called "pixels" (techno-speak for "picture elements"). The pixels collect photons and register their brightness for assembly into a completed image.

Figure 2.7. Meade Deep Sky Imager CCD Camera. Photograph by author.

One of the great advantages of the CCD chip is that it is far more light-sensitive than is film. An image that would take an hour to make on film can be recorded in just a few minutes or even seconds on a CCD. A CCD camera for astronomy is something of a different beast than those used for conventional photography. The CCD chip, like photographic film, does a much better job recording light when exposed to high levels of light for a brief period of time. Astrophotography involves exactly the opposite, exposing the chip or film to a very minute level of light for very long periods of time. Fortunately the speed at which the chip collects light improves dramatically as you reduce temperature. An astronomical CCD camera is equipped with a cooling system that chills the chip to a temperature well below zero degrees Celsius. The heat generated by the camera can register in the CCD pixels creating what is called "dark current," or false light. Cooling the chip to sub-freezing temperatures reduces the amount of dark current.

Another feature that increases light sensitivity is sacrificing color. Unless you have observatory class aperture, there is very little to be gained by using color sensors so most CCD cameras record light in shades of gray ranging between the extremes of white and black. In an 8-bit camera, CCD pixels measure shades of gray and report that data to a computer using a scale of 0 to 255. The "zero" value represents pure black while "255" represents pure white. Anything in between cor-responds to a shade of gray. Moving up the cost scale, cameras are available with 16-bit capability registering light on a scale of zero to 65,536. These cameras will produce images that are far smoother than the 8-bit camera with no large unnat-ural transitions between pixels. The image quality of the 16-bit camera is every bit as good as those taken on photographic emulsions. Eight-bit imagers tend to be somewhat grainy but if saving money is an issue, they are satisfactory because the human eye can only discern about forty shades of gray anyway.

Like with the 35-mm camera, buy as much quality as you can afford. That assures that you will only have to do it once. My first camera was the Meade Pictor 216XT. This is a 16-bit camera with a CCD pixel array measuring 335 elements by 256 ele-ments. It produces images that are super-smooth and when properly focused, nearly indistinguishable from a photographic print. It also is adaptable to color photography by installing a tricolor filter wheel. In this technique, separate exposures are obtained through red, green and blue filters, then stacked one on top of the other. This produces an image in what is called "RGB" color. I have not yet opted to do this because the filter wheel assembly is as costly as the camera itself. The Pictor 216 XT was also, at the time I purchased it, the only 16-bit imager available for under $1,000. Several smaller manufacturers are now also under that level with 16-bit capability and now even with color detectors. I have not had the opportunity to try out any of these cameras, but I was happy enough with my Meade that I eventually purchased their new Deep Sky Imager for color work. By the way, since all cameras use either the universal T-thread or standardized 1.25-inch draw tubes, you can mix and match telescopes and cameras. Just because you own a Meade telescope, you are not beholden to the Pictor line of CCD imagers. You can use a Santa Barbara Instruments Group camera and it will work just fine. My Meade CCD gets along just fine with my Celestron telescope since the interfaces are of uniform size. I cannot take advantage of some of the features in the Meade camera that are specifically designed for use with Meade's LX200 line of telescopes such as autofocus. Today I am using Meade's Deep Sky Imager, which allows deep sky images to be taken in color. It has a larger chip (wider field) than the Pictor 216 XT and costs less than half what the Pictor did in 1998.

CCD cameras have disadvantages as well. Each image must be carefully calibrated. This involves subtracting what is called a "dark field" and a "flat field." These techniques correct for any heat on the CCD that can generate false light and for any field curvature induced by the telescope. Taking one picture therefore involves actually taking three. You must first take a dark frame (usually done with the telescopes cover on) then a flat field frame (done with a white card over the telescope objective) prior to taking the actual picture. The two calibration frames are then deducted from the actual image to produce proper views. Removing this unwanted clutter is referred to as maximizing the "signal to noise ratio." Remember that signal is good and noise is bad. Compounding the prospective photographer's task is that the size of the chip is extremely small, only a few millimeters across. This means that at f/10, the field of view of the chip of the Pictor 216XT with my Celestron is a miniscule 5.5' × 4.1'. Using a telecompressor gets the field up to 8.7' × 6.5'. Remember that CCD imaging is an extremely difficult and demanding operation, so consider adding this to your equipment box only when your skills warrant it. Otherwise, you will spend a great deal of money only to get frustrated. Get good with camera and film then step up to a CCD imager. In today's market, CCD imagers start at about $150 for what is basically a web camera and can go up over $10,000.

Camera and CCD Mounts and Accessories

To get the most out of your camera equipment, it is necessary to have the proper supporting equipment. The devices used for 35-mm photography and for CCD imaging are very different from each other so we'll discuss them separately. In photography, 35-mm cameras cannot be directly connected to a telescope so adapter equipment will be necessary in order to make the mechanical connection. The minimum that you will need is a T-ring and a T-adapter. The T-ring is a metal ring with two threads. One screws into your camera and provides the interface for the T-adapter which screws into the other end. At the far end of the T-adapter is a universal thread ring that will mate to the rear cell of a Schmidt–Cassegrain telescope in place of its visual back. These two devices allow the camera to be mounted at the prime focus of the telescope and allow for photography at the telescopes normal f-ratio (for me, f/10). Photography using this type of setup is therefore called *prime focus* imaging. This type of setup allows nice, short-exposure images of the entire Moon or Sun, reasonable length images of star clusters or other bright objects in the deep sky and the shortest possible exposure time for any other type of object, unless a telecompressor is used. For imaging the planets or small objects, the magnification is generally inadequate to show any detail. For small targets, a different technique must be employed. We can gain extra magnification by introducing an eyepiece into the light path. This will require moving the camera several inches farther away from the telescope using an empty metal tube called a "tele-extender." After inserting an eyepiece directly into the visual back of the telescope, the tele-extender then screws over the outside of the visual back. The other end of the tele-extender screws directly into the T-ring on the camera. This has the effect of dramatically increasing the focal ratio of the telescope, sometimes into triple digits! Tele-extension photography can be extremely challenging because with so much power, it can be difficult to focus the telescope and like with the CCD imager, field of view is agonizingly small. This can make it difficult to even find your target, never mind focus on it and shoot. If you get good at it though, the results are well worth it.

Long-exposure astrophotography requires that the telescope be carefully guided throughout all the time the shutter is opened. If the target is allowed to drift, the result will be a badly streaked image. Modern clock drives are very good at tracking objects, but no motor drive is perfect so you will always need to correct in right ascension. With an extremely accurate polar alignment and a perfectly level telescope, you can eliminate the need to correct in declination. Still, very few of us are that good or that lucky so some minor declination correction will almost always be required as well. To make these corrections, you will first need a control device. This is a simple four-button (up, down, left, right) hand-held box that connects via a telephone-style jack to your drive base. It controls the right ascension correction by speeding up or slowing down the motor in response to your commands. For declination corrections, commands are sent to an external motor that inputs the north–south correction. The motor, normally purchased separately, connects to a power port on the telescope drive base. Using these motors draws a lot of power from your telescope battery, so be prepared to plug into either a larger DC battery or some source of AC power with a DC inverter. Most telescopes come with equipment to allow you to do the latter while DC/DC adapters (to plug into your car's cigarette lighter) are readily available.

Now that you have the means to make corrections, you need to be able to see what you are doing. With the camera shutter open, there is no means to look through the telescope to see the target. There are two ways around this problem. The old fashioned way is to mount a guide telescope on top of the main telescope.

This is usually a 2–3 inch refractor with a high-power eyepiece. The guide scope is mounted precisely in line with the main telescope so that by keeping the target object centered in the guide scope, you also keep it centered in the main scope. The advantage of this apparatus is that you can use the target itself to guide the telescope. This is a huge advantage if you are imaging a moving object, such as a close-by comet. The problem with the guide telescope is that if you are photographing a faint deep-sky object in your main telescope, it may prove too faint to see in the guide scope.

Most owners of Schmidt–Cassegrain telescopes prefer to use an off-axis guider for controlling the telescope. The off-axis guider connects directly to the rear cell of the telescope. Most of the light passes directly through the guider body to the camera. A small amount of light is redirected by a prism to an eyepiece at a 90-degree angle to the light path. This allows the observer to see a small amount of the field of view and pick a target to guide on. The advantage of the off-axis guider is that you can take advantage of the full light gathering power of your telescope to find a guide star. The problem with this system is that you cannot see the actual object you are trying to photograph. The prism does not extend that far into the light path. You must aim and focus the camera carefully, then hope that the off-axis guider is able to view a star onto which you can focus, center and guide the telescope.

A CCD camera adds to the aim, guide and focus difficulties because you cannot view through it at all. Attaining a proper focus is a difficult process of trial and error. This is very frustrating because image quality is very dependent upon a very precise focus. The range of focus position is literally only a few hundred *microns* wide. To help aim and focus, many amateurs now use a device called a "flip mirror." The flip mirror uses a fixed mirror that directs light to an eyepiece that allows the viewer to see exactly what the camera will see. When the target is centered in the eyepiece, the mirror is then retracted out of the way allowing the light to go to the camera. Once the camera is focused once, you can bring your eyepiece to focus. Then all you will need to do is center and focus in the flip mirror, and you will have a point and shoot CCD setup. The flip mirror can also be used for 35-mm photography provided it is large enough to allow the entire field of view to reach the camera. If the field of view is smaller than the camera frame (due to the obstruction of the flip mirror) then you will see the circular edge of the field of view on the frame, a phenomenon called *vignetting*. I use the larger of two flip mirror models offered by Meade. Some more modern CCD cameras will allow you to view real time video. This makes aiming and focusing easier, but it only works with the brightest of objects.

When you buy any photographic accessory, make absolutely certain that it is of sound construction so as to avoid any mechanical flexing caused by the weight of the camera. If the camera or CCD is allowed to bend the mount by even a few millimeters, then the light path will no longer fall into the center of the camera's field of view. This was a key selling point in selecting the Meade flip mirror. Other devices that I tried caused the camera to flex slightly off the telescope's optical centerline. T-rings and adapters and tele-extenders are less than $40 each but the flip mirror and illuminated reticule eyepiece can set you back close to $400 if you go for larger models.

Color Filters

When you view through your telescope, you are viewing the full range of the visible light spectrum. There are times however when seeing all the light is not necessarily the best thing. Certain features stand out when certain wavelengths of light are eliminated. Viewing Mars through a blue filter brings out details in the planet's atmosphere and clouds. This occurs because most of the planet's dominant red color cannot pass through the filter. A yellow filter will emphasize the presence of dust in the atmosphere. Mars's surface is best in a red or orange filter. An orange filter can enhance subtle differences in the atmospheres of Jupiter or Saturn. A neutral density filter reduces the brilliant glare of the Moon or Venus by simply reducing the surface brightness and total light. When less than about 17 arc seconds in diameter or close to the horizon, the disk of Venus can be nearly impossible to detect because it is lost in its own glare.

Filters come in two types for Schmidt–Cassegrain telescopes. Eyepiece filters screw directly into the eyepiece barrel. Make certain that when you buy eyepieces that they have threaded barrels; some cheaper units may not be threaded. These are simple filters that are easy to use. Eyepiece filters are also useful for tele-extension photography since an eyepiece is always involved. For prime focus photography, you will need a drop-in filter. These filters are inserted into the telescope's visual back before it is connected to the scope. The drop-in filter will also work for visual applications.

A set of six filters can typically be purchased for a modest price from astronomy equipment retailers for as little as $75. If you intend to do both visual and photographic observations, the drop-in set is probably best for you since the filter can be "dropped in" to a visual back, T-adapter or off-axis guider. A typical set will include red, blue, green, orange, yellow and a neutral density filter. If your telescope is a refractor or Newtonian reflector, you can only use eyepiece filters.

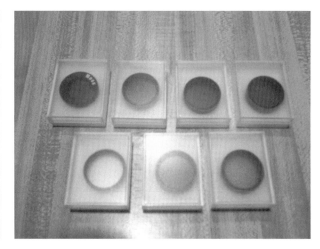

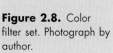

Figure 2.8. Color filter set. Photograph by author.

Spectroscope

At first glance this sounds as though I'm suggesting something as expensive as a luxury car. Some time ago that might have been true but not today. Spectroscopes are now available for amateur telescopes that thread into standard 1.25-inch eyepiece barrels for as little as $150. For many of the great mysteries of the heavens, the key to unlocking their secrets lies in their light. The light given off by a star, nebula or galaxy, split into the rainbow, shows us a great deal about what they are made of and what they are about.

A spectroscope for an amateur telescope consists of two parts, first a grating cell that threads into the barrel of your eyepiece. This is the actual prism-like device that splits the light of the star into its constituent components. The second is a projection screen that fits over the viewing end of the eyepiece and allows you to view the spectrum of the target. Some models will also allow you to thread the grating screen into the barrel of your CCD camera's 1.25-inch nose piece and give you the capability of imaging the spectra of the target. An imaging spectroscope is more properly called a *spectrograph*.

When you view the spectra of a star, you will notice that it is interrupted by bright lines and dark lines. The bright lines are called *emission lines*. When an electron loses energy and falls closer to the nucleus of its atom, it gives off a photon creating the bright line in the spectrum. When an electron gains energy, it absorbs a photon, creating a dark line called an *absorption line*. In later chapters, we'll talk about what these lines mean about the stars we seek to view.

Carrying Case

When you buy a telescope, it comes with the minimum accessories needed for you to begin operating. As months and years go by, your equipment stash grows by leaps and bounds as you have seen. Initially you tend to pile all of it into the same small cardboard box that your original equipment came in. I always stored my eyepieces and other equipment in the small boxes that they were shipped in and put those boxes inside the larger cardboard box that Celestron originally shipped my accessories in. Over the years, I've added two more eyepieces, a T-ring, T-adapter, tele-extender, off-axis guider, CCD camera and then a flip mirror assembly. So now I need something larger to carry all my stuff that will keep it all organized and at the same time protect it while in transport to the observing site.

What you need when you get to this stage is a large, sturdily built foam-lined carrying case for all of your equipment. A typical large-size equipment box is made of aluminum and will have prefabricated openings provided for standard-sized items such as eyepieces and star diagonals. Some may then offer larger openings for things like camera bodies and support equipment. A feature that is appearing in many cases today is the so-called "pluck-foam" case. This is a case that has no pre-cut openings in it. The interior is composed of hundreds of tiny squares of foam that you can "pluck" out of the case to create custom openings for your specific equipment needs. Whether you go with pre-cut or pluck foam, go with the largest case you can afford. They are designed to be portable so don't worry about getting something so big that you won't be able to move it. For example, the largest case offered by Orion in their aluminum line measures about 17 inches across by 12 inches wide by 5.5 inches deep and weighs less than ten pounds empty. If aluminum cases run a little steep on your budget, you may want to consider polyethylene (plastic) cases. These run cheaper, but remember that you get what you pay for. Think about that before picking a case to carry your expensive eyepiece collection around in. Camera stores sell these but usually at exorbitant prices. Astronomy houses offer better deals on large high-quality cases for as little as $45.

Laptop Computer

For the first time, you're probably asking why do I really need this before bothering to read why. A portable computer can do as many as three critical things for you. First, it can run powerful astronomy software in the field. Having a program such as *RedShift, Starry Night*, or *The Sky* can almost completely eliminate the need to carry paper star charts with you while you are out. The computer screen is also much easier to read in the dark. Using a quality program will allow you to see the stars on the screen exactly as they appear in the telescope. If something appears in the telescope that is not on the computer screen, you may instantly become famous. Secondly, some of these programs provide for telescope control as well if your scope's drive is compatible (my telescope predates Windows-based computers now). If you're using a Meade LX200 telescope, a laptop with Meade's Epoch IP2000 can drive the telescope with arc second accuracy to any target in the heavens. Some third-party software such as MaxIm DL can interface with many different types of telescopes. Both of these software programs can also control and download data from CCD cameras. If you have a CCD camera, then a laptop is an absolute must. Meade's Pictor line of CCD imagers comes with a Windows program called Pictorview XT, which serves to provide control of the camera, image downloading

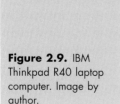

Figure 2.9. IBM Thinkpad R40 laptop computer. Image by author.

through either a serial or SCSI interface and also provides a limited range of image processing functions. The Deep Sky Imager uses Meade's AutoStar control suite. This software almost completely automates the process of acquiring and stacking images and takes a lot of the guesswork out of CCD imaging.

The computer you buy need not necessarily be the newest thing on the market. Unlike telescopes, computers become obsolete at speeds that make automobiles sound like a good investment. That means the price of a six-month old laptop can be less than two-thirds of what it was when it was new. Pay close attention to clearance sales as old models are moved out at stupid prices to make way for newer, faster models. Also pay close attention to brand name. The best quality equipment on the market today is in my experience the IBM Thinkpad series. The next thing is to make sure that your computer has the Big Three in abundance, or at least as much abundance as you can afford. These are processor speed, RAM, and hard drive space. Most modern programs require a high-speed processor to run them since astronomy programs perform millions of calculations per second. This is especially so if you are running an animation of an event such as an eclipse or occultation. Nearly all new computers today run at speeds over 2,000 MHz. If you are looking at something used, set a hard floor at about the 500 MHz level. This is the minimum speed you will require to run the most modern programs along with Windows 2000 or XP.

RAM (random access memory) is where program information is stored as you are working on it. You must have enough of it to run both your operating system and your application. If not, then the speed of the computer will be greatly diminished because it will try to make up the shortfall in RAM by creating what is called "virtual memory." This involves using part of the computer's hard disk as additional RAM. Virtual memory is extremely slow compared to RAM and will badly degrade performance if you have to begin using it. RAM is fairly cheap to buy and easy to install yourself (one of the few things you can upgrade on a laptop by yourself). Most new computers ship with 64 MB of RAM on board and are upgradeable to either 128 or 256 MB. RAM also comes in several different types that are not interchangeable. Make sure you buy what's right for your computer.

Hard drive space must be plentiful to store both large image processing and telescope control programs, astronomy and charting programs and to store your images. Image files can be very large all on their own, especially if you choose to save your data in the GIF, TIF or FTS formats. JPG files are smaller, but suffer some loss of image quality due to data loss. Do not buy any computer that has less than five gigabytes of hard drive space. This will be the minimum required to run a late edition of Windows plus a suite of astronomy, telescope control and image processing software, plus provide a margin of safety should it become necessary to employ any virtual memory. And to be completely honest, you will want to put other non-astronomy programs on it, if only to justify the expense to the IRS.

Finally make absolutely certain that you won't run out of power in the field. Make sure that if your computer manufacturer offers dual batteries or a heavy-duty battery that you buy it. If not, make certain that you can operate near an AC power source or buy an optional DC (cigarette lighter) power adapter. Most standard laptop batteries will not last more than an hour especially if you are running a CD drive with any of your software or if you have a CCD (Meade DSI or LPI) that draws power from your computer.

My first computer for field use was IBM's Thinkpad X20. This is a super-slim computer measuring barely one inch thick with the screen closed and weighs less than four pounds. I purchased it just as the X20 was being replaced by the more powerful X21 so I got it very inexpensively. The computer itself cost about $750 online. Being so thin, it does not have any internal floppy or CD-ROM drives, so I added a docking station to it that provided ports for those items as well as a serial port. The computer has a Intel Celeron processor running at 550 Mhz, a 5.4 Gb hard drive and has 64 MB RAM. It came with Windows 98 SE installed, which I later "upgraded" to Windows Me. The 12.1 inch screen provides only 800×600 maximum resolution, but that is more than plenty to support my needs. The only thing I needed to add after that was that DC adapter so I could run the computer from my car or portable power adapter. After losing the X-20 in a burglary, I replaced it with a Thinkpad R40 that runs Windows XP and has a far superior display. For those amateurs on a tight budget, a laptop seems to be a luxury of enormous expense, but once you have it and begin using it, it makes your time much more productive and enables you to do many more things than you could without it.

Portable DC Power Supply

Telescopes with drive motors, and CCD cameras and computers require power and lots of it. Telescopes and computers have internal batteries but not with enough amperage to last for a very long time under high electrical loads. As we have just mentioned, a laptop operating under standard batteries will not last much longer than two hours. Your duration will be even less if you are operating in extreme cold.

To be able to operate your equipment reliably during the night you will have to have an independent power supply. For many people, this can be your car battery. Many cars now come with a second 12-volt DC power plug in the trunk for operating things like computers, camping equipment and telescopes. I wanted to be able

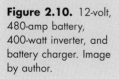

Figure 2.10. 12-volt, 480-amp battery, 400-watt inverter, and battery charger. Image by author.

to operate without my car, so I purchased a simple discount automobile battery. The battery supplies 480 cold-cranking amps of DC power, which is plenty to run my scope, CCD camera and laptop all night long on one charge. The telescope draws AC power from an inverter that is supplied with DC input from the battery. The CCD has no power source of its own and the computer battery is only good for a very short period of time. That time will be even shorter if the CCD needs to draw power from the computer USB port. So I use my computer's AC power adapter and plug it into the inverter.

Whether it's your car battery or something more portable, don't get caught without power. When you've got that once in a lifetime picture to take, there is nothing more disheartening than that flashing red light signaling the end of your power. You can buy a discount priced 12-volt car battery and a 400-watt inverter with battery clips for under $70 to supply all your AC power needs in the field. I would never trust such a battery to start my car, but it is more than adequate for field work. A charger to top off that battery from household current also costs about $30.

Star Charts

Just as you would not drive without a road map, so you should not try to navigate the skies without sky charts. Good sky charts are made all the more necessary by the fact that many of the objects you might be looking for, you will not be able to see with the unaided eye. Imagine trying to find an invisible road without a map. Or even with it. A good set of charts will be an invaluable guide to the treasures of the heavens.

You can either buy the old fashioned charts, on paper, or you can buy computer software for that laptop. Perhaps the best chart on paper to start out with is *Sky Atlas 2000.0* produced by Wil Tirion. This volume was the first set of charts produced for the 2000 epoch (charts are revised each fifty years to account for precession of Earth's axis and any proper motion of the stars). *Sky Atlas 2000.0* is available either in a large format version for use indoors or a smaller laminated field edition. The larger charts are clear and easy to read while the field version is more portable and will not get soggy in dew. This atlas displays the positions of over 10,000 stars down to eight magnitude, all the objects in the Messier catalog and dozens of other objects from the NGC catalog. If you are more advanced, consider an even more in-depth atlas such as the *Cambridge Star Atlas* or *Uraniometria 2000.0*. I currently own the *Sky Atlas 2000.0* Deluxe edition. Its large charts are easy to read and are ideal for viewing large sections of sky for planning purposes. The unlaminated version currently sells for about $60 while laminated charts cost about $120.

For outdoor purposes, I prefer something that can run on my laptop such as *RedShift 4*. My laptop is capable of multitasking both this program and the current version of Meade Autostar Suite in the Windows XP environment. I can point and click my way around the sky quickly and easily. The program interface is friendly (at least for me) and allows me to customize the view so that I can see the sky as it appears to the unaided eye (stars to magnitude +4.5 in northern Delaware) or through the telescope (stars to about magnitude +11 on a moonless night with clear conditions). The market is currently well saturated with astronomy software, most of which enjoys excellent reviews including *The Sky*, *Starry Night* and many others. Some software publishers offer various versions of their programs with differing capabilities and price tags. *Starry Night* is offered in *Beginner* ($34.95), *Backyard* ($58.95) and *Pro* ($149.95) editions so there is both a range of capability and price to suit just about anyone.

Personal Equipment

Remember that the most important piece of observing equipment you own is your own eyes. Keeping yourself comfortable in the field is critically important. If you are uncomfortable while you are observing then you will not be fully focused on what it is you are doing. For winter operations it is critical to keep warm without being too badly bogged down by heavy clothing and the like. Light down jackets keep one toasty warm without being bulky. If you must wear gloves (and sometimes you must) find gloves that are tight fitting and allow you a good feel while you are manipulating tiny eyepieces or other components. It would be a tragedy to drop that brand-new Nagler eyepiece and shatter it before it ever got into your telescope's drawtube because your were wearing thick bulky gloves. But at least your hands are warm! Remember that more than 90% of your body's heat is radiated away through your head. Keep it covered. A wool ski cap is excellent protection and keeps you comfortable. Down at the other end, keep those feet warm too. My grandmother once said that if your feet are cold, then everything is cold no matter what. Since your feet don't handle anything, use the heaviest sock you can possibly fit into your shoe and keep those tootsies nice and toasty. Personal heat packs are also an excellent idea. A little Hot Pocket can keep your hands and fingers comfortable for several hours not to mention prevent frostbite on the coldest nights.

Warm nights of course require light clothing such as shorts and t-shirts. Remember that there is one hazard associated with observing at night that is not present during the day . . . insects! The flying kind in particular can make life absolutely miserable for everyone and they seem to have more of an affinity for some people than for others. Personally once temperatures get above room temperature, I start sweating something awful so I usually carry a sweat towel. Mosquito repellant is an absolute must for summertime work. And don't forget a jacket. On a crystal clear night, if it is 80 degrees Fahrenheit at sundown, the temperature might be in the upper 50s by the onset of morning twilight if you've been out all night.

Don't forget food either. Carry plenty of liquids on a warm night to keep properly hydrated. Snacks are also good. A deli sandwich or something that can be kept cool will prevent hunger pangs from ruining an observing session. Stop at a Subway on the way out and grab something that will keep in the car for several hours in case you need it. Light juices are also good for refreshment and a bit of a sugar boost. Avoid sodas or other beverages with caffeine. Caffeine is a diuretic that will cause frequent trips to the bathroom, a major inconvenience when there is no bathroom available. Beer will do much the same in addition to the damaging effects of alcohol on your cognitive abilities and the physical functioning of your eyes (remember histotoxic hypoxia).

Setting Up Shop

Now that you have all this equipment together, please remember one thing. A fitness instructor gave me this advice about joining a gym a long time ago and it applies to astronomy or any other hobby as well. If it is frustrating or time consuming to get yourself and your equipment to the site, be it a gym for a workout or to your observing site with your telescope and equipment, you're not going to do it very often. It needs to be easy. For me, my eyepieces, CCD equipment and adapters are in one carrying case; my tripod is in a second box and the telescope in a third. I can get all of my equipment out of the house in only three trips consuming less that ten minutes. Setting up the Celestron is even easier. The tripod assembly requires only a few seconds and mounting and aligning the telescope is a breeze using my polar alignment finder, a feature that no decent telescope is sold without these days. From the time I first began moving to set up to observe to the time the telescope points at the first target is about fifteen minutes. If I am going to do serious photography, then a fourth trip will be required to bring out the power supply and the laptop. I will take more time to carefully polar align because precision will be critical. From first motion off the couch to first trip of the shutter consumes about thirty minutes. We'll discuss some of the finer details of getting organized, assembled and aligned in Chapter 4.

Make certain that you always have a flashlight that has both red and clear lenses. Use the red while observing to minimize degradation of your night vision and keep white light available as an option in case you need to fix a flat tire. Speaking of such emergencies two other great things to have are a cell phone and someone who knows where you are in case you get stuck someplace.

Observing Projects II – Comparison Shopping at a Star Party

Observing Project 2A – Commercially Produced Reflectors and Refractors

For the amateur seeking to find out exactly what is the right type of telescope for him to buy, there is no better place to comparison shop then the nearest star party. Here you will find telescopes of all sizes and shapes on display. Here what we will do is take a look at the differences between two common types of commercially produced telescopes, a refractor and a classic Newtonian reflector.

It should not be too hard at a well-attended star party to find these very common types of telescopes. The reflectors generally draw the larger crowds because they tend to be larger. The smaller refractors don't gather as much light and do not draw the same "oooohs" and "ahhhhhs" as the reflectors do.

Remember the things you read earlier on in this chapter about the strengths and weaknesses of both designs and evaluate them carefully against what your needs are. If you are looking for size and have no need of portability, then browse amongst the monster reflectors. Pay close attention to the quality of the image over the entire field of view. It is easy to just look at the middle of the field and be impressed. Make certain that there is not an unacceptable amount of "coma" or streaking of stars at the outer edge of the field of view. Spherical aberration causes this unpleasant effect and it is particularly evident in short focal-length telescopes. Also watch for how difficult it can be to view through the telescopes. Many reflectors require their owners to perform anatomically questionable acts to reach the eyepiece. Is he contorted into a pretzel? Does he need a ladder?

If reliability and simplicity is key, then browse amongst the people who have the refractors. What will strike you here is the sharper quality of the images. What you want to watch for here is the degree of color distortion created by the objective lens. This is the chromatic aberration that we discussed earlier in this chapter. If bright color fringes surround objects viewed in the scope, then the telescope's quality is likely suspect. Only the most expensive refractors are free of this effect, the so-called "fluorite" scopes, but it should not be excessive in any scope. Also remember that refractors may be relatively cheap and therefore come equipped with cheap accessories. Do not consider any scope for purchase that will not accept 1.25-inch eyepieces. All serious accessories are of this size and the most serious are of the 2-inch variety. These are likely beyond your ability and cost ceilings at this time anyway. But don't look at less than 1.25-inch eyepieces on any scope.

Remember to take that list with you to the star party and examine scopes that meet your needs, but since you're not spending money here make sure you take a good look through other types too. You may get a surprise and discover something that you want that was not on your original list. For example you may discover a love for computer driven telescopes that relieve you of the burden of having to navigate your own way around the heavens.

Observing Project 2B – Homebuilts and Dobsonians

If you are handy and enjoy building things yourself, you might find out something surprising. An amateur telescope maker with the right equipment and enough determination and patience can actually build a telescope of superior quality to what commercial telescope producers can make. Meade and Celestron will mix parts together to produce a telescope that will please most consumers but might still have minute optical errors. But the best telescopes are built not in major factories, but in people's basements.

Good astronomy clubs will provide resources on how to do this work and provide support to amateur telescope makers (ATMs). Delmarva Stargazers of Smyrna, DE has an annual mirror making seminar each year that is well attended from all over the northeast and even attracts attendees from as far away as the maritime provinces of Canada.

Figure 2.11. A large ATM-built Dobsonian at the Delmarva Stargazers star gaze star party. Image courtesy Delmarva Stargazers Yahoo Group.

The most popular type of telescope for a homebuilder to produce is the "Dobsonian" so named for the man who invented the box-shaped alt-azimuth mount that made it famous. Amateurs often call the Dobsonian scope a "light-bucket" because they offer huge apertures for very modest prices. They are light in weight, simple to operate and maintain and can offer superior quality for less money than commercial designs. Take a look through one of these and see for yourself.

Observing Project 2C – Wide Field Telescopes and Binoculars

Though I have talked up the importance of aperture, aperture, and aperture to this point, there are few things that are better to supplement your observing than a good wide field telescope or a pair of quality binoculars. Two popular brands of wide field telescopes are Bushnell's Voyager and the Astroscan. Both of these telescopes are designed with a short focal-length primary mirror of about 4.5-inch aperture. The Astroscan uses a primary mirror that has a parabolic curvature to it that helps reduce the exaggerated effects of spherical aberration and coma that occur when using such short focal lengths with mirrors. The scope is built into a red hard plastic case with a spherical bulb at the base. This bulb allows the telescope to be rotated in any direction on a cast iron tabletop base. The Astroscan is the ultimate in medium aperture portability. The Voyager, which is based on a design similar to Astroscan, uses a spherical mirror and so produces more curvature of field. But it is more modestly priced than Astroscan and makes a nice addition to your toolbox. Either scope can simply be pulled out and plopped on the table allowing observing to begin in minutes.

Binoculars also offer the advantage of portability while providing the image quality of a good refractor if they are well made. You also get to use both eyes, which can lend more of a sense of depth and a three-dimensional feel to your viewing. You will find many differing types of these at the star party, including some that are as large as 100 mm (4 inches). Many are sufficiently large so as to require a tripod of their own for satisfactory viewing. The thing to beware of with binoculars is that they come from many different manufacturers from many different parts of the world and can be of very different levels of quality. Try some out and you may find that a pair of binoculars just might make the perfect second telescope.

Observing Project 2D – Eyepieces and Other Accessories

The utility of a telescope can be greatly increased by the wide availability of accessories for it. As you make your way around the star party, take note of what the various amateurs around you are sticking in the eyepiece holders of their equipment. What they are using will largely depend on what they are viewing. Planetary watchers are more likely to have Plossl design eyepieces in their telescopes. Those observers watching the wonders of the deep sky will be using wide field designs like Erfles. Amateurs using the largest of scopes may well have the huge 2-inch eyepieces in their telescopes. Note that many of these telescopes need to be counterweighted to keep from tipping over with these huge oculars attached.

Remember that having a Barlow lens effectively doubles the number of eyepieces at your disposal. You will find several different types of lenses here. Most manufacturers of commercial telescope equipment offer at least one type of Barlow. The 2× type is the most common, but both Celestron and Meade offer 3× Barlow lenses as well. Orion makes a Barlow called the "Shorty" which as its name implies is very compact.

Most serious observers will not be caught dead without a good filter set. Most will have at least a basic set of red, blue, yellow, green and orange filters for planetary viewing. Most will also have as part of a basic set, a neutral filter to suppress the dazzling light of the moon. More advanced amateurs will have a wider variety of colors available and may employ various types of light pollution rejection filters.

There are also many accessories out there that while they do not directly affect viewing through the telescope can go a long way towards enhancing the viewing experience. Most equatorial mount telescopes have a drive motor to allow the scope to track an object as Earth rotates. Motors require power and there are plenty of creative ways to get that in the field. Generators and batteries abound out there. If your scope has an AC motor, then users of DC batteries will have to acquire an inverter to create the right type of current. The more electrical stuff you have, the more power you need. Many creative amateurs have used multiple batteries and inverters to create setups that would make the electric company jealous.

Other electricity guzzling accessories might include such items as a GPS driven scope, illuminated controls and setting circles, electric dew caps or heaters. Creative amateurs can find as many ways to consume electricity as they can find to create it.

Observing Project 2E – Personal Comfort Items

Note that as you make your way around the star party as the night wears on, it will get progressively cooler through the evening. Note what the astronomers around you are wearing to keep warm. Some will carry multiple layers of clothing which can accumulate as the night wears on, then can be peeled off later on after the Sun comes up and things get warm again. Others may opt for items like Mylar space blankets, which are super-lightweight and fold into very tiny areas.

Warmth is one basic need and food is another one. Note how your new friends (and it will amaze you how friendly amateur astronomers can be, a fact that makes this native New Yorker a bit nervous) carry food and beverages with them. Thermos bottles keep the coffee warm or others may have the amps to spare for electric heaters. Coolers come in many different sizes and shapes for storing various quantities of food. The Delmarva Stargazers always have hot dogs on the grill along with hot coffee.

Don't forget medical needs, which will grow as we grow older. Make sure you have a place to carry any required medications. For myself, I don't have that issue yet, but during hay fever season I will not be caught outside without my non-prescription allergy meds, a bottle of Sinex and a bottle of Visine.

Remember that everything you propose to carry into the field you will have to move by yourself to and from your car, so make sure that whatever it is you have that you can get it to and from your observing site without it causing so much grief you get frustrated with the whole thing.

CHAPTER THREE

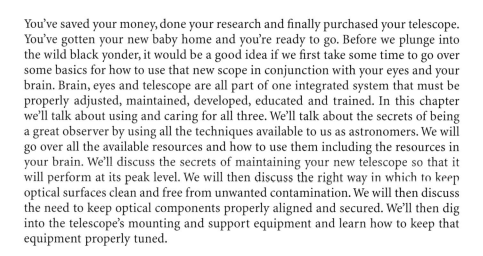

Putting the Integrated Observing System Together

You've saved your money, done your research and finally purchased your telescope. You've gotten your new baby home and you're ready to go. Before we plunge into the wild black yonder, it would be a good idea if we first take some time to go over some basics for how to use that new scope in conjunction with your eyes and your brain. Brain, eyes and telescope are all part of one integrated system that must be properly adjusted, maintained, developed, educated and trained. In this chapter we'll talk about using and caring for all three. We'll talk about the secrets of being a great observer by using all the techniques available to us as astronomers. We will go over all the available resources and how to use them including the resources in your brain. We'll discuss the secrets of maintaining your new telescope so that it will perform at its peak level. We will then discuss the right way in which to keep optical surfaces clean and free from unwanted contamination. We will then discuss the need to keep optical components properly aligned and secured. We'll then dig into the telescope's mounting and support equipment and learn how to keep that equipment properly tuned.

Training the Brain and Eyes

Remember that at the very beginning of this book I mentioned that one of the greatest misconceptions about astronomy is that it is a source of instant gratification. You cannot simply expect to look through the eyepiece of an amateur telescope, even a big one, or even an observatory class telescope and expect to see what shows up on those awesome photographs. All those beautiful galaxies that show up as swirling gorgeous pinwheels on photographs only show faint gray-green blobs in the eyepiece. So how can I see what is in those pictures?

Well the reality is that you cannot. The human eye is a wondrous instrument but it has limitations. Every image that falls on it is instantaneously transmitted to the brain without the opportunity to allow light to accumulate on the retina, so you can only see with what light is available in that instant. To see the detail that your telescope *can* show you, it's not just enough to mount eye to telescope. You must train the brain.

Your brain offers you several key tools that you can use to enhance the observing experience. The first is your *ability to learn*. You try something and it does not work. Do you try the same thing again? No, of course not! You try something else and get a different result. You try as many different variables on the same problem as you can think of until you create the result you are looking for. This is learning by *experience*. This can be frustrating over time because you may make an awful lot of mistakes before the proverbial light comes on. I can't begin to tell you how many rolls of film I've wasted over the years making bad exposures trying to figure it out. But in time I did. I've learned to make very beautiful wide field panoramas by properly adjusting for all the variables. You can also learn through *education*. You can train yourself by finding someone who knows how to do it and learning from them, you can save a lot of time and frustration, but you don't necessarily learn from your mistakes as much. Also be careful whom you choose as a teacher. Many legendary athletes were failures as coaches because they could not teach. Just because someone is a very good observer or astrophotographer does not mean that they are someone who is able to teach you anything at all. You should also always remember that if one really loves what he is doing then knowledge is both contagious and self-perpetuating. The more knowledge you *have*, the more you *want*. Seek out as much knowledge as you can about the objects you view. In the chapters ahead we will prime the brain by learning not only the science and physical characteristics of the objects we study, but also a bit about their history and mythology and about our explorations and studies and the things they have taught us.

The second key tool the brain possesses that is critical in the process of learning is the power of *insight*. The human brain creates insights by the gathering of *perceptions* and grouping them into meaningful wholes. A perception in turn is any sensual stimulation (sight, sound, touch, taste, smell) that has real meaning to the perceiver. If I look at two objects, say a potted plant that sits to the left of a table. The spatial relationship of the plant and table constitute a perception. If I stand on the opposite side of the table, the plant and table have the opposite spatial relationship. By combining the two perceptions together, you create an insight, the knowledge of what the true three-dimensional spatial relationship of the plant and the table is. Learning is enhanced when all senses are used to gather perceptions, but in astronomy we principally use only sight. Sound, smell, touch and taste are just not used at the telescope. Keeping your eyes healthy, rested and trained is therefore most critical to the process of gathering perceptions both in quality and in quantity. Studying and gaining knowledge is also crucial. Remember that knowledge speeds the creation of insights by allowing more orderly assemblage of perceptions by increasing the "real meaning" of each perception.

Insight and learning are augmented by something else that only humans can do and that is *imagine*. If you obtain a perception from observation of a moon of

Jupiter hovering just below one of its poles, you know it is not directly underneath Jupiter but must be in front of it or behind it. If you combine the perception at the eyepiece with information gathered through study, say the knowledge of which way Jupiter's axis is inclined, your imagination enables you to visualize a three-dimensional view of where that satellite actually lies. When you look at a faint galaxy and can only see the core, your imagination, combined with some education and study can show you where those spiral arms are so that you know where to look and exercise perhaps some averted vision. By no means should you see something that is not there but you should combine your insights and knowledge to allow you to *see* where before you could only *look*. In the chapters ahead we will use all these tools that you possess along with the equipment used to create those perceptions. Through exercise, practice and repetition you train your brain and get the maximum capability out of the entire observing system.

Unless you live in southern California or the desert southwest, you will not be able to get out and observe every night because Mother Nature will periodically shut down the sky. These nights are great for brain training. Pick up a book (like this one) or perhaps a copy of a current or even non-current astronomy magazine. Don't pick up anything too non-current though because astronomy is a dynamic science that is always changing. Information can become obsolete very quickly. The astronomy periodicals are well written documents that can be very educational. *Astronomy* magazine is written at a bit more of a basic level while *Sky & Telescope* is written at a somewhat more advanced level.

The eyes, like the brain also need to be trained. Remember that perfect vision is not required in the early twenty-first century. In Chapter 1, we discussed means of maintaining ocular health and various means of coping with ocular pathologies. We also discussed the techniques of dark adaptation and averted vision to best prepare the eye for the observing session and make use of the parts of the eye best suited for observing in minimal light. The eye works together with the brain to form and process images. As the muscles of your body grow stronger through exercise so does the eye. The more you view in dim light, the more your eye will be able to detect lower and lower levels of light. The simple principles of adaptation to changing environment and physical conditioning work here the same as they do in the gym. You will notice this as you observe more frequently. Your eyes will progressively be able to detect objects that are fainter as you learn more how to use the eyes and as their performance progressively improves. An aviation physician noting my deteriorating vision suggested to me that a simple way to gain optical strength is to train with an eye chart. As you practice you will be able to progressively read further and further down the chart. The more you practice the stronger you will get. I found that the weakening of my eyes stopped and even reversed itself slightly.

Preparing the Equipment

Like with the eye, the telescope itself must be properly prepared for use before observing begins. The most important thing that you need to do for your scope when it is set up is to ensure that it has time to become thermally stable before you begin to do any serious observations. If the air in the telescope is warmer than the outside air, it will distort the image to the point you may not be able to focus on anything. A closed-tube telescope like a catadioptric or a refractor may need as much as thirty minutes to cool down. An open-tube reflector will cool down much more rapidly.

The observing site itself must also be chosen with care. When selecting a place to observe, you must consider three criteria. First is the presence of extraneous light. Light pollution can quickly turn an observing session into an exercise in complete frustration. Even in the most light polluted areas, you can find a place to observe that has somewhat dark skies. Large parks make excellent observing sites because they can put a few miles between you and the "light domes" that envelop large and medium-sized cities. Secondly the site must have reasonably clear horizons so you can see down to the haze levels. If trees stick up too high in close to you, you will not be able to see much more than straight up. Most large parks have an open meadow where observers can set up not far from their cars. The third important thing is to avoid large expanses of asphalt. Dark pavement will radiate heat for many hours after the Sun goes below the horizon, creating turbulence in the air adjacent to the ground and creating unstable conditions and bad seeing. It is a good idea to scout out the site in daytime. Get familiar with the horizons and exactly where you will be setting up. Know which way north is and how you will align on it.

The telescope itself is not all you have to check on. You will also have a box full of cameras, eyepieces and filters. The most important thing you can do is make sure that it is all *organized*! You are going to have to be able to find absolutely *everything* in the dark, without the use of anything brighter than your red map light. All your eyepieces should be stored in order from longest focal length to shortest. Barlows should be stored together near the eyepieces. Filters should also be all together and stored in spectral order with blue or violet at one end and red and orange at the other end. The other side of your box should be dedicated to your cameras. Lenses are stored in order of focal length and other accessories should be close by such as cable releases. If you find yourself fumbling for equipment in the dark, you may as well not have it if you can't find it. Even worse, not being able to find equipment is an invitation to lose it.

A seemingly silly thing to do, but an exercise that can be very helpful is to bring all your things into a dark room and practice assembling it in the dark. This is a common military exercise that all military infantrymen must perform during basic training, learning to disassemble and reassemble their rifles in total darkness. Try it for yourself and practice setting up your equipment, go through some common changeouts of eyepieces and other equipment just the way you'll do it in the field. Go through it first though in full light. This will give you the chance to see how everything fits together, what is tight, what is loose and what is just right. You would think that precision-made equipment that is industry standard should all fit together but sometimes you can get surprised. The eyepieces for my 4.25-inch Bushnell do not fit my 8-inch Celestron in quite the same way, they are slightly loose. In reverse, the eyepieces of the Celestron have barrels that are a bit too long to fit the eyepiece holder of the Bushnell. The Bushnell eyepieces were not threaded to accept filters.

Cameras can also give you a surprise especially if they were used equipment. I bought a Minolta 35-mm camera for use with my Celestron and found that it did not work in quite the same way that my old one had. The attach point for the cable release was on the lens mount instead of being on the shutter button as it was on my old Yashica. I also found that the threads for the tripod receptacle were not deep enough to allow the bolt on the Celestron piggyback mount to fully tighten against the mount causing the camera to be very loose on the mount. I had to devise a spacer for the bottom of the camera to ensure it would not rotate on its bolt while I was taking pictures with it. When that did not work, I resorted to Velcro. So check out everything in daylight including how you plan to attach each accessory in all the various different ways. Once you are certain that everything fits right, then try everything out in the dark. When you are comfortable with how everything fits and your organizational plan, then its time to go out into the night and put it all to work.

Imaging Equipment

Cameras operate in a manner very different than the human eye does. Film forms a fixed image through a process called *emulsion*. The retina of the eye forms an image instantly and transmits the image to the eye continuously. Each time a photon strikes the retina, a photochemical reaction takes place. In emulsion, crystals of silver halide are photochemically changed by light in a similar manner to the way the retina operates. The energy present in each photon causes the film to register differently. Since blue light is much higher in energy than is red light, the crystals are capable of registering color by reacting differently to varying energy levels with the aid of an organic catalyst that transfers the color from the photon to the crystal. Moreover you can control how long the film is exposed to the light source so light can accumulate on film for extremely long periods. The longer the film is exposed the brighter the image appears and faint details that the eye cannot see begin to appear. The eye however can only see in real time.

Film does suffer from one important shortcoming known as *reciprocity failure*. When film is exposed too long, all of the silver halide crystals are exposed and the image begins to become foggy. There is therefore a practical limit to how long film can be exposed for. There is however a way to limit the effects of reciprocity failure. Prior to using film for astrophotography, you should put the film in the freezer for a few hours. This will "push" the film to a higher speed in effect and allow the film to record light faster before the effects of reciprocity failure set in. Amateurs used to use a dry ice cooled device called a *cold camera* to achieve this effect before CCD cameras came into common use.

CCD (charged-coupled device) cameras use a silicon chip covered with an array of light detectors that are extremely sensitive to light. Each detector records light in one of 256 shades or gray (8-bit) or 64,000 shades of gray (16-bit). The most sensitive models are grayscale cameras although now there are many models that can image the deep sky in color. Even grayscale cameras can take color images by stacking images taken through red, green and blue filters, respectively. Like with film, you can enhance the sensitivity of the chip by cooling it. Most cameras employ some type of cooling system to reduce the temperature of the chip. This reduces the amount of residual heat on the detector created by the electronics of the camera. This residual heat causes the light detector to see light where there is none. Cooling the chip to temperatures below freezing helps limit that effect. Some cameras use air flow (either passive or with the aid of a fan), some use water and some use an electronic cooling system like a Peltier-effect heat exchanger (heat is transferred from the imaging chip across an electrical conductor to another piece of silicon) and cooling fins or baffling. When using a CCD it is a good idea to ensure it is connected and powered on long before you actually intend to use it so it has time to cool. In addition to being subject to all the other imaging challenges that cameras have, CCD chips are extremely small, only a few dozen square millimeters of area is available so the field of view of the camera is rarely more than a few arc minutes wide. Finding your target is very tough unless you have either a flip mirror, which will allow you to see exactly what the camera sees or a camera that can show you the target in real time so you can see what the camera sees. The tiny field of view also makes precision guiding absolutely essential. The great advantage of the CCD is that they are incredibly fast. A CCD can record in a few minutes what might take an hour or more on film. You can also thread a spectroscope into the barrel of the camera to image stellar spectra.

CCD cameras are almost universally computer driven, so you must have a laptop computer that is capable of handling the camera and whatever other applications you are using in the field. Make sure that you spend some time training your brain to use the imaging software and the computer in the dark. There is little that is more frustrating than wasting time messing with your computer while the night slips away. The keys again to success are telescope alignment, software familiarity, focus, cooling and aiming.

The trick with long-exposure astrophotography of course is the fact that you are imaging a moving target. The observer must therefore ensure that the telescope that the camera is mounted to is perfectly polar-aligned. Most equatorial-mounted telescopes have a polar alignment finder to make precise alignment easy. Remember that the scope must not only be perfectly aligned but it must be perfectly level too. To that end, I carry a bubble-level with me everywhere. If the mount is not level, the target will begin to drift off the film in the north–south plane. Another problem is that no clock drive is exactly precise. A drive corrector will be needed to ensure that the target does not drift in the east–west plane. If you cannot deal with both of these errors, you will end up taking images of streaks instead of stars.

To make corrections you will need one of two things. The first is a guide scope, which is a high-powered telescope mounted in parallel with the main scope. You keep a target star centered in the guide scope while the main scope images the target. The other is an off-axis guider. This takes a small area from the field of view, not visible to the camera and allows you to guide on a star in that field. That is, if you can find one. Corrections are then input via a hand-held device that enables you to drive the telescope along either its declination or right ascension axes to keep the target centered on your crosshairs. You do not have to be perfect, but you will have to be very patient and be able to catch errors while they are still very small.

Getting your Bearings in the Sky

The sky, like Earth, is a sphere and so it can be divided up with a coordinate system in a manner similar to the way Earth is. The system used to define position on the celestial sphere is somewhat different than is the one used on Earth. The north–south plane of the sphere is defined by the value we call *declination* and it is directly analogous to latitude. The celestial sphere has an equator that is parallel to Earth's equator. Declination, like latitude is measured in degrees, minutes and seconds. Because lines of declination always run parallel to each other at all locations on the sky, a degree of declination at the pole spans the same angular distance as one at the equator.

The east–west plane is somewhat different. Rather than measuring in degrees, the east–west plane is measured in time units that we call *right ascension.* Since there are twenty-four hours in a day, the sky is divided into twenty-four hours of right ascension. The zero hour line is at the point of the vernal equinox, where the Sun crosses the equator going north. Therefore the six-hour mark is where the summer solstice occurs, the autumnal equinox is at the twelve-hour mark and the winter solstice is at the eighteen-hour point. Each hour is divided into sixty minutes and each minute is further divided into sixty seconds. Units of right ascension are much more coarse than are degrees of declination, at least at the equator. At the equator, an hour of right ascension equals fifteen degrees of arc, but as you move farther north and the hour lines begin to converge, the angular distance separating the lines closes. At declination +40, an hour of right ascension is equal to only 13 degrees of arc. The separation of right ascension lines decreases steadily until it reaches zero at the poles.

Telescope Maintenance

Cleaning precision optics is not the same as cleaning household windows or mirrors. Remember that the lenses and mirrors of your telescope are tasked with collecting miniscule amounts of light, then amplifying, magnifying and focusing that light into a viewable image. The slightest damage to the glass surface of your objective lens or mirror can ruin it as a viable scientific instrument. It is simply not acceptable to pick up a bottle of Windex and a paper towel and try to wipe the lens/mirror surface clean. An aviation mechanic who I was assisting in an inspection once chastised me about "using the right tool for the right job." There are tools you use on cars and tools you use on airplanes." The same is true for cleaning a telescope. There is a right way to clean a mirror in your bathroom. That is by no means the right way to clean the mirror of your telescope. Let's discuss how it should be done.

Any optical component that is exposed to the open air will eventually become contaminated by dust no matter how diligent you are about keeping the dust caps in place. After all, you must remove the cap if you wish to do any observing with the telescope. During the time that the scope is in use, surfaces are becoming contaminated. This can happen a lot faster than you might think. If you are an allergy sufferer, you no doubt have reflected many times on how your automobile seemed to change color overnight from black to green. The greenish tint is caused by pollen settling on the vehicle overnight. If your scope is outdoors on a high pollen night, the same thing could be happening to your lenses. If you work in a dusty area (something you should try to avoid just on general principles) you will have the same concern about contamination. Then there is the issue of liquid contamination. You've spent the whole night trying to find that faint dwarf galaxy that you read about in *Sky & Telescope*, then as fast as you found it, it is gone because the objective of your telescope has become fogged with dew.

In addition to the optical components discussed in the previous chapter, a good addition to any equipment setup is a good optics cleaning kit. Both Celestron and Meade offer cleaning kits, as should any camera store worth its salt. A typical kit will contain a box of special tissues for lens cleaning, a bottle of lens cleaning fluid and a can of compressed air. When cleaning your telescope, remember that you have two major objectives. Remove the dirt and avoid scratching any of the surfaces as you do so. Given a choice between a small amount of dust and a single scratch on my objective, I will take the dust any day. Contrary to popular opinion, objectives must become very dirty before optical performance begins to become impaired. It only takes one little scratch to render a precision scientific instrument all but useless. When cleaning your telescope, you should always start with the least intrusive method possible. That means that if you can remove the contamination without touching the telescope, so much the better. For dust removal therefore, your best friend is that can of compressed air. As long as the dust is not bound to the glass by some other contaminant such as tree sap, the dust should blow right off. If the dust resists, then it is time for step two. Prepare the lens cleaning fluid according to the directions that are provided with it; it should be diluted. Apply a tiny amount (using an eye dropper) to the area to be cleaned. Use the corner of a piece of lens cleaning tissue and by just barely allowing it to touch the top of that drop of fluid, clean the surface by dragging the fluid droplet across the lens. Do not if at all possible allow the tissue itself to touch your telescope. Only once you are absolutely certain that there is NO dust or pollen on your telescope, should you allow the tissue to touch the glass. Lens cleaning tissue is designed in such a way that the fibers of the paper are woven in a pattern that will not scratch the glass. If the tissue catches a dust grain however, that grain can be dragged along the glass surface leaving minute scratches behind. When the lens is spotlessly clean, you may then use the lens cleaning tissue to dry the optical surface. Dealing with sticky contaminants like the aforementioned tree sap requires a bit more care. You may apply a slightly more liberal amount of lens cleaning fluid to the contaminated area and allow it to break up the sap deposit. Then if there is no dust on the lens surface, use the lens cleaning tissue to dry the area. Take great care not to use too much pressure on the glass surface.

Certain types of telescopes require more care than others. Traditional Newtonian or Cassegrain reflectors have optical tubes that are exposed to the elements whenever they are in use and as such, the primary and secondary mirrors are continuously exposed to contamination when the covers are removed. These surfaces are exquisitely difficult to reach and clean and for that reason the popularity of these scopes fell off sharply when the closed-tube Schmidt–Cassegrain design appeared in the early 1970s. Refractors and Schmidt or Maksutov Cassegrain telescopes have tubes which are closed so interior surfaces are almost always contamination free. Just always remember to keep the dust cover on the visual back or focus tube. The objective lens of the refractor or the corrector of the catadioptric telescope is exposed to the elements and must be properly cared for.

Eyepieces are also finely machined optical components of your telescope and must be cared for in the same way as other optical surfaces. For removing dust, again use the compressed air wherever possible. The can will usually come with an extension to allow you to gain access to cramped areas near the edges of the lens. Don't forget to clean both sides of the eyepiece since there is an exposed glass surface inside the tube as well. Some pieces of equipment may be more difficult to clean than others. A Barlow lens for example may have the lens buried in the middle of a long tube where only a dust grain can get at it. Don't do what one friend did and place a piece of tissue on the end of a long toothpick. He got the lens clean, but only after the toothpick scratched the Barlow into retirement.

By the way, also take care to remember your finder scope as well. I've met more than one person around an astronomy club who does a great job caring for their main scope only to have a finder that is nearly useless because of the dust that has built up on it. Clean it with the same attention and care that you clean your main telescope with because it is an important part of the optical system. A telescope that cannot be aimed is completely useless. Use the compressed air can to clean the objective lens and the fixed eyepiece. If you need to resort to fluid and tissue cleaning, remember that if you rub too hard in addition to scratching the lenses you may also push the finder scope out of alignment. This will cause you to have to spend much time later realigning the finder scope, time you would rather spend doing other things.

As the evening wears on, the temperature invariably will begin to cool and things around you may begin to get a little soggy. Dew will form on any surface when its temperature cools below the *dew point*. Dew point is the temperature to which a parcel of air must be cooled in order to saturate it with only the existing moisture. The glass components of the telescope will cool faster than the air so even though the effects of the cooling temperature will not be immediately seen in the weather (fog), you will see it on the objective of your telescope. Different types of telescopes are affected in different ways. Open-tube reflectors are actually well protected because the glass elements are deep inside the tube of the telescope where they will cool much more slowly. Refractors are more susceptible since the lens is much closer to the outside air. Still a typical refractor is slow to dew because they usually have a tube extension that shields the lens. Catadioptric telescopes are the most vulnerable because the corrector plate is at the very end of the tube where it has no protection at all. The exposed corrector will often dew over very quickly at the beginning of a session. Most catadioptric owners (including yours truly) have added a dew cap to their collection. A dew cap is a plastic or metal tube that is designed to fit over the front end of the telescope to help retain some heat near the corrector plate and prevent the formation of dew. Once it has formed however, there is only one way to safely get rid of it. Falling temperatures put dew on your lens; only rising temperature can get rid of it. If dew is an issue for you, bring a battery powered hair dryer with you. Blowing some hot air over the corrector plate will quickly dissipate the dew. Always remember to do this after the end of an observing session. If the telescope lens is dew covered after bringing it inside, turn that hairdryer on the lens until it is dry. Do NOT ever cover a wet corrector or lens. The result of that will possibly be an accumulation of fungus growing in the greenhouse-like environment underneath the lens cover.

Caring for precision optical surfaces is not at all difficult, but it must be done correctly. Clean lenses and mirrors allow the sharpest possible views of the heavens. Just remember that if you're nervous about it, don't be afraid to get it cleaned professionally. If you can't do that, a slightly dirty lens or mirror is still far better than a scratched one.

The Art of Collimation

Refractor owners can take a break here, though if you're thinking of joining the ranks of reflector owners, please read on. In order for a reflector of any type to produce a sharp image it is absolutely required that its mirrors always be in precise alignment. Before a telescope is shipped the manufacturer will usually perform a check using a laser to ensure that the mirrors are properly aligned with each other to produce the best image quality possible. Over the course of time, the secondary mirror may work its way out of alignment and need to be adjusted. This can happen as a result of bumping or jostling of the telescope or through mechanical flexure caused by repeated heating and cooling of the telescope. You might ask why the manufacturer does not design the scope so that it is inflexible one the mirrors are in alignment. The answer to that is because there is no substance known to our science that does not expand and contract to some degree in response to changing temperatures. The tolerances we are talking about are infinitesimally small. Remember that the spherical aberration that rendered the Hubble Space Telescope myopic is the result of an error in mirror shape of only a few angstroms; we're discussing *billionths* of a millimeter. In addition, changing temperature conditions may cause the mounting and telescope tube to flex, grow and shrink. The secondary mirror therefore is always equipped with an adjustable mounting to allow its alignment to be corrected periodically.

Though it is possible to precisely collimate a telescope these days with a laser, it is fairly easy to do without high technology. It just takes some patience. When you unpack the scope, one of the first things you should do is remove the covers and look through the eyepiece holder without an eyepiece. What you should see is a perfect reflection of your eye in the center of the secondary mirror. If your eye's reflection is dead center in the mirror, then you have a perfectly collimated mirror. If not, then you will need to adjust it. At the front of the telescope on the secondary support you will find three small screws with either Phillips or Allen heads. Your telescope will usually be supplied with the right screwdriver to adjust these. Turning the screws will move the secondary mirror. As you look through the eyepiece holder or visual back, turn one of the screws *slightly* and observe the result. The adjustment is very coarse, especially if you consider that even a few degrees of rotation of the screws will change the collimation completely. Carefully adjust the screws one at a time until the image of your eye is perfectly centered in the field of view. This technique will allow the telescope to function well enough to let you do a more precise collimation under the stars.

Once you are outside, center and focus a moderately bright star at low power. Move the focuser gradually out of the focus position. As you do so, watch what happens to the star. As its image blurs and expands, it forms the shape of a doughnut. The hole in the center is the shadow of the secondary mirror and is called the *airy disk*. Using the appropriate screwdriver if necessary, collimate the mirror until the airy disk is precisely centered in the blurred star image. Refocus the telescope and notice the improvement in image sharpness. Repeat the procedure then with a high power eyepiece in place. You may discover that the alignment was not so perfect as you thought with low power. With the airy disk centered using the high-power eyepiece, you have now achieved perfect collimation. Many observers like to collimate now with the aide of a laser aide. Generally the technique involves shining the laser into the eyepiece holder and watching where the dot falls on the primary mirror. It works and makes money for the laser sellers, but I must admit that I prefer doing it the old fashioned way. Owning a Schmidt–Cassegrain is helpful to me here since the optics are much more ruggedly mounted than on typical Newtonian reflectors. In sixteen years, I have only needed to collimate the telescope twice. On both occasions, it was after being transported over a long distance and in rough conditions, something I do not generally do to the telescope. Still I check my scope's collimation regularly because if it is off, the scope's performance will be too.

Mountings

The most optically perfect telescope money can buy will not do you the slightest bit of good if it sits on a mounting that is poor or poorly maintained. It is important to ensure that all the connections are tight and the telescopes axes and locks are properly maintained. Clock driven telescopes may have internal batteries that must be properly cared for if they are to deliver maximum performance. It is as important to take care of your telescope's mounting as it is to take care of the telescope itself.

The heart of the mounting is the telescope's tripod. Take the time to ensure that the legs operate smoothly and that all of the connections are tight. The tripod should not exhibit even the slightest amount of wobble when it is erected. Verify this with the tripod's legs fully extended and the tripod fully loaded with the telescope and all of its accessories. If you are using a German style equatorial mount, make certain the counterweights are properly set. What I liked about my first mounting is that it has only one hinge to look after and it is made of solid cast iron. This type of mounting is called an *isostatic* mount or "wedgepod." It combines the concepts of a tripod with an equatorial wedge. Two legs of the tripod serve as the wedge, connected by the cast iron hinge. The telescope bolts directly to them. The two legs are then supported at the appropriate angle by the adjustable length third leg. I used this mounting for sixteen years and never had to do any maintenance on it of any kind. The only drawback of this mount is that I find it more cumbersome to polar align than a standard Celestron with tripod and equatorial wedge. I must physically move the entire mount around while looking for the pole. If I need to adjust the third leg, I cannot do that and see through the telescope at the same time. The wedge/tripod arrangement lets you do this with the aid of two simple adjustment knobs. It allows the telescope's polar axis to be pointed at the pole with great ease of efficiency. The wedge and tripod arrangement greets the user with an intimidating array of knobs, nuts and bolts to keep tightened. It only takes one to work its way loose to turn a joyous observing session into a frustrating evening. Make certain you spend some time with your tripod and your telescope's other related support equipment. I decided for advanced astrophotography to get such a wedge and tripod for myself.

My telescope itself consists of the optical tube, a fork mount that the telescope swings through for storage and the drive base. In 1986, internal batteries were not common. I bought mine from a dealer who specialized in innovations such as these. Roger W. Tuthill of Mountainside, NJ was one of the first to specialize in retrofitting new telescopes with internal batteries. Tuthill devised a means of placing two lead acid batteries inside the drive housing of a Celestron telescope allowing it to be taken away from any source of power for up to eight hours at a time. A DC voltage regulator and an AC inverter turn the battery power into household current for the standard telescope drive motor. Today, all the major manufacturers offer this feature standard, but this was a revolution in telescope portability in 1986. Batteries need to be cared for just as telescopes do. Nickel-cadmium batteries in particular need to be looked after in a careful way. Believe it or not, for a "ni-cad" to work at its best, you need to be a little nasty to it. The first time you use the battery, don't be concerned about running it until its dead. Kill it. Run your scope until the battery will not run the scope anymore. Don't be afraid about running the battery so low that you won't be able to charge it again. When you unload a ni-cad battery, the voltage "bounces back" sufficiently to allow it to accept charge. By depleting the battery like this you help to extend its life by exercising all of its internal cells completely. Users who only use batteries for a short period of time before powering down or connecting to external power are actually harming the prospect of getting a long life from their batteries. Some telescopes that were among the first to have internal power provided by the manufacturer used size AA lead acid batteries. If you have one of these and you are planning to store your telescope for any length of time, make sure you remove the batteries and store them somewhere safe. Batteries left in the telescope or any other user will eventually leak as they age and ruin the contacts inside the battery compartment. There may be other accessories with your telescope that employ battery power. LED illuminated finders use small calculator size batteries. These small batteries have very short lifetimes if you leave them running. Be gentle with these batteries. Find and center your target, then turn the light off.

Speaking of finder scopes, remember that just as they deserve the same care as your main telescope does, so does its mounting deserve to be treated in the same way as the main scope's. The scope is usually mounted in a bracket with three centering screws set at 120-degree intervals. Tightening or loosening the screws can adjust the aim of the scope to keep it in line with the main telescope. Loosening one screw always means tightening two others. If you do not, the scope will be loose in the bracket and slip out of alignment. Check it regularly and keep it precisely aligned. There are few things more frustrating to deal with than being unable to see your target because the finder is out of whack.

As a general rule, keep everything clean. Heavy-duty mountings have heavily reinforced brackets that may be heavily ribbed. These ribs make excellent hiding places for dust and dirt. Any such nook or cranny makes a great place for corrosion or rust to crop up. Don't let it happen. Be aggressive about keeping every part of your telescope clean, not just the glass surfaces. Keep all of the bolts tight and lubricated. Make certain that batteries are treated the way they're supposed to be. Use tender loving care for those small ones, but remember some batteries like it when you're a bit nasty to them.

Cameras and Support Equipment

Cameras are usually pretty easy to deal with. The best advice to give with photographic equipment is to keep all lens covers in place and the camera itself inside its protective case when it is not in use. Camera lenses do occasionally need to be cleaned and when you do so, you should give them the same level of detail as you do your telescope lenses. For loose dust, use compressed air to get the surfaces clean unless more forceful techniques are required. Pay attention to the camera's viewfinder mirror too. Since you will be frequently be removing the lens of the camera, you will be exposing the interior of the camera to dust and dirt far more frequently than the average user will. If the mirror becomes dirty or smeared, it will inhibit the ability of your telescope and eyes to sight faint objects through it. Faint nebulae may even disappear behind that smear. Take care to ensure that the interior workings of the camera also remain dust free. Similar advice must also apply to users of off-axis guiders. Also remember to make sure that your camera's battery is kept fresh. If you leave it for several weeks, then pull it out into freezing weather, that battery you forgot about is going to turn on you.

T-adapters, T-rings and tele-extenders do not require a great deal of care because they are simply empty tubes or rings. Still they can collect dust and once they are integrated with your camera or CCD, they can scatter that dust on eyepieces and camera innards. Not only must your lenses be clean, but also so must anything that potentially comes in contact or close proximity to them. You don't have to be as formal about cleaning such non-optical components. Instead of using your supply of compressed air, just blow in the tube to get the dust loose.

CCD cameras are very delicate pieces of equipment. The only optical component is a small piece of optically flat glass that covers the CCD chip, which is opaque to infrared light. Of course, this must be kept meticulously clean using the same techniques that apply to any optical component. Make certain that the camera body is always stored in a dust free environment or as close to one as is reasonably possible. This will ensure that the data ports for the camera are kept clean as well as the camera's innards. It is also important to ensure that the camera's internal software is kept updated. This software is written into a type of memory that is designed to be very secure. Many software professionals refer to this type of hardware as *firmware*. Your camera's firmware is much like the BIOS of your personal computer. It contains the instructions that tell the camera how to start, cool, operate, image and transmit. Occasionally CCD manufacturers will come up with an improvement in the operation of the camera that requires a firmware update. You should be familiar with how to update the camera's internal firmware. Usually this will involve downloading a small file to your computer, then uploading it to the camera via a cable connection. Your operating manual will explain how it works. If you can't make heads or tails of the manual, then call the company's technical support office. If they don't have a technical support office, then you should not have bought a camera from that company in the first place. When you are finished with your camera, always put it back in its storage box or case. I was not kidding when I said that these cameras are delicate pieces of complex equipment. The only place where the camera belongs when you are not using it is in a nice cozy foam-lined box in its own custom-fitted cutout. Then make sure the box lid is closed nice and tight and keep that dust where it belongs . . . outside.

Many CCD cameras run from an outside power supply. This will not be through the telescope's battery because the camera consumes far too much amperage. Some cameras are powered through the computer's USB port. If so, then make sure whatever external power supply you are backing up the computer with has enough juice to support anything else attached to it.

Computers and Software

Mechanically, computers should be considered every bit as delicate as any optical component of your telescope. Take extra care of the LCD flat panel display of your laptop. These so-called active matrix screens are without a doubt, the majority of the value of your computer. Treat this screen with the same reverence reserved for your telescope's objective lens. Try to avoid sudden jars or bumps while handling it and when not in use, make certain it is stored inside a padded carrying bag designed for laptop computers.

From a software perspective, protect your data. Your acquired image files should always either be backed up or copied onto CD or some other non-erasable media. The first time you crash Windows and have to reformat your drive (therefore losing all the data on it) you will understand why I can't put enough emphasis on this point. A good way to protect yourself against operating system crashes is to pay regular visits to your system's Windows Update site. If you have ever worked in manufacturing, you understand the importance of making sure that your product is as perfect as it can be. In his post-astronaut memoir, Michael Collins points out that the Saturn V that launched him and his crewmates on the Apollo 11 mission had six million parts. With even 99.9% reliability, that meant that some six thousand failures could be expected. Each time a new version of Windows is released, it contains a mind-numbing average of 50,000 *flaws*! Most of these are relatively minor problems that deal with the operating systems interaction with certain third-party software. Microsoft addresses each issue as it arises and a corrective "patch" is posted on the Update site. Occasionally a large number of issues are corrected at once in a single large release called a *service pack*. A service pack will change your version number so be aware of any such changes. Other issues may involve security vulnerabilities that might allow someone on the Internet a chance to sabotage your computer. Windows also provides automatic access to updated drivers for your hardware. This can be critically important to you if you change or upgrade operating systems. If you convert a Windows machine to Linux or something else, your software drivers will not work with the new operating system. Even if you upgrade from one version of Windows to another, you will need new drivers. For example, I learned in upgrading from Windows Me to Windows XP that my video card driver does not get along with Windows XP. Keep your drivers up to date and keep Windows up to date.

Windows XP comes in two types of programs; the personal version intended for use on home PC's and the Professional version intended for use in the network environment. The Home version is oriented towards home users who run games, the Internet and many more basic functions. The Professional version is designed for heavy-duty networking where either workgroups or large servers are involved. Basically you will find three versions of Windows on the market for use on personal computers. If you've found a real dirt-cheap computer, you might find it is still running Windows 3.1. This is a 16-bit operating system that dominated the operating system market during the first half of the 1990s and made a multibillionaire out of Bill Gates. Many versions of CCD imaging software supplied with older generation cameras will still operate on Windows 3.1. The problem is that such a computer may not have nearly enough RAM to work with large image files or to upgrade to a later version of Windows. I would only suggest one thing if you purchase such a computer. Whoever owns it must be able to provide you with the original Windows floppy disks to reinstall the operating system. If something corrupts the operating system, you must be able to reinstall it and Microsoft will not be able to help you since support for Windows 3 versions ended many years ago.

In 1995, Microsoft introduced the first Windows 4 version, marketed as "Windows 95." This program (for you computer geeks, version 4.00.95) was the first of six in a product line that was built from 1995 to 2001. Always remember that whether you are running Windows 95[8], 95A, 95B (OSR 2), Windows 98, 98 SE, or Windows Me (millennium edition) they are all basically the same operating system. Each succeeding version simply takes all the patches that were made available to fix problems in the previous version and re-releases them in a single concise integrated software package. Succeeding versions also introduced some new functionality and updated drivers for new hardware as it is marketed. Because of the massive expansion of computer peripherals over the past seven years and the massive size of upgraded utilities such as Internet Explorer, the size of Windows itself has exploded. Windows 3.1 occupied less than 10 megabytes of hard disk space. Windows 95 was three times the size of Windows 3.1. Windows 98 was exponentially more massive and was the first Windows version not to be offered in a floppy disk version. Windows 98 ate up from 175 to 260 megabytes of disk space. Windows Me was the first operating system to consume a gigabyte (1,000 mb) and Windows XP Home Edition is a whopping 1.5 gigabytes. If you have bought an old laptop with limited disk space, upgrading to a later version of Windows may not be practical. The minimum operating system you should consider for use is

---

[8] Marketing names are different from program version numbers. For the record, Windows 95 Service Pack 1 is 4.00.95A, Windows 95 Service Pack 2 is 4.00.95B, Windows 98 is 4.10.1998, Windows 98 Second Edition is 4.10.2222, Windows Me is 4.90.3333.

Windows 95. That will at least run a 32-bit planetarium program and a CCD camera utility such as Pictor View. Also remember that Microsoft will not support any version of Windows that is more than three years old. Microsoft for example as of this writing has ended support for all versions of Windows older than XP and no longer provides security updates for anything older than Windows XP Service Pack 1.

Microsoft's flagship operating system today is the Windows XP (version 5.1.2600) product line that debuted in September 2001. For myself, I could not get my hands on it fast enough. The program is built around the system kernel designed for Windows NT and its follow-on, Windows 2000. By building a personal use system around the new kernel, Microsoft finally managed to rid its PC operating systems of their notorious instability. There have been drawbacks however. Some programs intended for use with Windows personal operating systems do not provide support for the Windows NT kernel and as such will not function with Windows XP. If you have a pet favorite image processing, camera control or planetarium program, make certain that it supports Windows XP before you upgrade your operating system. If the program you use predates Windows XP, check and see if it supports Windows 2000 or Windows NT. If it specifically does not support Windows NT or Windows 2000, then it probably will not work under Windows XP.

Your applications themselves require frequent updating. Make sure that whatever brand of planetarium software you buy, it is well supported by the manufacturer and offers frequent updates from a good website. A slick software program, particularly a planetarium program is worthless if it cannot be kept up to date. As new comets and asteroids are discovered, planetarium software quickly becomes obsolete. Even stars, generally considered to be eternally unchanging do in fact change from time to time. In recent years, the star Beta Scorpii has exhibited an unusual brightening. Usually the three stars that make up the Scorpion's head are of nearly uniform brightness. During the past three years, Beta Scorpii has increased in brightness by nearly half a magnitude from its normal magnitude 2.3. As much as I have enjoyed using *RedShift* in its various versions over the years, I will likely look at a different brand next time since *RedShift* does a very poor job of keeping its databases updated. Camera control programs are also often updated. Remember that it is quite normal for software programs to be released chock full of errors. In addition to correcting for their own mistakes, all programs must be continually revised to account for constant changes in the operating system environments. Remember that since the first Windows 95 appeared, there have been *seven* different Windows operating systems. Each time a new Windows appears, your applications must be upgraded to operate properly in the new system. Nearly every major camera control program regularly updates their software for these reasons as well as to introduce improvements and new functionality. Do not underestimate the value of improved or updated computer control. It can vastly extend the ability of hardware to perform beyond the standards for which it was originally designed. By the time Voyager 2 reached Neptune in 1989 more than twelve years after it was launched, its software had been rewritten so many times that it had become a spacecraft with capabilities far beyond what had been originally designed into it. That growth allowed Voyager 2 to overcome difficulties and problems that would have and probably should have crippled the mission. Improvements in the camera control software for my Pictor 216 XT allow for use of differing image formats, cooling and connection to more modern PCs. NASA has made available to Meade its "Drizzle" technology to allow long exposure to be taken using the Deep Sky Imager with an alt-azimuth mount without an expensive field derotater. Stay up to date and don't accept any program that cannot be kept so.

Power Supplies

Depending on how much equipment you are carrying into the field, you may have several different power sources. Many small rechargeable batteries are of the nickel-cadmium type. Remember that to get the best endurance and voltage out of a ni-cad, you need to be a little nasty to it. Make certain that you run all your rechargeable nickel-cadmium batteries down to exhaustion. Remember that dirty or rusty contacts do not transmit electricity so keep them clean periodically. Cold temperatures can also cause voltage loss. To preserve the charge of the batteries, keep them in a warm place as much as possible. Sub-freezing temperatures can cause you as much as a full volt out of the battery's charge. Lead acid batteries in particular need to be kept out of the cold.

To get the best performance from your telescope, brain and eyes everything must always be kept in as close to pristine condition as possible. Keep all your lenses and eyepieces meticulously clean and scratch free. Make certain that your mountings are kept tight and secure so as to minimize vibrations. Make sure that your computer hardware and software are both properly maintained and that your batteries are ready to deliver the juice. Lets head out into the field and get to work.

Observing Projects III – Caring for your Brain and your Stuff

Observing Project 3A – Training for the Brain

As we said earlier, unless you live in San Diego or the desert southwest, Mother Nature will limit how many days per year you can observe. Here in Delaware, skies are only completely clear about one-third of the time and there are stretches during spring and summer where skies are unobservable for weeks at a time without a break. Skies here are also very heavily light polluted between Philadelphia and Baltimore. Smaller cities in between like Wilmington and Dover make matters even worse. And of course there is that blasted Moon!

For the nights when you can't get outside or the light pollution is just too much, consider some of the industry's better periodicals for getting up-to-date information on everything that is going on that you cannot see. *Sky & Telescope* and *Astronomy* magazines are both published on a monthly basis and are excellent resources for detailing current happenings in the sky and timely articles on new discoveries and reviews of new products. The cover price can be pretty expensive but the subscription rate can save you a lot of money.

Both magazines also offer the publication of many quality books on a wide variety of subjects. From lunar to solar observing, to the planets to the deep sky, there are no limitations on the things you can get good information on observing. Books are also offered on subjects like astronomical computing, telescope making and maintenance and the exploration of the solar system by NASA's intrepid robotic probes.

Spend some time on the Internet too. Appendix C at the back of this book lists on-line resources that were used to fact-check the preparation of this book. They can provide some fascinating bits of information that when the weather clears can be used to enhance your observing experiences further. A little knowledge can go a long way.

Observing Project 3B – Assembly Check

Take all your things into a brightly lighted room and practice assembling and aligning your telescope. With all the lights on, set up the tripod, the equatorial wedge if it is a part of your mount, then put the telescope on it. Then lay out your accessories and go through a dry run of placing eyepieces and other accessories on the telescope, removing them and storing them. Try mounting cameras and other equipment. Set up your computer and get it running. Do everything you will do when you are out in the field. Then when you are done, tear it down as you would in the field.

When you have done all of this and feel comfortable, do it all again in the dark. After all, you will be observing in the dark, so you should feel comfortable with assembling your equipment in the dark. By doing this at home you know you will be comfortable, dry, and if you lose anything you know its not too far away and you can just turn the light on if needed. You will not have that option when you are out in the field so get comfortable at home. Time yourself and see how long it takes. This is important because if you are doing a project that is time-sensitive, such as a Messier Marathon, you have to know how far in advance you need to be ready.

Observing Project 3C – Loading the Car

Now this *really* sounds silly, but it's important. Astronomy is, for most of us still a hobby. If it becomes too much hard work, we will cease to do it because all the hard work detracts from the joy of it too much. One of the most frustrating things you will encounter is difficulty getting all your stuff to and from the car. Time yourself and figure out how long it takes to get all your stuff from its storage location into the car or out to the back yard observing site. If it is taking you more than fifteen minutes, you really need to organize better. The more work it takes to get yourself going, the more infrequently you are going to put yourself through it. Also consider how heavy the load is in addition to how voluminous. You may not have to make a lot of trips but even two or three can still be too many if you are moving so much weight that you throw your back out. That's why one of my telescopes is a tabletop model that I can be set up and observing through in less than two minutes flat. For the Celestron, I can be in the car, or mounted in the back yard in less than ten minutes. That includes the time to carry out the tripod and wedge, telescope, table, eyepiece case and computer. Polar alignment time is not included in that, but that does not take me very long either.

Observing Project 3D – Organization

Ever feel the total helplessness that goes with being disorganized and unprepared? It's a feeling I often had out at the telescope when I would carry my eyepieces out with me in the original box that Celestron had shipped them in. That's all well and fine when all I had was one Plossl and one Orthoscopic. When you own six eyepieces, a Barlow, a telecompressor, an LPR filter, a filter set, a CCD camera and a 35-mm SLR, that box gets a little crowded and a bit disorganized. Want to be the bad guy at a star party? Try turning on your white flashlight to look for something in that box. You will become very unpopular very rapidly.

Organization is the key to success. You must be able to find any tool at all at once and in complete darkness. The only way you can possibly succeed is to know where everything is in advance. That means that every piece of equipment goes back in the box in exactly the same way. Spend some time and brainpower figuring out the best way to organize your box so that everything is in a logical order, stores easily and can be found by touch in the dark. Then try it in the dark to prove to yourself that your plan actually works.

CHAPTER FOUR

First Night Out

Finally with your studying done and your mind primed, your eyes trained and all your equipment cleaned, prepped and ready to go, its time to head on out into the field to let your new telescope see "first light." There's a lot that goes into a night of productive observing, much more than just throwing the scope together and looking around, although that certainly can be fun. A serious night of observing can require some serious advance planning. You must plan where to go, when to arrive there, how you will get your stuff there, how you will set up your equipment and so on. You must also carefully plan on what you are going to do with your precious time. Sometimes the plan is based on personal constraints or desires (how long can you stand the cold) and sometimes it must be based on the happenings in the sky. If you are out to watch an event such as an asteroid occultation, your entire life for that evening is going to be based around those few seconds when that rock hides that star. You must be ready when that moment comes. Planning to observe a few select targets in the deep sky requires a different kind of planning. Faint galaxies require the presence of a dark sky so your planning is not for a particular moment but for a range of hours within a range of days when the Moon is not in the sky. Perhaps the most demanding planning task will be on the night when you dare attempt the Messier Marathon. This is a one-night run where you attempt to observe every single object on Charles Messier's famous list. There is a narrow window during March and April where it is possible to see every object on the list in a dark sky. The marathon requires not only selection of the right night where no Moon is visible during the window of opportunity, but careful planning of that night almost minute by minute from the end of evening twilight to the beginning of morning twilight.

The first night out should be designed around giving you a careful opportunity to set up your scope properly while there is still some ambient light, then taking a good two to four hours to view a selected series of targets which will allow you to sample the full range of your scope's capabilities. A perfect night might include a night where the Moon will be visible in a dark sky for a short period of time after twilight ends or rises later leaving a few hours of dark sky time before moonrise. It should include two or three bright planets, then a handful of deep-sky objects. These can include nebulae to test your scope's ability to pick out those faint glows from the background, galaxies that will test your scope's ability to resolve fine detail in faint objects and double stars in several combinations of relative brightness. To begin the planning process, use either an astronomy software program such as the ones we've discussed in prior chapters or a simple guide to events of a desired evening. Such a guide fits on one page that is published in each January issue of *Sky & Telescope*. Let's plan out one particular evening that would seem to fit a perfect first night perfectly.

The Events of one Night

Lets take a look at the events of the night of April 18, 2003. On that night, the sun sets at 7:44 PM Local Daylight Time[9]. Evening twilight ends one hour and thirty-seven minutes later at 9:21 PM. Four minutes later, Mercury sets in a dark sky. This is a rare event for northern hemisphere observers. In Mercury's six apparitions in 2003 (three morning and three evening), this is the only one in which it will climb high enough to set (or rise) in a dark sky. For the next sixty minutes, until 10:25 PM local time, the sky will be dark. At that time, the Moon rises, flooding the sky with its magnitude −11.9 brilliance. The moon is just two days past full and still 89% illuminated. It is also at a very close perigee (closest to Earth) so its disk is swollen to 32′ 56″ in diameter, about as large as it can ever appear. During this dark sky period, there are two other planets in the sky. Saturn is in the west about 30 degrees above the horizon and almost due west. The planet is well past its outstanding opposition of the previous December, but is still a fine telescopic target at magnitude +0.2, sporting a 17″ disk. The planet's rings during this time are open nearly as wide as they can ever appear, tilted open more than 26 degrees. This is an angle great enough to allow you to view the entire ring system as the far end rises up into view behind the planet. As twilight ends, Jupiter stands nearly at its highest for the night, some 68 degrees above the horizon, directly due south. Jupiter is magnitude −2.0 and its disk is 38″ across and is accompanied by its fine contingent of moons. During the dark sky period, there are several deep sky objects available for viewing to test out that new scope. Right after twilight ends, the Great Nebula in Orion (M42) is just high enough to permit crisp viewing in the southwest about 25 degrees high. Just above and to the right of Jupiter is the bright open cluster M44. Also called the "Beehive," this cluster is one of the finest open clusters in the sky. Standing about 40 degrees up in the northeast is the galaxy M51.

---

[9] The times are drawn from calculations using both *RedShift* 4 and *Sky & Telescope's 2003 Sky Almanac*. *RedShift* is based on my position in Northern Delaware while *Sky & Telescope's* are based on a location near Peoria, Illinois that I have then corrected for my location using instructions provided by *Sky & Telescope* in the almanac.

This famous ninth magnitude spiral galaxy is one of the sky's great showpieces and a fine test of your new telescope's ability to resolve detail. Also visible, climbing in the east is the constellation Leo that has one of the sky's great double stars. Gamma Leonis consists of two components nearly equal in color and brightness. The stars are closely spaced, but should be easy for the 8-inch scope. In the sky then, we have the chance to view in about a three-hour period, three planets, a bright nebula, an open cluster, a showpiece galaxy, a bright double star and finally the Moon to finish the evening. This short shopping list should provide a fine first test for your telescope. Let's get organized then for the evening ahead.

Where to go and Getting your Stuff there

In a perfect world, you reside in an area with a perfectly dark sky and there are no trees to obscure the view. Humidity is always low, insect life is non-existent and the temperature rarely strays from the low 70s.

Unfortunately, not many of us get to live in San Diego. For me, living in the mid-Atlantic area, winters are cold and windy while summers are oppressively hot and filled with many swarms of various sized flying bugs. Weather patterns are frequently stormy and when weather systems stall (as always seems to occur whenever an eclipse or other monumental event occurs) the sky stays cloudy for as long as seven to ten consecutive nights. Worse than that, the prospect of finding a dark sky area is daunting at best. Between Boston and Washington D.C., the skies are almost hopelessly floodlit in most areas as lights from major cities like New York, Baltimore and Philadelphia and lesser cities such as Wilmington, Atlantic City and Trenton fill in the gaps. This does not mean that finding a dark sky site is impossible and in fact if you intend to pursue this hobby with any degree of seriousness, then this needs to be your first endeavor.

This is where finding a good astronomy club can be vital. Remember that not only must the site you select be dark but it must also be safe. After joining a local astronomy club after I moved to Delaware, I learned that they had access to a dark sky site at Tuckahoe State Park, not far across the border in Maryland. I could also set up shop about fifty feet from my house where trees would not interfere though streetlights make deep sky viewing very difficult. The dark sky site is on a baseball field in the center of the park and is dark enough to draw large crowds annually to the club's two large annual star parties. Thus I have sites and opportunities for doing both casual stargazing and also for doing deep sky work under as close to ideal conditions as one can find on the east coast. The next challenge then becomes getting all that equipment to the site. Here it becomes vitally important to ensure that all your equipment is stored in an organized way.

As I elaborated on in Chapter 2, if I am going out to do serious photography, I can get all my stuff out to the car in only three trips. The telescope comes first in its storage case. The tripod comes second, then the eyepiece case and personal backpack and laptop on the third trip. It all fits comfortably in the back of my Mercury Mountaineer with some room to spare after the back seat is folded out of the way. All the optical components ride in foam lined cases to protect them from the bumps and jiggles of the road. The laptop is protected by the padded case that it normally travels in. With all my stuff loaded, its time to head out.

Leaving about an hour and fifteen minutes before sunset (6:30 PM) gives me time to get to a local convenience store to pick up my midnight snacks. A cold roast beef sandwich and a bottle of water and/or fruit juice will set me up for the night. If I need some munchies, I'll grab a bag of potato chips for something extra. Then it's on to the park. Arriving there about at sunset, I will pull my car onto the observing site and open the back. Pulling out the tripod, the first great challenge will be to find a spot that is as close to perfectly level as possible. Once level and with the tripod oriented roughly north, I will then go about getting the wedge installed and polar aligned. Since the pole star will not yet be visible, I will use a lensatic compass to set up the initial north alignment of the tripod and wedge. The trick to using a compass is to take great care to account for *magnetic variation*. Variation is the error that occurs in compass indications as a result of the difference in location between true north (Earth's rotational pole) and magnetic north. The magnetic north pole is actually located in extreme northwestern Canada. At Tuckahoe State Park, the value of magnetic variation is about 11 degrees west. From my position, the needle of the compass points about eleven degrees west of where true north actually is. The telescope therefore must be set up not on the indication of magnetic north, but eleven degrees to the east of magnetic north. Working with care, it is possible to point the telescope within a few arc minutes of exactly due north. The telescope is now roughly aligned along the east–west axis, now it must be aligned north and south. The equatorial wedge is adjusted by means of a knob that drives a jackscrew, raising or lowering the tilt plate of the wedge. A scale indicates latitude on the side of the wedge. Once this is set, I am now within a fraction of a degree on both the north–south and east–west axis for alignment.

Mounting the telescope is next. The Celestron Super C8 Plus has one particular design flaw that is somewhat annoying. The two declination circles are adjustable, which they should never be. They occasionally need to be reset to ensure that when the telescope is pointed straight up with respect to its base that the circles read 90 degrees. To do this, I will place the telescope on a flat surface and check that it is level with a bubble level. Then I will raise the scope to a vertical position and use the level to check that it is straight and vertical. The setting circles can then be adjusted to 90 degrees. This is a pain in the proverbial neck that should not be necessary, but occasionally must be done. With this small maintenance task out of the way, the telescope is then ready to be mounted on the tripod. Three heavy-duty bolts secure the twenty-three pound telescope, fork arm and drive base assembly to the wedge tilt plate. The first bolt is threaded on the drive base then fits into a slot on the top of the tilt plate. After tightening the bolt, the other two bolts then can be screwed in. I must of course take great care to avoid moving the carefully aligned tripod during this process.

Finder scope alignment is another task that is best accomplished under daylight conditions. I will carefully point the telescope at the pointed top of a stationary object like a house or structure and carefully center that target in the main scope. Do not use a treetop or something else that sways in the wind. Make sure that the target you select is at least 200 yards distant. At this minimum distance, the focus of the finder and the telescope on your target is about the same as it is on the stars. Finder alignment is much easier to do during the day when I can follow the lines of the alignment target to find the point at the top of that house. I will then switch to a high-power eyepiece and center it again in the main scope. When that is done, I will then carefully use the three setting screws in the finder scope bracket to carefully center the target on the finder crosshairs. Proper finder alignment on my scope is super-critical because not only do I need it to find objects in the sky but to accomplish the precision polar alignment since on the Celestron the finder also serves as the polar alignment telescope as well. Once aligned, I will install the batteries in the reticule illuminator LED and place the LED and battery pack in its receptacle on the finder scope. Normally I keep the batteries out of the illuminator since they tend to die quickly if the LED is inadvertently left on, something that a weak design and a weaker memory makes very easy to do.

The next job is to set up the power supply battery. This is a simple 12-volt DC automobile battery. Batteries of this type are great for astronomy because they supply a huge amount of amperage. The batteries are intended for use by high electrical draw items like car starters so there is plenty of power available for a 400-watt AC inverter. Because the astronomy equipment is relatively low amperage, the battery can sustain my equipment through a night of work on a single charge. I set the battery on the ground near the telescope and clip the inverter to the terminals. This will allow me to power the computer from the battery and the computer then can distribute power to my CCD. I will then set up a portable table (which normally is always left in the car) to set up the computer. The computer is plugged into the inverter and then started. By now, it is about forty-five minutes after sunset and it is time to begin observing.

The last critical task to be performed is the precision polar alignment. As soon as Polaris begins to come into view, I will rotate the telescope to 90 degrees declination and sight through the finder scope. In addition to a crosshair sight, the scope also has a dual ring to allow for precision positioning of Polaris. A template provided with the telescope shows exactly where Polaris should appear on the ring. With the telescope already aligned within a degree of the correct position, only some minor repositioning of the mount is required to gain a precise alignment. With this accomplished, it is not necessary to guide the scope during long-exposure photography on the declination axis. The only corrections that will be required will be in right ascension to account for the occasional variations in the accuracy of the telescope's gears. My scope uses a Byers designed worm gear that for the time was the most accurate clock drive ever produced for a telescope. Even by the standards of today's GPS computer driven telescopes, it is still pretty good. With precise polar alignment, I can take photographs of up to several minutes without any guiding at all.

First Light

Make the first object you look at a bright star high in the sky. As darkness falls on April 18, 2003 the bright orange star Arcturus is riding high up in the eastern sky. I will center it carefully in the finder scope crosshairs and then check for it in the main scope. I want the star to be exactly centered in the field of the main telescope. If it is, that means the finder is indeed perfectly aligned. The next step is to make certain that the telescope focuses properly and provides a sharp image. It is likely at this point that the star is slightly out of focus. Turn the focusing knob to move the scope farther out of focus. The image of the star will enlarge and blur. As it does so in a reflector design, the star image will assume the shape of a doughnut, a fading ring of light with a hole in the middle. If your telescope is a classic Newtonian design, the star will have dual spikes surrounding it. These phenomena are optical shadows created by the secondary mirror and the support rails for that mirror if there is no corrector plate. Focus the telescope until the star is at as precise a pinpoint and as bright as it can appear. If you have a Newtonian reflector, the dual spikes (called diffraction spikes) will merge into single spikes at the point of perfect focus. Once you have had success with a bright star, find a fainter one. Focusing Arcturus at magnitude −0.3 is easy. Try a second or third magnitude star in the darkening sky and try that.

The first target of the night hovering low in the west is Mercury, the elusive innermost planet. When viewing Mercury through my old 2-inch telescope, I just barely was able to muster enough magnification to discern the planet's tiny crescent face. Tonight, Mercury is about 8″ across and 31% illuminated. Using the 26-mm Plossl, I can clearly see the crescent phase during the fleeting moments when the viewing is sharp. The challenge is finding the planet before it sinks so close to the horizon that it can't be viewed clearly. Because Mercury is now a crescent, it's only magnitude +0.6 and a challenge to locate, never mind focus on and observe. At forty-five minutes after sunset, the planet is hanging just barely nine degrees above the western horizon. As the planet sinks nearer and nearer to the horizon, the sky grows darker, but the viewing gets worse as Mercury's light has to travel through more and more of the atmosphere in order to reach my telescope. As Mercury sinks from sight, it is time to turn somewhere else.

Twilight ends at 9:25 PM local time and now with the sky fully dark, it is time to test the light gathering ability of our scope. The treasures of the deep sky will be available to us for the next hour until the Moon shows up and washes it all away. A great place to start is the Great Orion Nebula, heading down fast in the west. Now about 18 degrees up in the sky, there is still time to get a crisp view of it. From my dark sky position, the fourth magnitude glow of the nebula is clearly visible and I can sight on it directly, without having to hop from star to star to find it. From an urban sky, start at the left star in Orion's belt, and then go directly south. As you do so, the nebula will drift into view from the south edge of your field[10]. In my old 2-inch telescope, the nebula appears as a featureless patch of gray. Buried within the gray are two faint stars that make up the two bright corners of the famous Trapezium. These are four newborn stars that are just emerging from their birthplace in Orion's stellar nursery. The little Tasco could only show two of them in a light-polluted sky. The Celestron clearly shows all four stars, but the grandeur of the view is in the nebula itself. In an 8-inch telescope, it is no longer gray and no longer featureless. The nebula has a clear greenish tint in visible light[11]. The

---

[10] If your finder is the type that allows you to view directly through it, south is at the top. If it has a right-angle viewer as mine does, then south is at the bottom, but east and west are still reversed.

[11] Many photographs of M42 show it as pinkish in color. This is because your eyes are very sensitive to the higher energy light while film is much more sensitive to the red low-energy end of the spectrum.

featureless blob now is brighter in some spots and less so in others. The nebula has spikes and tendrils and differing textures. Knots and kinks appear in the gasses as they can be clearly imagined spreading out into deep space, driven out by the light energy of the stars that have formed from within it. This is the object that for many is the one that truly fires their desire to pursue the hobby more seriously, sort of a second-stage catalyst, just as Saturn and its rings provided the initial first boost into the heavens.

Speaking of Saturn, the ringed wonder is right near by in Taurus. Though it has faded from the −0.5 magnitude luster that it displayed back in December, it is still an eye-catching +0.2 magnitude and 17″ in diameter. The planet's magnificent ring system spans 225% of the planet's diameter at any given time so right now the rings span a total of 38″ from east to west tip. The planet was some 40 degrees above the horizon at sunset and 30 degrees high as twilight ended. As the clock ticks towards 9:45 PM, Saturn has slipped to only 26 degrees high, so we want to take a look at it before it descends towards the horizon murk. My old Tasco showed Saturn itself as featureless butterscotch colored ball surrounded by its trademark rings. On dark sky nights, the little 2-inch refractor might also show Titan, the planet's largest moon glimmering at magnitude +8.4. To see Titan required a dark sky, good seeing and perfect focus. Through the Celestron, at 76x, the rings and planet appear much more three dimensional in appearance to me, thanks to a much higher quality optical system. The 3D effect is enhanced by the wide tilt of the planet's rings, right now about 26 degrees to our line of sight. This is very nearly the maximum opening that they can ever display. I can also see the planet's shadow peeking out from behind the planet, casting it over the rings. The planet itself at this magnification is visibly oblate, or "out of round." Saturn spins on its axis in less than half the time that Earth requires. The centrifugal force created by that rapid rotation causes the planet to bulge at its equator visibly. The superior resolution of the Celestron also brings into view some of the details in the planet's ring system. The Cassini division between the white "B" ring and the darker "A" ring is easily visible. This is a feature that I could never see with the little Tasco. On nights when Saturn is higher and the sky calm and clear, I believe that I have even been able to distinctly see the planet's inner "C" ring. This feature, called the "crepe" ring by astronomers, is so called because of the difficulty of seeing it. Saturn also has more company. Titan stands out clearly tonight, it is close to the planet just 128 seconds to Saturn's west. But Saturn also has other moons that were hopelessly beyond the range of the Tasco. In the Celestron, Rhea jumps into view at magnitude +9.4 between Saturn and Titan. Tethys and Dione are on the east side of the planet, in closer than Rhea at magnitudes +10.2 and +10.5 respectively. On perfect evenings, I have also been able to see Enceleadus which at magnitude +11.7 is very difficult and tonight it is too close to Saturn itself for easy viewing. Mimas is the closest of the planets eight large moons (larger than 500 km) is very near the limit of the Celestron at +13.0. Out beyond Titan is Iapetus, tonight at magnitude +11.1 and Hyperion at magnitude +14.1. Iapetus is a true celestial oddball. One side of the moon is very bright and the other side appears to be nearly completely black. When the moon appears to Saturn's west, it becomes as bright as magnitude +10 while when it ranges to the planet's east side, it fades to +14. Tonight, Iapetus is within my range making for a total of five companions (Titan, Rhea, Tethys, Dione, Iapetus) that I can see accompanying the ringed planet tonight. As Enceleadus moves farther from the planet, that will make six moons that are within the range of my 8-inch telescope. Though I have tried for years, I have never been able to say that I have seen Mimas. Hyperion is beyond the reach of an 8-inch telescope and Phoebe at magnitude +16.6 is visible only with research-sized telescopes.

As the clock ticks to 10:00 PM local time, let's spend some time with the other two deep sky objects on our list of first night targets, starting with the fainter one, the galaxy M51 in Ursa Major. The galaxy can be easily found just under the end of the handle of the Big Dipper[12]. Once located in the field of view of the telescope, the galaxy clearly appears as two distinct parts. What you are seeing are the core of M51 itself and the center of its companion, a small elliptical galaxy that M51 seems to be in the process of absorbing. Averted vision may allow you to see that the galaxy has two distinctive spiral arms, with the companion galaxy seeming to hang off the end of one of them. The galaxy is magnitude +8.9 total, though portions of the two cores appear to be brighter. The galaxy is an easy find for the Celestron, but was a source of endless frustration for the Tasco, with which under light-polluted skies, I could never find it.

M44 is a small association of several dozen stars traveling together through the Milky Way. The old 2-inch refractor showed M44, also known as the "Beehive Cluster" as an association of about two dozen stars of varying brightness. The Celestron reveals that the Beehive is not a group of dozens of stars, but of hundreds all traveling together across the heavens. The cluster itself is bright enough to be seen with the unaided eye in the constellation Cancer, shining at magnitude +3.9. At about the 10:00 PM time frame, the cluster is high in the southwest hanging about 55 degrees high. To the unaided eye the Beehive appears to span more than a degree of sky (70 arc minutes) or more than twice the size of a full moon.

Just to the east of the Beehive, barely 75 minutes away is the brilliant beacon of Jupiter. The solar system's largest planet was at opposition to the sun in February, so the planet has faded somewhat from its largest and brightest standing. Still, Jupiter is slightly brighter than magnitude −2.0 and its disk spans 38″ in diameter. In the old Tasco, Jupiter appears as a bland yellowish ball with a generous bulge at its equator. Jupiter rotates even faster than Saturn does and the action of centrifugal force bulges out the planet's waistline by nearly 11%. The old Tasco also clearly shows the planet's four Galilean moons as star-like points of light. Each of the moons measures less than 1.5″ in diameter, too small for the Tasco to resolve. In the Celestron, the superior optics of the telescope clearly shows the various belts and bands of the planet's atmosphere. If the Great Red Spot is in view, it will appear as a slightly off-colored oval. In recent years, the Spot has faded considerably in intensity to more of a pale beige color.

By 10:00 PM, Leo the Lion is at its peak in the south, some 60–70 degrees high in the southern sky. Leo holds one of the skies favorite double stars, Gamma Leonis. Also called "Algeiba," this star makes a good, but relatively easy test of your new telescope's ability to split close double stars. Algeiba consists of two components. The primary star shines yellow at magnitude +2.0, while the secondary star at

---

[12] The Big Dipper itself is not a constellation, but a formation of stars called an "asterism," that is, a constellation within a constellation. The Big Dipper makes up the body and tail of Ursa Major, the "Big Bear."

magnitude +3.1 has more of an orange color. The two stars are relatively close to one another, separated by only about six arc seconds. This was theoretically in range of the Tasco, but I could never split this pair with the little department store refractor. When I first turned the Celestron on this pair, I could separate them easily into two yellowish components. These stars are both aging giants that are evolving off the main sequence into helium-burning red giants. The Gamma Leonis pair represents a look at the future of our Sun about five billion years from now.

Finally by 10:25 PM, the darkness of the sky is shattered as a nearly full moon heaves itself over the eastern horizon. The Moon is just past full and still almost 90% illuminated. The Moon as it rises appears overwhelmingly huge. In part, this is due to atmospheric refraction close to the horizon and also in part due to the fact that the Moon is near the perigee point in its orbit, the point where it is closest to Earth. The Moon is shining tonight at magnitude −11.9 and with that light, all of the deep sky objects that we have looked at to date are now washed out of the sky or severely muted by moonlight scattered across the sky. The Moon is a good subject for the Celestron, not for its light gathering capability, but for its ability to produce crisper images at high power than the Tasco could. Using magnifications in excess of 100x on the Tasco usually produced images that are badly blurred because the scope cannot gather sufficient light to produce a sharp image at that kind of power. The Celestron can use up to four times as much power effectively because it has four times the aperture. One feature that is strongly recommended is the use of a Moon filter. The Moon is bright enough in a small telescope; in a large instrument its brilliance can be absolutely overwhelming. A Moon filter is a neutral colored filter that reduces lunar light by as much as 87%. This will allow increased contrast between lunar features and give relief from the overwhelming glare. This particular phase is not really the best time for viewing the Moon. When the Moon is nearly full, there are few shadows on the lunar disk to show relief of lunar features. Without shadows, the lunar surface appears relatively flat, whereas when it is at first or last quarter phase, the shadows of the lunar mountains and craters dramatically show the roughness of the lunar surface.

The first night out is a good night for some simple forms of astrophotography. My telescope came with a simple piggyback bracket. The bottom end of the bracket screws into the top of the telescope's rear cell just to the right of the finder mount. The top end has a standard thread that will allow any 35-mm camera to be attached directly to the top of the mount. This will allow the telescope clock drive to drive the camera as well. You can produce some startling results with just your camera. By this time of night in April, Ursa Major is just beginning to arc overhead. Attach your camera and its cable release to the telescope. Set the exposure time knob of the camera to "B". Frame your target and set the camera's focus to infinity. The typical 35-mm camera lens has a field of view of about 35 degrees of sky. This should be sufficient to cover the entire Big Dipper asterism.[13] Then all you have to do is open the shutter and lock the cable release. Experiment with several exposure times. Try exposures times of one minute, five minutes, ten and longer. The darker your sky is the longer you can expose the film. One minute should be

---

[13] The Northern Cross (part of Cygnus the Swan) and the Teapot (Sagittarius the Archer) are other examples of bright asterisms.

sufficient to reveal all the second and third magnitude stars that make up the Big Dipper pattern. By five minutes you will be able to see stars in the image far below the naked eye limit. If the sky is dark, you should be able to reach eighth magnitude. There is a practical limit to how long you can leave the shutter open. If there is any light pollution in your sky, as the exposure lengthens, the background of your image begins to brighten and block out the fainter stars in your field. Eventually the light pollution will fog out your entire image. That's why you try several exposure lengths to find out what the maximum time is that you can leave the shutter open. After the Big Dipper area, try again with Leo, high on the meridian, Gemini, still high in the west and Virgo, rising up the southeast.

By eleven in the evening, you've been outside now for about three and a half hours and its time perhaps to call it a night. Remember that it is now dark out, so as you disassemble your equipment, make absolutely certain that you do not forget anything or leave anything behind. Remember not to cover any optical component that is covered in dew. This is important! The corrector plate of Maksutov or Schmidt–Cassegrain type telescopes is particularly vulnerable to dew accumulation since there is little protection afforded for the glass surface, so that battery powered hair dryer would be a good thing to carry with you. If you cover the objective while it is wet, you will create a perfect breeding ground for fungus or other contamination to grow and fester on your lenses. Dry it first, or get it in a dry place and let it dry out before covering the lenses. Pack carefully as well, make certain that everything is back in its foam-lined cases or padded bags. Remember that you have been up for a long time and you may want to have something with caffeine in it for the ride home. Just remember that caffeine will keep you awake, but it has an unfortunate side effect. Know where the pit stops are if you have a long ride home.

The Day After

After getting home from work the next day, spend a little time with your equipment and see how it endured in the open. If your observing site is in an open dusty area (the infield of a baseball diamond), make certain that no dirt or grit has blown into any of your eyepieces or telescope optics. If so, clean the affected surface carefully using the compressed air in your optics cleaning kit. At the risk of belaboring previous points, it must be firmly stated again that the optical components of your oculars and telescope will not tolerate even the tiniest scratches in glass surfaces. If you do not know what to do or do not have the proper equipment for the job, it is better to leave dirt or dust in place on the lens and if it becomes intolerable, have it professionally cleaned.

Honestly evaluate how the telescope performed the night before. Are the images as bright as you might have expected. Did the telescope see down to its magnitude limit? How was the resolution? Could you split double stars close to the scope's Dawes limit?

One of the most important things you need to do in observing is to keep a log of everything you see. Important things to note are included in this excerpt from my log from the night of April 18, 2003.

Observing Projects IV – The First Night out

Observing Project 4A – The Observing Log

Observing Log	April 18,2003						
Object	Time	Alt	Azimuth	Mag	Size	Seeing	Remarks
Mercury	20:29	8	275	0.6	8.0"	2	31% Illuminated, Poor seeing
M42	21:29	18	240	2.4		3	Beautiful rich detail. All Trap. stars visible
Saturn	21:45	26	280	0.2	17.1"	4	Titan, Rhea, Tethys, Dionne, Rings Open 26 Deg
M51	22:00	80	360	8.9		4	Bright cores, arms visible with averted vision
M44	22:10	50	260	3.9	1' 10"	4	Bright open cluster
Jupiter	22:15	55	255	–2	38"	4	4 moons, no red spot tonight
Gamma Leonis	22:25	60	220	2.0/3.1		4	Yellow/yellow binary 6.0" split
Moon	22:50	10	130	–11.9	31' 50"	3	Nearly full moon, 89% illuminated.

An observing log taken in the field should be simple and concise. You are taking it as you observe so keep it small and brief. It should be just enough to jog your memory from the night before as you read it. As I go through this log, even years after I took these notes, it helps me to remember key things I saw in the sky that night. This can be extra helpful when dealing with objects that change over time, like comets and planets. Note the vital statistics, such as *what* you saw, *when* you saw it, *where* was it in the sky, *how bright and how big* it was, how good the *seeing* (1–5, with 5 being best) was and anything noteworthy. Each one of these lines was typed in Excel in the field in just a few seconds.

Observing Project 4B – Joining an Astronomy Club

Picking a good club provides several very important benefits to you as an amateur astronomer. To list just a few, first among them would be the companionship of those who enjoy our hobby and science as you do. Among those people you will find some surprising talents. In my own club, I have already benefited from the friendship of one amateur who is an electrician who helped repair my telescope when its internal inverter failed. He got me back up and running with a few passes of his soldering iron. Another helped locate us a reasonable dark sky observing site with clear horizons and dry ground to observe from. Others have lent their time to teach others and us about the areas of astronomy they are most passionate about. Others have allowed us to share in their innovations, their large telescopes and their hot beverage supplies.

Most important to remember is one simple fact that can make the club experience a joy or a frustration. Astronomy clubs come in two distinct types, those that stargaze and those that talk. A club that goes out and observes on a regular basis will encourage you to go out and do it yourself. The more you do it the better you will get at it! Clubs that do nothing but talk will eventually bore you to death. The thrill of the first night out will bring you back for seconds if you read the first four chapters and came out with reasonable expectations. But without new challenges, and someone to teach you how to accomplish them, you will quickly become bored. Find a club that goes out and observes, with members who own a wide variety of telescopes and even better, one that sponsors at least one annual weekend star party. *Sky & Telescope* maintains an excellent listing of clubs on its website. The database can be sorted by state and will allow you to easily find a club near you. Go to some meetings and find out what that club is about.

Now we have trained out eyes, purchased our equipment, learned to set up and care for it and demonstrated that it will perform as it should. Now it's time to take that equipment and begin to unravel the mysteries of deep space. But before pressing into the deepest parts of the cosmos, it would be perhaps better to sharpen our observing skills by starting a bit closer to home.

CHAPTER FIVE

Mysteries of the Moon

Figure 5.1. Ten-day old Moon. Photograph by author using 35-mm SLR and Celestron 8 at prime focus.

Professional astronomers hate it as much as romantic poets love it. It floods the sky with horrific quantities of unwanted light that washes out the grandeur of the deep sky for two weeks out of every month. It forever presents the same face to us, never allowing us to see its other hidden half. It is bland and it is boring and it never changes over billions of years, never mind in the span of a human lifetime.

"It" is our Moon.

The Moon is Earth's stalwart companion, loyally following it through space completing one circle around Earth each 27 days. It begins to appear in the west shortly after sunset each month, gradually growing fatter and brighter and higher in the sky each night until its full brilliance washes out the night sky of any FFTs (faint, fuzzy things) for days at a time. Then it begins to slim and fade as it transitions into the morning sky waning to half lit, then to a crescent and finally disappears in the east leaving behind a few precious nights of dark sky before starting the cycle all over again.

But the Moon is actually anything but a boring place and not quite so static as one might believe. The Moon is a geographical record, frozen in time, of the earliest days of the solar system and of what Earth might have been like. It will fill your eyepieces with amazing stories of meteorite impacts of astonishing power and lava flows that filled entire "seas." Parts of it have been battered to a point where there is no flat spot to be found anywhere and other parts of the lunar surface are as flat as prairie country. Different areas of the Moon have been exposed to very different environments. In this chapter, we will explore the various different elements of the lunar surface and use it as a practice ground for honing our observing skills under reasonably bright conditions. After all when the Moon is in the sky, no one will be spending much time looking into deep space, so it makes an excellent time to see the stories the Moon has to tell.

Before setting out to explore the mysteries of the Moon, we should first dispel some myths about the Moon, the first being the statement that the Moon is a moon. In fact it is not. The Moon is not in any way a true satellite of Earth. The Moon in fact fails the most basic test of whether or not an object is a satellite or a planet. As an object orbits a parent body, it "falls" towards that body and only that body. In a stable orbit, the falling motion is counterbalanced by forward speed. An object in a perfectly circular orbit moves forward at such a speed that the curve of the attracting body causes the surface to fall away from the satellite as fast as the satellite is falling, thus the object remains at a constant altitude as it circles its parent. When in orbit of a body, the orbiting moon falls only towards its parent planet. Jupiter's moons for example fall only towards Jupiter, never towards anything else. The Moon on the other hand, never falls only towards Earth, but rather falls towards the Sun. The Moon and Earth revolve together around a common center of gravity (called a *barycenter*) that resides approximately 1,600 kilometers below Earth's surface. Earth and the Moon revolve together around the barycenter once each 27.1 days and both fall together towards the Sun. So the Moon is not a satellite of Earth, but rather the two comprise a dual planetary system moving together around the Sun. Except for Pluto and Charon, there is no other pair of large bodies in the solar system so closely matched in size and mass as Earth and the Moon and none are so large. Both bodies have important effects upon the other.

The most important effect of the Moon's influence on Earth is ocean tides. The Moon's gravity pulls on Earth's oceans, causing them to rise when the Moon is overhead and recede when the Moon is at the horizon. Earth's gravity also exerts tidal forces on the Moon. Earth's pull is strong enough that it holds the Moon with sufficient force that the Moon is tidally locked to Earth. As a result of this, the Moon rotates around its axis in the same amount of time it takes to complete on revolution around the common center of gravity between the two bodies. This means that the Moon presents the same face towards Earth theoretically at all times. There are three minor variants in this motion that allow observers on Earth to see more of the Moon than would ordinarily be possible. The first is the fact that the Moon does not always orbit Earth at the same speed. When it is at its closest to Earth, it is at a distance of about 356,400 kilometers and travels faster than normal. When it is at its farthest from Earth, it is at a distance of over 406,400 kilometers and travels a bit slower than normal. The rotation rate of the Moon however is constant. So the slight changes in velocity of the Moon along its orbit allow observers on Earth to catch a glimpse slightly around each limb. As the Moon accelerates and decelerates in its orbit, it appears to nod back and forth, showing us alternately a peak around its western limb, then a peak around its eastern limb. The Moon's axis is also tilted slightly with respect to its orbit, about five degrees. The Moon's orbit is also inclined slightly with respect to Earth's equator. These two facts allow us to alternately peer over the Moon's north pole and under the Moon's south pole. All together, an Earth bound observer can actually see up to 59% of the Moon's surface over time. So the Moon does not always show those of us on Earth exactly the same face all the time. The combined effects of these motions and changing viewpoints are collectively called *libration*. Those areas of the Moon that we only get to see part time holds some rather interesting terrain, which we'll get a chance to explore later in this chapter.

Another common misconception about the Moon is that it's pretty much the same all over, a homogenous ball of rock. That assumption falls apart upon cursory examination of the Moon's near face. Though the Moon is not nearly as diverse as is Earth, it is not uniform either by any means. Bright highlands areas around the circumference of the lunar disk alternate with the dark lowlands called *maria* or "seas" dominating the areas in the center of the lunar disk. A bright rocky material that is primarily *anorthosite* dominates the highland areas. This is a brightly colored mineral, which is high in aluminum content. These areas of the Moon are far more heavily cratered than are the seas because these areas have not had their geological past erased by volcanism. The low-lying *maria* are primarily made of miles deep accumulation of *basalt*. This type of rock forms in layers when volcanic lava cools in pre-existing depressions of the lunar surface. Basalts are very dark in color. Under pressure, when it cools, it creates a black colored glassy material called *obsidian*. By a simple observation we can determine that the low-lying *maria* are much younger than the heavily cratered areas nearer to the lunar limbs. The Moon was obviously very heavily bombarded throughout its early history, leaving the polar areas and the eastern and western limbs obliterated with craters. The same bombardment had to occur in the *maria* areas, but subsequent volcanic activity wiped out the craters and replaced it with a smooth surface, which is not so heavily cratered. The current surface formed at a time when the planets had swept the solar system relatively clean of the vast amounts of dust and rock that so decimated not only the Moon's surface but also that of all the inner planets.

A far more interesting question about the Moon is how did those *maria* come to be? Only a relatively small area of the Moon, about 17% of its total surface area, is of this type. The Moon's far side is known to be almost completely free of the volcanic seas. Are the low depressions where the seas formed an artifact of the Moon's creation? Could they be a "birth mark" of some kind? That in turn calls into question just how the Moon itself came into being. There are many theories that seek to explain how the Moon came into existence and came to orbit Earth in such an orderly manner. Some believe that the Moon may have formed independently and been captured after drifting too close to Earth. Others believe that the Moon formed as a part of Earth that broke away. We can learn a great deal about both theories by studying the surface features of the Moon. And by looking at many of the same decisions that were faced by those who were planning the Surveyor and Apollo flights of the 1960s and 1970s, we can learn for ourselves how these theories of lunar creation gained strength and encountered resistance.

Picking Landing Sites

During the 1960s and 1970s, four major programs were mounted by the United States to conduct up-close studies of the Moon. Three unmanned programs: Ranger, Lunar Orbiter and Surveyor paved the way for the explorations of Apollo. The Ranger probes were the first to teach us to fly the trans-lunar highway to reach and impact the Moon's surface. They returned hundreds of up-close television images of the surface in selected areas. The Lunar Orbiter created the first comprehensive maps of the lunar surface including the far side of the Moon and conducted more detailed surveys of prospective landing sites for the forthcoming Surveyor landing craft. These small craft were sent to relatively safe landing sites that were candidates for Apollo landings. One of the Surveyor sites was also the target of an Apollo landing three years later. Later on in the observing projects of this chapter we will study and search for the regions of each of the Apollo landing sites and study more closely why they were selected.

When Apollo was in planning, the initial objective was simply to get there and get back safely so science took something of a back seat in the mission planning. NASA selected landing sites for the first two missions that were relatively safe and flat. Apollo 11 was targeted into the Sea of Tranquility, an ancient lava plain just east of the center of the lunar disk. The landing site was flat as a dime and the approach paths were devoid of rising terrain and offered the crew the easiest possible approach. The landing site was also close to the equator, which assured that the rotation of the Moon during the time on the surface would not carry the lunar module too far away from the orbital plane of the command module. The equatorial location also made the task of planning the trans lunar trajectory much easier and left the option of a free return trajectory open to the crew for a much longer period of time[14].

In the same vain, Apollo 12 was targeted into Oceanus Procellarum, about 1360 km west of where Apollo 11 landed. Like the Sea of Tranquility, the Procellarum landing site offered a simple orbit to reach, the safety of free return and a clear approach to a flat landing zone. Apollo 12 also landed within 600 feet of the Surveyor 3 spacecraft, which had preceded it to the Moon by three years. The Apollo 12 crew was able to bring back several pieces of the Surveyor probe and a much larger quantity of rock. But the rock was relatively young, volcanic in nature, confined to a narrow range of age and containing no surprises and no clues as to the origin of the Moon. To unlock the secret of the Moon's origins, future

---

[14] A "free return" allows a spacecraft outbound from Earth to fly around the Moon and return directly to Earth in theory without any large engine firings. This could save a crew's lives in the event that a spacecraft's engine became incapacitated en route to the Moon, an event that did in fact occur on Apollo 13 in April 1970.

expeditions would have to be able to get at the older rock on the lunar highlands. So NASA would have to become bolder in selecting landing sites.

Apollo 13 was the first flight to be targeted for the lunar highlands, aiming for a site called Fra Mauro. This is a highland area located between the Apollo 11 and 12 landing sites. Apollo 13 of course never made it to its planned landing site, but nine months later Apollo 14 did. Though the rocks found here were older, they did not contain any geological evidence that would tie them to the early history of the Moon. For the good booty, NASA needed to look at more exotic locations and also needed to bring in some extra brainpower.

Apollo 15 was the first mission to land an extended distance from the lunar equator, touching down on the plains to the north of Mount Hadley. It also required an approach that was far steeper than any that had ever been flown. The so-called Hadley Basin offered geologists a look at some of the most ancient terrain on the Moon and some of the widest variety of rocks and soil within a narrow area. The 15,000 foot high Mount Hadley rose to the south of the lunar module while to the west ran the deep Hadley Rille, which channeled huge amounts of lava during the Moon's active periods billions of years ago. The program also brought in prestigious geologists such as Eugene Shoemaker, Lee Silver and Farouk El-Baz to assist the astronauts in training. The geologists were not favorably received at first. Upon meeting El-Baz, one astronaut reportedly asked if the program was "all out of American scientists." Undaunted and wildly enthusiastic, El-Baz worked closely with Apollo 15 command module pilot Al Worden in identifying key features from his perch in lunar orbit. El-Baz had worked previously with Apollo 13's Ken Mattingly before he was scrubbed from the mission and Apollo 14's Stu Roosa. El-Baz' passion was contagious, especially in Worden's case. Worden's wildly enthusiastic analysis of the Taurus Littrow valley resulted in the area being selected as the landing site for the final Apollo mission. Shoemaker worked in the field with David Scott and Jim Irwin on identifying geological features in the rocks that would yield critical clues to the Moon's ancient history and origins. Shoemaker's work paid off on Apollo 15's first EVA when Scott exclaimed, "I think we found what we came for." What Scott found is the sample that is today called the "Genesis Rock," the oldest sample ever brought back from the Moon, estimated at 4.6 billion years old. Scott and Irwin also brought back a "deep core" sample showing 42 separate layers of soil, which had been undisturbed for 500 million years. Scott and Irwin also studied Hadley Rille, where exposed bedrock in the far face of the canyon showed the astronauts more than 3 billion years of the Moon's history.

Apollo 16 landed in the Descartes highland regions of the Moon. Some at NASA had pressed for Tycho as a landing site, but it was dismissed at far too dangerous despite the opportunity for scientific return. Descartes offered the opportunity for the first time to sample actual volcanic rocks. The basalts brought back from the *maria* on Apollo 11 and 12 revealed that the Moon's interior was once at least partly molten and that practically meant that the Moon was once geologically active. The Apollo 11 and 12 samples unfortunately covered only the very narrow range of the Moon's history when the volcanic seas were formed. Descartes offered the opportunity to find the geological record of the Moon's early volcanic history. In that regard, Apollo 16 proved to be an enormous disappointment. Most of the rocks were found to be *brecchias,* rocks that were formed from small stones and soil fused together by heat and pressure from catastrophic impacts. The final disappointment proved to be an enormous boulder called "House Rock" which was a breccia made up of individual fragments that were as large as six feet long. It was later theorized that the rocks that littered the area around the landing site were formed in the cataclysmic impact that formed the Imbrium basin some one thousand kilometers away.

The final mission of the program, Apollo 17, was targeted at the Taurus Littrow valley after it was exposed as an area of interest during the flight of Apollo 15. The final landing site was in such a position that reaching it required the first night launch of a manned space vehicle in history. Apollo 17 was also the only mission of the program to carry a trained geologist on board. The addition of Harrison Schmidt to the crew in place of Joe Engle made the mission among the most efficient of all the Apollo missions in terms of missions objectives completed and knowledge and science gathered. Schmidt and mission commander Gene Cernan explored two large massifs . . . giant landslide areas in two of the longest traverses of the program. Schmidt would recover rock fragments from the South Massif more than 4.6 billion years old. Cernan and Schmidt spent some 77 hours on the Moon, 22 of them outside and traveled more than thirty kilometers on the surface.

Several other fascinating areas were under consideration for use as Apollo landing sites. One of which is the Moon's most spectacular signature crater, Tycho. Some scientists pushed the huge rayed crater, which gives the Moon the appearance at times of a peeled orange, as a landing site for Apollo 16. The Marius Hills complex was touted as a potential landing site for Apollo 18 and the area around the huge crater Copernicus was tentatively chosen as the landing site for Apollo 19. Both missions were cancelled shortly after Richard Nixon became president in the aftermath of budget cuts and Nixon's sudden weak knees after the near disaster of Apollo 13. But they make fascinating targets for exploration through the eyepiece of our telescopes. With well-trained eyes, we can discover their beauties and secrets. In the words of Farouk El-Baz to the crew of Apollo 13 at their first meeting "Anyone can look, but few really *see*." El Baz wanted to create his own brain within the mind of Al Worden on Apollo 15. We will now learn his lessons. The Moon is where we will first learn to *see*.

Origins of the Moon

There are several theories that might explain how the Moon was created. Three have gained popularity over the years but all have faults. Any theory in order to gain acceptance and be plausible must be able to answer all open questions without raising any new ones. No existing theory currently is capable of doing that.

Any theory of lunar creation must be capable of explaining several issues that no existing theory can do with complete certainty. The first issue that must be explained is the close relationship in size between Earth and the Moon. This is unique among major bodies in the solar system and so must be explained by an event that is rather rare. Secondly is the fact that the samples returned from the Moon exhibit the same oxygen isotope composition as those examined on Earth. This means that the Moon formed in the same environment as Earth did. Rock samples examined on Mars have very different oxygen isotopes compositions for example. By inference, this means that the Moon and Earth formed in close proximity. Third is the fact that the Moon orbits Earth in a stable, nearly circular, prograde path. Fourth, such a theory must explain the dramatic difference in density between the two bodies. The density of Earth is 5.5 grams per cubic centimeter, the Moon is only 3.3 grams per cc. This is because the Moon is almost completely lacking in iron, while Earth is the most iron-rich body in the solar system. Finally if a theory proposes that the Moon is spun off from the primordial Earth, it must account for the lack of an impact scar somewhere on Earth's surface.

One of the earliest theories is that the Moon condensed out of the solar system's primordial debris cloud in the vicinity of Earth. But if this were true, then the Moon would be made of most of the same material that Earth is and should exhibit similar composition. But it does not, the Moon is almost a completely iron-free zone. The Moon's low density relative to Earth, the lack of any iron in the rocks and the fact that the Moon had no measurable magnetic field all illustrate this fact. The strength of this theory is that it readily explains the stable nature of the Earth–Moon system in a stable circular orbit and the Moon's tidal lock to Earth. It also explains the similarity in oxygen isotopes.

A second theory proposes that the Moon formed separately in the inner solar system and wandered too close to Earth and was captured by its gravity. But this theory fails to pass even a *prima fascia* test. The Sun is by far the dominant gravitational force in the inner solar system. For an object to be captured in a prograde orbit (counterclockwise as viewed from above Earth's north pole), it would have to approach from behind. As it neared Earth, it would be accelerated and flung back out of Earth's gravitational field into a solar orbit[15]. This "slingshot" effect is frequently used to boost deep space probes into the outer solar system. To be captured by a planet, an object must approach from the front, so the gravity of the capturing planet would act to slow the would-be Moon down. But this would in turn create an orbit that is both highly elliptical and retrograde in nature. This theory also cannot explain, except by chance, the Moon's lack of iron. It also does not explain why lunar samples contain similar oxygen isotopes. If the Moon formed elsewhere, it should exhibit differences in this measure in the same way that Mars does.

The leading theory today was formulated some 25 years ago. It suggests that late in Earth's formation, a large planetary sized body impacted Earth. The impact dislodged a massive amount of mantle material from both planets with sufficient velocity to enter Earth orbit. These materials then coalesced into the Moon. The theory satisfies the oxygen isotope matter because the Moon would have been formed from material, most of which came from Earth. Since the impact occurred late in Earth's formation, most of Earth's iron would have sunken into the core; this also explains why the Moon lacks iron. It also explains the Moon's neat and orderly orbit. It also accommodates the unique nature of the Earth–Moon system because such an impact event would be so rare. It is plausible that no other planet has

---

[15] The dominance of the Sun in this manner was clearly illustrated recently by an object that Earth's gravity captured briefly but after the object made several orbits of Earth, it drifted back out into a solar orbit again. What some thought might be a new "moon" of Earth was later identified as the third stage of the Saturn V that was used to launch Apollo 12.

undergone such an impact and that may be why no other planet has such a large moon. But this theory too leaves open items to be answered. First is what became of the impacting planet? To dislodge enough material from Earth to create the Moon, the impacting planet had to have been nearly the size of Mars! If so, then where did all that material go? And if the impactor had an iron core of its own, then where is the iron? Proponents of the impact theory postulate that any iron could have been merged into Earth's core and computer models do support this idea. But what became of the rest of the impactor? If it settled onto Earth's surface we should then be able to detect large quantities of rock that are not native to Earth (oxygen isotopes again). If it remained in space, it should have formed a ring around Earth. There is none to be found. Had it been coalesced into the Moon, we would see the differences in oxygen isotopes in the Moon's rocks. The final issue with the impact theory is the lack of a plausible explanation for an impact site. Though billions of years of erosion would have reduced the size of the wound, it would still be clearly evident. It is not. Some have suggested the Pacific basin as an impact site, but the depths of the ocean and composition of the seabed are no different than they are in the Atlantic or Indian Oceans. Despite its shortcomings however the impactor theory continues to hold up as the best current model on how the Moon formed.

Visual Lunar Phenomena

There are many phenomena visible involving the Moon that can be very easily enjoyed by eyesight without the use of any optical aid. The most common and obvious are the Moon's changing phases, which result from our changing perspective relative to the Moon–Sun line as the Moon circles Earth each month. When the Moon sits between Earth and the Sun, it shows us its dark side and we cannot see it at all. As the next week progresses, the Moon emerges from twilight appearing as a thin crescent and begins to reveal a bit more of its lighted side each day until it is half illuminated at "first quarter" about seven days after new moon. The Moon continues to grow brighter and fuller reaching a point directly opposite the Sun in the sky about 14 days after new moon. Now the Moon's disk is completely illuminated. During the next fourteen days, the process reverses itself and the Moon fades away again.

When the Moon is close to new, a beautiful effect takes place, which is called "Earthshine." In Earthshine, light striking Earth is reflected back towards the Moon. This causes the Moon's night hemisphere facing Earth to glow with a faint blue light. This is sometimes poetically called the "Auld moon in the new moon's arms." Earthshine can vary in intensity depending upon how much cloud cover is present on Earth, what type of surface is directed towards the Moon, oceans reflecting back more light than land. As the Moon waxes towards the first quarter, Earthshine becomes less and less apparent each night until it can no longer be seen, then gradually reappears after the last quarter. When the Moon is a thin crescent, Earth is nearly full, so a great deal of Earth's surface reflects a large amount of light to the Moon. As the Moon waxes, Earth wanes providing less and less light, so the Earthshine fades as the Moon reaches the first quarter.

Occasionally the Moon's path through space will take it in front of several bright stars. The Moon always remains within about five degrees of the ecliptic. Within this ten-degree wide band through which the Moon might travel are the first magnitude stars Aldebberan, Regulus, Antares, and Spica. The passage of the Moon across another object is called an *occultation*. When the Moon's path carries it in front of a bright star, it will usually do so as part of a long series of such events as the Moon's orbital plane drifts in front of the star, then away from it again. The Moon may occult one of these stars or more each revolution around Earth in a repeating cycle that can last for many months. The Moon can also pass through the Beehive, Hyades and Pleiades star clusters in the same way creating many occultations in the same night. An occultation of a star is said to be the most instantaneous sight that a human eye can behold. Since all of the planets orbit near the ecliptic plane as well, the Moon may also occult planets too. Long series of occulations can occur with Jupiter and Saturn because they move so slowly, but the inner planets move too quickly to remain in the Moon's path for a long period of time. Occultations of Mars occur infrequently, Venus rarely and Mercury almost never. Occultations can have important scientific value as well. Antares was found to have a companion that was unseen until the Moon occulted it. With the blinding glare of the parent star hidden, Antares bluish green companion peeked into view.

During its travels, while occultations of a given object are rare, *conjunctions* of the Moon with all the aforementioned objects occur every month and often can create startlingly beautiful patterns in the evening sky. Since the Moon travels around the zodiac once a month every month, there will always be a chance to use the Moon to guide you to something else. Many times for example, I've used a very young or a very old moon to help find Mercury when it was buried low in the twilight.

Periodically the Moon's path will carry it either directly across the face of the Sun, causing the Moon to cast its shadow on Earth's surface, or the Moon will travel through the shadow of Earth. This causes an event called an *eclipse*. These events are extremely rare for any given spot on Earth. Solar eclipses actually occur more frequently than do lunar eclipses but they can only be seen by the fortunate few who will fall under the Moon's tiny shadow. A solar eclipse that is at least partial must occur twice each year and there can be as many as five. A lunar eclipse cannot occur more than three times in a year and there do not have to be any. The reason why is that the Moon's shadow has a target to hit that is nearly 8,000 miles across, but the diameter of Earth's shadow at the distance of the Moon is about 4,500 to 5,000 miles across. But far more people can see lunar eclipses because anyone standing on the night side of Earth can see one when it occurs while solar eclipses are visible only to the lucky few who are in the path of the Moon's shadow. Lunar eclipses can occur in three types *penumbral, partial* and *total*. In a penumbral eclipse, the Moon only passes through the outer part of Earth's shadow, where the Sun's light is only partially blocked. The effects of a penumbral eclipse therefore are only very subtle and in fact if the Moon is not completely immersed in the penumbra may not be noticeable at all. In a partial eclipse, part of the Moon enters the dark umbral shadow where the Sun is completely hidden. The Moon appears as though a bite has been taken out of it. In a total eclipse, the Moon is completely immersed in the umbra which may cause a variety of beautiful lighting effects to occur, or may cause the Moon to vanish entirely. The coppery red color of the Moon is caused by light being refracted into the shadow by Earth's atmosphere. Aside from their beauty, eclipses can have important scientific value to scientists who study climate and climate change because the amount of light transmitted into Earth's shadow tells a great deal about the amount of dust or other contamination in the atmosphere.

Eclipses are not random events, but rather occur in families of similar events. A family of eclipses is called a *saros*. When an eclipse occurs on a given date, then exactly eighteen years, ten days and eight hours later, Earth, the Moon and the Sun will all return to almost exactly the same relative position and a nearly identical eclipse will occur. The only differences is that because of those extra eight hours the eclipse will occur 120 degrees further west around Earth and the Moon will pass about 200 miles further north or south through Earth's shadow than it did in the preceding event. So if a total lunar eclipse were to occur at midnight in London, then eighteen years and ten days later, another eclipse of nearly identical circumstances will take place at midnight for viewers on the U.S. west coast. Eighteen years and ten days after that, another similar eclipse will occur in eastern Asia. Fifty-four years and thirty-one days after the first event, another eclipse will occur at the longitude of London at about midnight, though the Moon's track across Earth's shadow will have shifted south by about 600 miles. This resonance of three saros is called an *exligmos*. The north or south drift is caused by the fact that the Moon's *synodic* (new moon to new moon) period is not exactly the same as its *Draconian* period (node to node). The nodes of the Moon's orbit drift slowly over time and the alignment at the next eclipse in a saros is not precise. This means that a saros cannot continue indefinitely. An eclipse series begins with the Moon penetrating the upper or lower (depending on whether the Moon is nearing the ascending or descending node of its orbit) part of Earth's shadow beginning a series of penumbral, then partial eclipses that become deeper and deeper with each of approximately seven to ten successive events until they become total. The Moon will travel further and further north or south until it begins to clear the opposite side of Earth's shadow and then eclipses become partial, then penumbral again until the Moon passes clear of Earth's shadow completely, ending the saros series. Currently there are 41 lunar saros series in progress. This means that within any period of 18 years and ten days, there will be 41 lunar eclipses then the entire cycle starts over again. Astronomers catalog and number saros series sequentially. Eclipse series that occur with the Moon nearing its ascending node are given even numbers and those with the Moon at its descending node are given odd numbers. The oldest current series in progress is Saros 109, which will have the last of its 73 events on August 18, 2016. Saros 109 began with a penumbral eclipse at the extreme south end of Earth's shadow on June 17, 718. I did not leave a number out. This series of eclipses is about to conclude a run of 1,298.1 years! Imagine a Broadway show running that long! A lunar eclipse saros must have a minimum of 69 events (1,226 years) and the maximum is 89 (1,586.6 years). The newest series is Saros 149, which began with a slight penumbral eclipse on the south edge of Earth's shadow on June 13, 1984 and had a second penumbral eclipse on June 24, 2002. This series will have 70 more events before ending with a penumbral eclipse on the extreme north side of Earth's shadow on July 20, 3246, a run of 1,262.1 years!

The Lunar 100

In the April, 2004 issue of *Sky & Telescope* author John Wood set out a list of 100 objects on the Moon that characterize the geological history and makeup of the Moon. Wood set out to create for the Moon a list of objects that are organized from easy to see and observe and increasing in difficulty with increasing number on the list. The first object for example is the Moon itself and is designated "L1." The second phenomenon is Earthshine (L2). The list then progresses to general surface characteristics such as the dichotomy between the highlands and the maria (L3) and then the Moon's most prominent mountain range, the Apennines (L4).

In creating the Lunar 100, Wood set out to create a lunar equivalent to Charles Messier's famous 110-member list of deep sky wonders. To observe all 110 objects on the Messier list is regarded as a right of passage for amateur astronomers. Wood set out to create a similar challenge for lunar observers. When Messier created his list of objects, he was attempting to create a clearinghouse of objects that could be mistaken for comets, which was Messier's primary interest. He did not intend to create an educational list for amateurs. Wood's list filled with features intended to illustrate important aspects of the Moon's geological history. In some ways the Lunar 100 poses an even more difficult observing challenge than does the Messier catalog. It is possible for example to view all 110 Messier objects in one night, but it is a physical impossibility to observe the Lunar 100 in a single night or even a single month. As libration rotates some objects into view, it hides others, or may present targets into our line of sight during lunar night.

Let's break out the telescope now and go for a visit to the Moon. The Moon's many different landforms tell the hidden story of its past and its origins. When the Moon is bright and prominent there's not much else to see in the night sky, so let's have a look.

Observing Projects V – The Stories of the Moon

Observing Project 5A – Tycho (L6): The Great Rayed Impact Crater

Tycho (L6) is one of the Moon's most beautiful signature craters. It sits in the far southern reaches of the Moon's visible face and is most famous for its characteristic "rays." Though no Apollo mission landed here, many at NASA called for a landing at this spectacular landmark. Such a landing never occurred because of the high risk inherent in landing on such rough terrain and because the trajectory required would preclude a free return trajectory. In 1968, the unmanned Surveyor 7 did land some 30 miles north of the crater rim in a bright ejecta blanket. Tycho is a relatively young crater, possibly less than 500 million years old. It is approximately 102 kilometers across and remarkably deep. The floor of the crater is approximately 4,800 meters below the rim. Tycho is the classic example of an

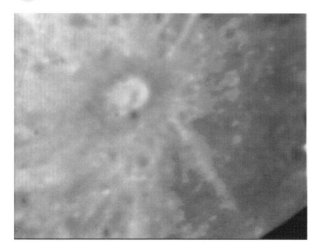

Figure 5.2. Tycho and its ray system. Image by author using Celestron C8 and a low-resolution video eyepiece.

impact crater. For many years it had been argued whether the Moon's craters were volcanic in nature or created by impacts. There seems very little doubt in the case of Tycho. As you look in on Tycho, notice that the crater floor is rough and it is bright. The material here is primarily anorthosite, the aluminum-rich material that makes up most of the lunar highlands. The same is true for the rays that emanate from the crater. That bright material is also anorthosite indicating that it is primarily surface material that had been blasted out of the impact site by whatever rocked the Moon's surface within the last 500 million years.

The walls of Tycho display the youthful nature of the crater. The walls are rough and terraced, suggesting that no eroding processes have taken place to age the walls. Also you should take note of the mountain formation on the crater floor. Tycho has a triple peaked mountain that rises from the crater floor, which towers nearly 5,000 feet above the crater floor. This feature is also indicative of an impact formation. A volcanic eruption would be unlikely to leave any peaks standing on the crater floor, never mind three of them. Can you bring enough magnification to bear on the crater to see all three? The triple peak should be easy in a telescope of 8 inches although I have also been able to see it clearly in scopes as small as 4 inches.

The best time to study Tycho, or any other lunar feature for that matter, is when the Sun is relatively low on the lunar horizon. Then the surrounding peaks and valleys create long shadows that cast the area in dramatic relief, showing just how rough the area in and around Tycho. If you study the area at local noon, the Sun casts very short shadows and thus even the roughest lunar surface features can appear flat. Local sunrise at Tycho occurs when the Moon is approximately eight days old. So look for Tycho when the Moon is about a day past first quarter. For the next forty-eight hours, the Sun is less than 20 degrees above the local horizon. Look again if you are a night owl when the Moon is about a day before last quarter (20 day old Moon) and Tycho is within about forty-eight hours of sunset.

Observing Project 5B – Plato – A Volcanic Blast?

Located well to the north of the lunar equator is the Moon's Alps mountain range. Many of the mountains here tower over 15,000 feet above the Imbrium basin to the south that the mountains surround. In the middle of these bright highlands is a huge dark floored crater that rivals Tycho in size. Oddly enough, Plato is not listed in Charles Wood's Lunar 100. The monstrous crater is 99 kilometers across and its floor is some 2,200 meters below the rim, about half as deep as the floor of Tycho. While Tycho's floor is bright and rough, Plato's is dark and flat suggesting that some part of this crater's history involves volcanic activity rather than being purely an impact event. The rock that makes up the floor of the crater is the same type of basalt that forms the floor of the nearby Imbrium basin and other lunar seas. While Tycho is very young, Plato is very old, between some 3.2 and 3.8 billion years old. It was formed during the Moon's early geologically active period. Exactly what process formed the monstrous crater is unclear but what is obvious is that whatever cataclysm created Plato bore deep enough to flood the massive crater with a layer of lava thousands of feet deep. Another issue of interest as we look at Plato is that even after billions of years, the crater floor is still almost perfectly flat. Assuming that the Moon has been geologically dead for billions of years, what force of nature then has been keeping the crater floor smooth for billions of years?

Plato is an anomaly in that it is so different from the surrounding terrain, the event that created it having crushed the mountain area around it. Many of the mountains surrounding the crater formed long before Plato itself did. The crater is some 3.8 billion years old, but the mountains surrounding it were thrust up hundreds of millions of years before during the Moon's formative period. What cataclysmic disaster blew this enormous hole out of the lunar surface, leveling mountains that stood some three or four miles tall? One of the things that are not commonly found on the lunar surface is a volcanic mountain such as those found

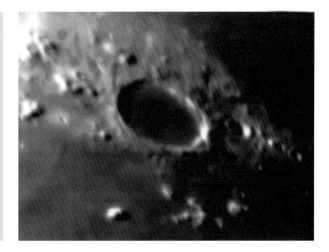

Figure 5.3. Crater Plato. Celestron Super C8 Plus and Meade DSI CCD. Photograph by author. The DSI's higher resolution chip provides a much sharper image.

on Earth or Io. But this is as close to a major volcanic feature that can be clearly viewed on the Moon's visible side. Did a volcanic catastrophe blow a mountain apart and leave this crater behind? Or might it be an impact crater that penetrated so deep as to flood the crater with lava? Nearly four billion years ago, the Moon had an active interior that would support either type of activity. When Tycho formed within the last half billion years, the Moon was already geologically dead, thus is was not possible to flood the area with lava.

Plato sits at approximately the same lunar longitude as does Tycho so your best opportunities to view it occur at about the same time. Local sunrise occurs when the Moon is around eight days old and the viewing geometry is very favorable for the next three days. Sunset occurs at about day 23 or about one day after last quarter, so the best time to look begins about three days prior to sunset.

Observing Project 5C – Star Hopping Practice, The Apollo 11 Landing Site

Selecting a landing site for the first lunar landing, mission planners for the flight were primarily interested in one thing, safety. The objective was to find a relatively flat spot with a paucity of craters or mountains that would allow for an approach to take place over a broad stretch of flat terrain. The logical choice would be an approach to the western side of a flat lava plain, since the Apollo missions orbited from east to west (retrograde). For the first landing, the planners chose the western side of the Sea of Tranquility. This particular spot also gave the advantage of a landing site near the lunar equator, which meant even as the Moon rotated underneath the orbit of the command module, the CM would not drift away from the landing site over several days. The low-latitude landing also made life much easier for the trajectory planners who were always adamant about maintaining the option for a free return in the event something went wrong.

Figure 5.4. Sea of Tranquility. Can you find Apollo 11's landing site? NASA Lunar Orbiter IV image.

In this project, we will attempt to zero in on the landing site of Apollo 11 using a technique that will become very valuable to you in later efforts to explore the deep sky. The technique is called "star-hopping." Star hopping involves beginning at a bright or distinct feature that is easily found, then from it, zeroing in on a fainter object by following a trail of other features. The features that we'll use are listed in the Lunar 100 and we'll travel from point to point in ascending order from easy to find to hard to find. You will not by any means be able to see the actual landing site on the surface but to identify the point of landing is exciting because it will build your confidence in your ability to navigate at the telescope. You will also have the gratification of looking at the spot where one of the most important events in the history of the humanity took place.

The mission of Apollo 11 was timed such that it would land at the target site just after local sunrise, about a day or so afterwards. This was done for the same reason that we like to observe lunar features when they are near the lunar terminator. The lunar shadows at that time would give the crew a good degree of depth perception to help them better judge their height and forward speed during the final moments of their manually controlled descent to the surface. With the Sun about ten degrees above the local horizon, shadow lengths are ideal. You will want to have about the same lighting conditions as the crew did. These conditions typically occur when the Moon is waxing and about five days old. You can also have success when the Moon is waning and sunset is nearing at Tranquility Base when the Moon is about twenty days old, but for the novice, this is far easier at sunrise.

Start out by locating the Sea of Tranquility (Figure 5.4). As the Moon waxes, two *mare* come into view near the Moon's east limb. Mare Crisium is the first to appear, then Mare Fecunditis appears to its southwest. Then over the next two days, sunrise occurs over Mare Tranquillitatis. Since the landing site is on the extreme western edge of the mare you must wait for the terminator to expose the entire area. Once you have identified the Sea of Tranquility, search along its southwestern edge for two twin craters in a northwest to southeast line nearly touching each other. Each of these two impact craters is approximately 30 km across. The northwest crater in this remarkable formation is called *Ritter* and its counterpart to the southeast is called *Sabine*. The two craters are listed together in the Lunar 100 as L38. With good shadow relief, these two craters should be easy to see with even low magnification. Now things get harder. Use the highest magnification eyepiece that the telescope and seeing conditions can bring to sharp focus and center on Sabine. The two craters are easily seen at the lower left in Figure 5.4 near the edge of the Sea of Tranquility. In the image, north is up and east is right. Remember to account for the orientation change in your telescope. If you are using a refractor or Cassegrain design with a star diagonal, north is up, but east is left. In a Newtonian, north is down.

From Sabine, look very carefully off towards the east. You are using very high magnification now because the next thing you are looking for is a string of three tiny craters each of which is less than three miles across. At high noon, you could never see them because without shadows to provide contrast they are just too small to be viewed. Shadow relief makes seeing them possible with steady air and effort. The westernmost crater in the string is called *Aldrin* and it is about 55 km east of Sabine. About another 49 km east of Aldrin is an even smaller crater called *Collins* and about 40 km east of that crater is the largest of the string called *Armstrong*.

These three craters are named for the three astronauts who crewed the mission that landed just to the south of this spot and are collectively listed in the Lunar 100 as L90. The challenge is to find Collins. From Collins, scan directly south to a small bright crater called *Moltke* which is about 6 km across and some 88 km south of Collins near an outcropping of highlands. The crater is easily visible because of the contrast it makes with the surrounding basalt plains. If you imagine a line connecting Collins to Moltke, exactly halfway down this line, then about one-third of the length of the line to the west is the spot where Apollo 11 set down on July 20, 1969. If you cannot find Collins, then the landing site can also be found by drawing a line between Aldrin and Moltke. The site is on this line about one quarter of the distance from Moltke to Aldrin.

Later on you will apply this technique of using a bright object or an easy to find feature to navigate to more obscure objects and zero in on your treasure. Before computer driven telescopes flooded the market, star hopping was how amateurs learned their way around the sky and made even its faintest treasures as familiar as the features of any road map. Here we used the brightest features of the Moon to zero in on conspicuous features, then to find obscure ones on our way to one of the most historic places in the solar system. There are two important things also to remember when viewing the Moon and looking for fine details. First you should always use a neutral filter to reduce glare. A good Moon filter will cut the total light from the Moon by almost 90%. Secondly, find the highest power eyepiece that you can possibly use under the existing condition, put it in the scope and crank it up! There is an abundance of light available from the Moon to make crisp images possible where it otherwise would be a waste. So this is one of the places where when it comes to using high power, its time to go for it.

Observing Project 5D – Geological Forces at Work I – The Marius Hills (L42)

The Moon as we see it today is geologically and volcanically dead. It is completely motionless through to the center of its core. Unlike Earth which is geologically very alive with shifting plates and a molten interior generating currents of molten iron and shifting powerful magnetic fields, the Moon is completely still. The Moon was not however always this way and certain features on its surface indicate with certainty that the Moon was once very active with a shifting surface and burgeoning fluid mantle.

Near the Moon's western limb, well to the south and the west of the lunar Alps, surrounding the crater Marius is a complex area of bulging hills which ring the crater on three sides known as the Marius Hills. In the Lunar 100, the hills are cataloged as L42. NASA strongly considered that area for the landing of Apollo 18 during the period of time before the Nixon Administration, when there still was an Apollo 18. The same types of disturbances created the Marius Hills in the lunar mantle that created the Badlands of North Dakota here on Earth. Magma pushing up from the liquid mantle below created bulges in the crust but could not push close enough to the surface to break through before the fluid rock hardened beneath the surface. These formations are called "volcanic domes" and are signs of the presence of an active fluid mantle long in the lunar past.

To find the Marius Hills, a good place to start is actually far away at a familiar place, the crater Copernicus, designated L5 in the Lunar 100. Look about 530 km to the west into Oceanus Procellarum, where another prominent crater stands largely by itself. This crater is called *Kepler* and it is easily identified by a prominent ray system. From Kepler, go about 400 kilometers to the west-northwest across Oceanus Procellarum to find another solitary crater. We can make such large jumps across the surface here because large craters are in short supply in this smooth area of volcanic basalts. This crater is Marius and it too is easy to find because it largely stands by itself. Surrounding the north, west and south sides of this crater are the Marius Hills. The best observing time for the Marius Hills complex is when the Moon is approximately nine to ten days old and again when the Moon is about twenty-four days old just prior to local sunset.

The hills are softly rounded and dome shaped, suggesting that they were driven up from below. Note as you look through the area the absence of any sharp peaks such as what you saw when looking around Plato. The area surrounding the Marius region is largely flat which tells us there was not any other major mountain building force ongoing in that area. Marius stands alone, conspicuously sticking up in the middle of one of the Moon's largest volcanic seas. It is likely whatever force excavated Oceanus Procellarum left the crust beneath weakened to future volcanism. The once active mantle pushed magma up towards a weak spot driving up the hills. They are relatively shallow, only reaching heights of about 5,000 feet or so. The crater itself is about 40 km across and only about 5,000 feet deep. The walls of the crater support the hills to the north and west, but never caved in under the mountain-building processes. But in the northern hills, magma did break through to the surface. To the north is a very difficult to see channel called Rima Marius. At some point in the mountain-building process, it seems likely that magma did break through the surface and was channeled away by into the surrounding lava plain through this channel. Rima Marius is approximately 290 kilometers long and a maximum of 2 kilometers wide. More typically, the rille runs about 500 to 1000 meters wide.

So the Moon was once volcanically active with dynamic volcanic mountain-building forces, just like Earth. Did the Moon also once have active and moving crustal plates? Fault lines would indicate the presence of such plates. Does the Moon have any?

Observing Project 5E – Geological Forces at Work II – The Straight Wall (L15)

It is called "California's Terrible Wound." The San Andreas Fault extends across hundreds of miles through the mountains of southern California where the Pacific and North American continental plates run together and the Pacific plate is forced beneath the surface and into the mantle of Earth. As the plates rub against each other, unimaginable amounts of energy are building up, waiting to be released. This scenario is played out on fault lines all over Earth. Did these processes ever take place on the Moon?

Located near the Moon's central meridian is the *Straight Wall* which is in the Lunar 100 as L15. This is the most obvious example on the Moon's near side of a fault line. If you use a map to locate Mare Nubium, the Straight Wall pops into view very easily, most especially at low Sun angles. The wall runs for approximately 155 km across the eastern edge of Mare Nubium. Look for the Wall beginning about eight days after new moon and again as sunset approaches around twenty-two days after new moon. The long shadows created by the low Sun at these times gives rise to a stunning illusion. The wall truly looks like a wall but in fact this so-called "wall" is sloped only seven degrees from horizontal.

We don't know enough at this point about how the Moon's crust is divided and whether or not more of these fault lines exist below the visible surface. It is a fascinating point of geology to note that the location of the fault is near the "shoreline" of Mare Nubium and highlands begin to the east. Most of Earth's faults can be found in similar geological confluences, such as the Pacific's "ring of fire" that runs from the tip of South America, up the Andes and the U.S. west coast, up Canada and Alaska, then all the way down the coast of Asia into the islands of Indonesia. The Moon has been dead geologically for billions of years, but features like the Straight Wall serve as fossilized reminders that the Moon may once have been as geologically dynamic as Earth is today.

Observing Project 5F – Peaking Around the Rim I – The South Pole–Aitken Basin (L98)

As the Moon moves around Earth in its monthly orbit, changes in its orbital speed and the inclination of its orbit and tilt of its axis cause it to appear to nod back and forth and up and down giving us tantalizing glimpses of what lies around the rim on its far side. As the Moon's south pole tips into view, we gain a glimpse at what has become one of the most fascinating places in the universe over these past ten years.

Scientists had long speculated about the possibility that some craters on the Moon may have been formed by comet impacts over the millennia. Since comets are made mostly of water, large amounts of water ice may have been deposited on the lunar surface during those impacts. Exposed to the Sun and to the effects of daytime temperatures well above the boiling point of water, the ices would have boiled off and escaped to space eons ago. But at the south pole, many crater bottoms reside in permanent shadow and the temperatures there are always around −130 degrees Celsius. Any ice deposited in these areas could then remain frozen leaving perhaps millions of gallons of frozen water available for use by future human explorers either for consumption or for distillation into the components of rocket fuel. In the mid-1990s, the NASA/DOD Clementine spacecraft turned speculation into the very exciting possibility that lunar ice might be for real when its onboard spectrograph signaled the presence of hydrogen in those permanently shadowed craters. NASA's late 1990s follow-on mission, Lunar Prospector, also confirmed the presence of hydrogen and attempted to detect the presence of water by crashing itself at the end of its mission into one of those craters. Researchers hoped that the probe's impact would throw up a cloud of vaporized water that could in turn be detected by instruments on Earth. No such cloud appeared, but the evidence for water on the Moon remains compelling enough to warrant future investigation. NASA's upcoming Lunar Observer probe will further these investigation in advance of the return of human exploration of the lunar surface during the next ten to fifteen years.

The South Pole–Aitken Basin is L98 on the Lunar 100 list. It is placed so high on the list because its position on the lunar rim makes it difficult to see under the best of conditions and requires a favorable libration to bring it into view at all. The view is foreshortened by the viewing angle, so viewing into the crater bottoms is impossible. Since sunlight cannot reach the crater bottoms either, there is nothing to see anyway in the bottoms of these craters. Since the viewing angle is so shallow, we get a much truer representation here of just how rough the Moon really is. Here you can see that the Moon has been battered and tortured by meteor impacts to a point where it is considerably "out of round." And as the Moon revolves around Earth, each month the south pole will show us a slightly different perspective which makes it worth returning here every month because during no two months will it ever look exactly the same. NASA will undoubtedly be looking at this area as a future landing site as it begins to formulate its return to the Moon over the next decade. Here is your chance to get an early preview of what might be one of the most exciting space discoveries ever.

Observing Project 5G – Peaking Around the Rim II – Mare Orientale

Many of the Moon's most fascinating geological features lay hidden just out of view around the Moon's far side. The far side of the Moon is dramatically different in appearance from the near side in that it is almost completely lacking in lava filled *maria* that are so evident on the Moon's near side. But there is one fascinating exception, Mare Orientale. Normally this amazing feature is hidden from our view on the Moon's far side, but when favorable western librations occur, about half of the Orientale basin peaks into view around the Moon's western limb.

Mare Orientale is distinguished by a series of concentric rings that were created by the shockwaves that accompanied impact or eruption that carved out the basin. When Orientale librates into view, we get a good foreshortened look at the twin rims of peaks that surround the lava plain at the center. Orientale had teased astronomers and geologists for centuries until the Lunar Orbiter probes of the 1960s got our first good direct look at the entire area, confirming that the rings ran for the entire circumference of the basin. This tells us that the force that created them was first massive, and second relatively recent. The basin shows little evidence any erosive force has taken place over time other than the flow of lava through the basin.

The best time to look for Mare Orientale is when the Moon is full. At this time, it is local sunrise at the basin and the rims will show you their greatest shadow relief. Sunset locally occurs just before new moon, but then the Moon will be low in twilight and not worth observing. Then you need the benefit of a favorable libration to bring the rings into view around the Moon's western limb. Though the Moon does rotate this area into view each month, it does not always occur when the Moon's western limb is illuminated. Nor does it always occur when lighting conditions are at their best. Good opportunities to view Mare Orientale are thus few and far between. Since viewing is best at full moon anyway and you will not be doing much else, make sure you note when viewing is good for Mare Orientale and take a good look at one of the Moon's few hidden treasures that actually occasionally peaks into our view.

The Moon is at first glance boring and brilliant, obscuring all else in the heavens when it is present. But on close inspection, it can be an endless treasure trove of eye candy telling the story of the chaotic early years of the solar system. It is a wonderful geology laboratory frozen in time, little changed in nearly four billion years. It is also an invaluable tool for you to hone your observing skills, by finding fine details and being able to navigate from feature to feature working in bright light before you move on to the deep sky when only very faint levels of light will be available. So go and visit the Moon as you begin to unlock the mysteries of the heavens. There's far more here than meets the eye.

CHAPTER SIX

Secrets of the Sun

Figure 6.1. The Sun on Dec. 27, 2002. Celestron Super C8 Plus and 35-mm SLR at prime focus. Image by author.

Its brilliance and heat make possible all that we know on Earth. Its luminosity lights our days and its heat warms our world. As technology has advanced we've learned to harness its power to create clean and inexhaustible supplies of energy for both spacecraft and terrestrial applications. Our Sun and the uncountable billions of other stars in the universe are far more though than just that. They are astonishing nuclear furnaces that within them create from atomic fusion nearly everything that we are made of. The nuclear reactions of the Sun and other stars create from simple hydrogen every other element of the universe. The calcium in our bones, the oxygen we breathe, the iron in our blood were all forged in the nuclear hearts of the stars. Carl Sagan, the greatest publicist our science has ever had, said it best in his book *Cosmos* when he said, "We are all made of star stuff."

Though the Sun is the dominant force in the solar system and seemingly all-powerful and mighty in its own neighborhood, in the grand scheme of things the Sun is actually something of a dullard. From our point of view, the Sun is overwhelmingly bright, shining at an *apparent magnitude* of −26.7 because we are so close to it. But if a common measuring stick is used, then the Sun does not seem so bright after all. That common measuring stick is called *absolute magnitude*. This is the measure of how bright a star would appear at a standardized distance, which is defined as 10 parsecs (32.6 light years).[16] At that distance, the Sun would only appear to shine at about magnitude +4.8, faintly visible to the unaided eye. By comparison at the same distance, Sirius, which appears to be the brightest star in our night sky at magnitude −1.5 from our perspective 8.8 light years away, would shine at +1.4. Spica appears to shine at about magnitude +1 from 270 light years away but at 32.6 light years would shine at magnitude −3.5 (about as bright as Venus). Deneb, at the head of Cygnus, appears to shine at magnitude +1.25 from 3,200 light years distance has an absolute magnitude of −8.7!

Though nowhere near the high end of the market in total luminosity, the Sun pretty much lives in the stellar mainstream. In 1913, the astronomer Henry Norris Russell plotted a graph, which took into account a star's total luminosity and spectral type. Spectral type is based mostly upon temperature (but also partly on spectrographic qualities) and is the horizontal scale on the graph. Cooler stars are plotted to the right on a scale that runs starting with the hottest stars classified as "O". The original modern scale then runs through B, A, F, G, K and finally the coolest stars are spectral type "M." Each of these classes is then subdivided by number, 0 through 9. Lower numbers are the hottest and higher numbers are cooler stars. Luminosity is graded on the vertical scale of the graph by absolute magnitude. Spectral class usually also corresponds to a star's color. Cooler "M" type stars are red. The hotter "O" and "B" types are blue. The Sun is spectral type G2. By combining these two values (G2 and absolute magnitude +4.8), we can plot the Sun as a data point on the graph. If we were to plot the spectral type and luminosity of a thousand stars at random, we would find that a huge majority of them would occupy a narrow band that runs from the bottom left of the chart (cool, faint, red stars) to the upper right of the chart (hot, bright, blue stars). This band is called the "main sequence." The one thing that all main sequence stars have in common is that they are all in the hydrogen fusion portion of their lifetimes. Some years before Russell's work became known, it was learned the astronomer Ejnar Hertzsprung had performed the same analysis and independently discovered similar patterns. Today, the charting system they independently created shares the names of both men, the Hertzsprung–Russell diagram. In recent years, new spectral types have been added to the classic Hertzsprung–Russell diagram to accommodate new types of stars that have been discovered. S-type stars are cool red giants with peculiar oxide absorption bands that typical M-types do not have. The C-type class has been created for very low luminosity stars cooler than M types that astronomers call "carbon stars." For the mysterious brown dwarf types,

---

[16] The term "parsec" is not techno-babble out of Star Trek, but an actual expression of distance. It is shorthand for "parallax second." It is the distance at which a star would appear to shift by one arc second on the sky due to the motion of Earth around the Sun. This distance is equal to 3.26 light years.

spectral types "L" and "T" have been added to the right of "M." Our Sun resides not only within the main sequence, but sits pretty much right in the middle of it in terms of overall luminosity and temperature. Our Sun is about as average as average can be.

Life and Death of the Sun

The Sun, like all other stars, was born from a cloud of interstellar dust and gas that began to condense under gravity. As the pressure increased at the center of the cloud, temperatures rose as a result of a release of gravitational energy and when sufficient mass, temperature and pressure had been built up, nuclear fusion began to take place at the core of the newborn Sun. From the time the first contraction of the interstellar cloud takes place to the first fusion takes approximately 30 million years for a star the mass of the Sun. Material not pulled into the core of the newborn star may make smaller stars, brown dwarfs or planets. In the simplest and most efficient form of fusion, two hydrogen atoms are fused together to make one helium atom. The energy released as a result of this reaction causes the star to emit radiation across the entire electromagnetic spectrum, including the visible range. Depending upon how much hydrogen is available to a given star, it will burn somewhere along the main sequence. Ironically, the larger the star, the hotter it will burn and the exponentially faster it will run through it fuel. The Sun is estimated to have approximately a ten billion year supply of hydrogen to fuse during its main-sequence lifetime. Massive, hot blue stars like Rigel will only live for a few hundred million years before exhausting their fuel, while tiny, cool red stars like Proxima Centauri can burn for many times longer than the Sun can ever hope to live without ever changing.

The Sun has been converting hydrogen into helium for some five billion years and will continue to do so for about five billion more. In the meantime the Sun is converting billions of tons of hydrogen each second into helium and that helium is accumulating in the core of the Sun. At present the pressure in the Sun is insufficient to fuse helium into heavier elements so it sits in the Sun's core accumulating. The energy radiating outward from the core precisely balances against the mass of the Sun trying to collapse inward under gravity, maintaining the Sun in a steady state. When the Sun's supply of hydrogen begins to run out, the energy output will fall off and the core of the Sun will collapse under gravity. As the core collapses, gravitational energy will cause the core to again begin fusing hydrogen, but only in a narrow ring around the outer edge of the still collapsing core. The renewed fusion will then begin to push the outer layers of the Sun away into space, enormously increasing its surface area while temperatures at the surface cool dramatically. The combination of increased surface area will be roughly offset by cooling temperature causing the Sun's total luminosity to remain constant.

Meanwhile at the center of the Sun, inside the shell of fusing hydrogen, the helium rich core is still collapsing and still increasing in temperature. When core temperatures reach about 100 million degrees Celsius, the helium begins to fuse into beryllium, then carbon and oxygen. This reaction is not nearly as efficient as was the hydrogen fusion so as helium becomes the Sun's dominant source of output; the Sun will cool and become red in color. The outer layers of the Sun will escape into space forming a shell that reflects light from the dying core called a "planetary nebula." Eventually when most of the helium has been fused into heavier elements, fusion ceases and all that is left is the solid collapsed core. The Sun lacks sufficient mass to fuse carbon and oxygen into heavier elements. More massive stars may have multiple shells surrounding their cores performing fusion of hydrogen, helium, beryllium, carbon, oxygen, silicon, sulfur and finally a core of iron. What will happen then, we'll discuss in a later chapter. For the Sun, its continually collapsing core will create pressures so extreme that even atoms cannot stand up to it. Electrons will be crushed into the nuclei of their atoms creating heavy neutrons (called "degenerate matter") and emitting white light from the Earth-sized remnant. The degenerate core, consisting entirely of neutrons will become so dense that a teaspoon of it will weigh a ton! The dead Sun will continue to glow in this way for many billions of years before extinguishing into a dark heap of degenerate matter.

While the Sun will end its life in an astronomical whimper, for today it is the great source of power and life in the inner solar system and the greatest natural nuclear physics laboratory available to us. Let's take a closer look at our amazing Sun and unlock some of its secrets.

Safe Solar Viewing

The Sun is a treasure trove for observational astronomy but unlike any other object in the universe, observing the Sun is by definition hazardous to your health and you must make sure you use proper precautions and proper equipment. You must make sure that you use a safe and properly made solar filter. There are many types that are in use, some of which block solar light at the objective and others that block light at the eyepiece. If your telescope is an older design which blocks or shunts light away at the eyepiece end of the telescope, please do yourself and your eyes a favor now and *throw it away!* Eyepiece solar filters are extremely dangerous because the objective of the telescope amplifies the light many times over as it brings the light to focus on the glass of that filter. Many of these glass filters have shattered as a result, causing serious eye injury to the observer. The only filters considered safe for use today are objective filters.

Objective solar filters are made by every major manufacturer for their own telescopes and can also be purchased from many popular on-line astronomy stores. Filters generally come in two types today, glass type and Mylar-type. Glass filters offer the advantage of a more natural color, but require great care to avoid scratching them. Scratches could badly distort the view and let unsafe amounts of light through. Mylar filters have the advantage of having redundant layers of material. A typical Mylar filter will use two sheets of Mylar with the inner side of each sheet coated in aluminum. Since the aluminum filter media is not exposed to the elements or to your hands, you cannot scratch it without punching completely through one of the Mylar sheets, which is a lot tougher than you think. There is a disadvantage to the use of such filters and that is that they yield an unnatural color to the Sun. Mylar filters block infrared radiation almost completely but this also biases the red end of the spectrum. As a result, Mylar filters tend to create an image of the Sun that is an unnatural blue in color. But this is little more than an aesthetic inconvenience because the Sun is basically one color anyway. In most of my photographs of the Sun, such as the one above, I remove the color information from the image anyway. The grayscale is actually more natural in appearance and sharper in quality.

When using any solar filter, make sure you inspect it carefully for damage before attempting to use it on a telescope. I will check my Mylar filter for pinholes by carefully scanning the Sun across the entire area of the filter before attempting to use it on the telescope. This is time consuming, but worth the investment of some time before risking my observing eye on that filter. I then will place the filter over the objective of the telescope and secure it there with some tape. Any tape will do just as added insurance that a gust of wind will not rip the filter off the telescope while I am looking through it. My particular type of filter fits over the outside of the corrector cell and does not plug in to the interior of the cell. The wind can thus get under the filter rim and pull it off in theory. The filter fit is very tight, but I'm not willing to take risks on it.

Aiming the telescope is also something that must be done with care. I don't have an additional filter for my finder scope so I cannot use that. Don't even think for a moment that because the finder is small, it cannot hurt you. A glimpse of the Sun through the finder scope will blind you just as fast as the main scope will. So I must find the Sun with both dust caps left on the finder. Aim your telescope by pointing it roughly at the Sun, then watching the shadow on the ground. As you get closer to the Sun, your scope's shadow becomes less elongated and when the aim is perfect, the silhouette of your telescope on the ground should be a perfect circle. Now perfection is usually beyond the reach of most of us, but this should get you close enough to see the Sun's glare in the eyepiece and from there you can find it for yourself.

Even without a telescope, it's very easy to observe the Sun with something as simple as two pieces of cardboard or even a shoebox. The simplest thing to do is punch a pinhole in one side of the shoe box and tape or glue a white piece of paper to the inside of the other end of the box. By pointing the pinhole at the Sun, you will project an image of the Sun's disk on the white paper. You can do the same thing with two rigidly aligned pieces of white cardboard. Either way, it's a cheap, easy to make and perfectly safe way for large numbers of people to safely view the Sun.

Visual Solar Phenomena

When looking at the Moon, we learned the importance of looking for the fine detail in craters and other features that stand out very plainly in sight. The Sun's outer gas layers are arrayed in three layers. The Sun's outermost visible gas layer, which we see when we view it through a telescope, is called the *photosphere*. The other two layers, the *chromosphere* and the *corona* are invisible under normal circumstances. Looking at the photosphere requires attention to detail because at first glance, it looks homogenous. But on close inspection, the Sun begins to reveal a

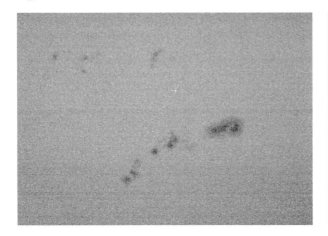

Figure 6.2. Close up of sunspot group on Dec. 27, 2002. Celestron C8 and 35-mm SLR with eyepiece projection. Photo by author.

lot of detail. Its face is not really smooth as you might think at first but rather looks more like a plate of white rice (or yellow or blue or whatever color your filter renders the Sun). The Sun's face is completely covered in these so-called *faculae*. The faculae are areas of upwelling gas in areas of rising convection. The gas rises to the surface, cools and falls back into the Sun. The Sun's entire surface is covered in these surging pockets of hot gas causing its surface to perpetually rise and fall in mountains and valleys of 10,000 degree Celsius gas. Through any telescope with a proper solar filter the mottled surface of the Sun easily reveals itself. Remember that when you look at these seemingly tiny pockets of gas, each little cell you see can easily swallow Earth whole. Another amazing fact about the gases is their astonishing age. Each time a pocket of gas wells to the surface it represents the culmination of a journey of some fifty thousand years from the time it began to rise up from the center of the Sun.

The Sun is powerfully magnetic and currents of magnetic activity are perpetually coursing through the photosphere. At times, many magnetic currents become bundled together in close proximity. This causes the local magnetic field to increase in strength by thousands of times what is normal for the solar surface. The strong magnetic field inhibits the upward movement of gas creating a "cool" area on the surface of the Sun. The blocking action of the magnetic field creates a depression on the surface of the Sun that can be several hundred kilometers deep. The average temperature in the depression is approximately 4,200 degrees Kelvin[17], about 1,500 degrees cooler than the average temperature in the photosphere. Because the area is cooler, it does not glow as brightly as the surrounding surface area and in fact the contrast is so great that the depressed area appears black. This is an optical illusion. If viewed in isolation, the surface area of a Sunspot is actually extremely bright. The actual area of depression in the photosphere is the darkest area of a sunspot and is called the *umbra*. A lighter colored outer area called the *penumbra* surrounds the umbra. The lighter color is caused by faculae

---

[17] The Kelvin scale is preferred for use in most physics applications. One degree Kelvin equals one degree Celsius. The zero point of the Kelvin scale is set at absolute zero, the theoretical point where all atomic motion stops. Zero K equals −273° C.

encroaching upon and overriding the outer edge of the sunspot, which is still visible below the overriding gasses. The average sunspot is approximately twice the span of Earth. Spots can form by themselves on the Sun but often appear in pairs. When two sunspots appear in close proximity they are usually areas of opposite magnetic polarity, like the two ends of a magnet. Powerful magnetic currents can be conducted between the spots and this in turn can cause powerful solar eruptions called *solar flares*. The Sun can throw off massive amounts of hot gasses into space along with enormous streams of electrically charged particles. When these particles interact with Earth's magnetic field, they cause the beautiful high-altitude light displays called *aurora borealis* or "northern lights" (in the southern hemisphere, this is called *aurora australis* or "southern lights"). sunspots can also occur in large groups of 100 or more in a long string that can persist across the Sun for many weeks. Often times such a group will form and rotate out of view around the Sun's far side and then reappear about two weeks later. Sunspots also only appear between a range of latitudes between about 40 degrees and 7 degrees north and south of the Sun's equator. Sunspots can never be seen in the polar regions. The reason for this is that the Sun does not rotate as a solid body. The Sun rotates at the equator once each every 25 days, but requires 29 days for a rotation at the poles. This amplifies the Sun's magnetic field at the equatorial latitudes.

Sunspot activity is cyclical, peaking on an eleven-year cycle that is extremely regular. From the peak, sunspots will fall off to a minimum over a three-year period then activity will gradually build again to a new peak over the next eight years. At peak times of activity, such as occurred in 2002, sunspots can number as many as 150 per day. When the sunspot cycle ebbs to a minimum, the Sun can be completely devoid of any spots for many weeks at a time. As activity begins to increase again, sunspots will first appear in the mid latitudes at about 35–40 degrees from the equator and then as activity increases, moves towards the equatorial regions. As each cycle is completed, the Sun completely reverses its magnetic field. At the end of the next eleven-year cycle, the field reverses again. This twenty-two-year pattern is called the "Hale cycle."

As the Moon travels around Earth, at least twice each year it will block at least part of the Sun creating a solar eclipse. If any part of the photosphere is left exposed then a partial eclipse will be the result. Even if the Moon passes centrally across the Sun's face, it may not completely cover the Sun. It is one of the most amazing coincidences in nature that the Sun is simultaneously 400 times larger than the Moon and 400 times farther away. When the Moon is near the perigee point of its orbit, the Moon appears slightly larger than the Sun and a *total* eclipse occurs. If the Moon is farther away from Earth near the apogee point of its orbit, then an *annular* eclipse occurs with the Moon leaving a ring of photosphere visible behind it. Other than the unusual "ring of fire" effect, there is no significant difference between a partial eclipse and an annular eclipse. It is the total eclipse that is most special. When the brilliant photosphere is completely hidden, the beautiful outer atmosphere of the Sun comes into view. The rim of the Moon is surrounded by the pinkish *chromosphere*. The name literally means, "sphere of color." After gases cool during the climb through the photosphere, they begin to heat up again during the ascent through the chromosphere. Temperatures will rise from about 5,700 K in the photosphere back up to around 10,000 K in the chromosphere. As the chromosphere becomes visible, one might also see the appearance of solar *prominences* peaking out of the pink gases. Prominences may or may not be associated with sunspot activity. Those that are associated with sunspots are called "active" and will persist for several hours. Active prominences can be among the most spectacular features on the Sun and sometimes will form dramatic arcs of gas between sunspots. Prominences that form away from sunspot activity are called "quiescent." These prominences may extend for tens of thousands of kilometers into space and persist for months. Gases erupting from a quiescent prominence will eventually peak out in altitude and fall back into the photosphere and can be seen as a sort of gaseous "rain" falling back out of the outer atmosphere. The Sun's outer atmosphere is called the *corona* and is the one of the great signature sights of astronomy. The sight of the corona during a total eclipse is one of those visions that turn people on to a lifetime of astronomy. The corona is divided into two major components. The high-energy "K" corona consists of high-energy electrons streaming from the Sun in various streamers and plumes. As the photosphere generates active regions beneath it, the corona changes in shape and texture to the eye as various plumes and streamers grow and shrink. The high-energy gasses in the corona can reach temperatures of nearly one million degrees K. The outer part of the corona, called the "F" corona, glows in a softer and more consistent white light that is scattered by slow moving dust particles around the Sun. The beauty of the corona is the thrill that brings amateur and professional astronomers alike from around the world to stand in the narrow shadow of the Moon.

Solar eclipses occur in saros series just as those of the Moon do. Currently there are 39 solar saros series in progress. It is ironic how the laws of probability work and don't work with eclipse tracks. During the last ten years, the same town in west Africa experienced two total solar eclipses within 18 (June, 2001 and December, 2002) months but the entire mainland United States is in an eclipse drought that began with the total eclipse of 1979 and will not end until 2024 (not counting annular eclipses of 1984, 1994 and 2017). Solar saros series are numbered just as lunar series are, though the odd–even node relationship is reversed from lunar eclipses. Odd numbers are reserved for ascending node events while descending node events get the even numbers. A solar saros progresses much the same a lunar one does. For example, the youngest currently running series is Saros 155, which began its run of 71 events with the first of eight partial eclipses on June 17, 1928. The series will then go on to produce 56 central eclipses, of which 33 will be total, 20 will be annular and three others will be a mix of both. The series then ends with seven partial eclipses, the last of which will occur on July 24, 3190 ending a run of 1262.1 years!

Observing Projects VI – The Power and Beauty of the Sun

Observing Project 6A – Tracking Solar Activity

The activity level of the Sun generates far more than just sunspots. As the Sun becomes more and more active, it affects our lives more and more particularly as we become more and more dependent on technology. Satellites route more and more of our television, phone calls, cellular communications, Internet, air and sea navigation services and information. Long-range communications between airlines and air traffic control can be disrupted. As the Sun grows more and more active, the potential for service disruptions grows. Our dependency on space-based technology has grown to such an extent that it has required the birth of a whole new science, *astrometeorology*, or literally "space weather." Here is your chance to be the weatherman.

As often as possible, keep an eye on sunspots. As the Sun's activity level increases, watch for sunspots to increase in terms of both quantity and quality. As activity increases sunspots will also begin to appear in groups. They will also become more persistent. The key things to watch for are sunspots that are forming in pairs, or in pairs of groups in close proximity. These are prime places for solar flares to occur or for large prominences to be thrown into space. If one should occur while it is rotating past Earth in space, enormous amounts of high-energy radiation and gas will be thrown at us in what as known as a *coronal mass ejection*. These create prime conditions for aurora to form and by keeping an eye on the activity of the Sun you may be among the first to know when an aurora is imminent, along with the various negative aspects of strong solar activity. Astronomers now monitor the activity of the Sun very closely because an energetic Sun is a threat to many billions of dollars of commercial and military space-based assets. The low-orbiting space shuttle and International Space Station are not at risk because they operate well beneath the protective shield of Earth's magnetic field but high-orbiting geostationary communications and weather satellites are endangered when coronal mass ejections strike.

Watch sunspots as they form, grow, spawn groups, then shrink and die or rotate out of view. If a large sunspot group rotates out of your view, make sure you note when it disappeared and make a note of what, if anything, appears from around the Sun's opposite limb in about twelve days. If a sunspot is particularly energetic, it may continue to persist during the fortnight's journey behind the far side of the Sun. Remember that even if the Sun is not active, large sunspots can at times appear. In late 2003, despite being deep into a post solar maximum decline, two massive sunspots nearly 100,000 miles across appeared on the Sun and maintained their strength for more than 10 days until rotating out of view. To the surprise of many solar observers, twelve days later the two massive sunspot groups returned and traveled again across the front face of the Sun before disappearing around the far side to die.

If you become an avid solar observer and the weather is bad, or if you're one of those who is just a bit apprehensive about peering at the sun's blinding glare you can follow the Sun on the Internet at the SOHO web site. The Solar and Heliospheric Observatory images the Sun through eight different filters, including a coronagraph (camera which creates an artificial solar eclipse). One camera images in normal visible light and allows viewing of sunspot activity no matter what the weather on Earth. SOHO is positioned in space at what is called a "Lagrangian point" where the gravity of the Sun and Earth balance each other out. The spacecraft slowly circles this point about 1.3 million miles sunward of Earth. From this position, SOHO serves as our sunward sentinel, guarding Earth against the ravages of the Sun. We'll talk more about SOHO a bit later on in this chapter.

Observing Project 6B – Total Solar Eclipse

At least once in every amateur astronomer's lifetime, he or she will take the journey to see a total solar eclipse. It is one of nature's most rare and most beautiful spectacles. Traveling to see an eclipse requires careful forethought and planning. It can involve travel to some of the world's most remote (Antarctica 2003) or most dangerous (west central Africa 2001 and 2002) places. The average location on Earth gets to view a total solar eclipse about once every 300 years, so the odds are against your having home field advantage for an eclipse.

Another important consideration is the quality of the event in question. How wide is the eclipse track and how long will totality last? If you will only travel to see one eclipse in your lifetime, should it be the thirty-second eclipse next year or the seven-minute eclipse in three years? The choices are not always easy. The duration of the event will depend upon the size of the Moon. When the Moon is near perigee, it will produce maximum duration eclipses because the Moon is maximum size. But when the Moon is at its mean distance, it is just barely large enough in size to cover the entire Sun and thus totality may only last a few seconds. Any farther away than this and the Moon will be too small to cover the entire Sun, causing an annular eclipse to occur. Some events are hybrid in nature where at the beginning and end of the eclipse track, it is annular, but the curve of Earth brings its surface far enough towards the Moon to allow the shadow to reach the surface and then the eclipse becomes total along the center portion of the track. But such totalities are very short in duration.

You must also carefully consider the weather where you are going. An eclipse track can be many thousands of miles long, so there will often be plenty of choices. You must carefully evaluate the quality of the weather at your planned viewing location. A desert or arid location provides the best chance of clear weather. Autumn events are far more likely to produce clear skies than are springtime or summer events. Choose carefully, as a trip to a remote eclipse site will easily consume several thousand dollars both in travel costs and shipping for your equipment. It would be the understatement of the century to state that cloudy weather would be a disappointment.

Speaking of equipment, plan to take along as much as you can carry to the observing site. You will certainly need your telescope, its associated solar protection, field power and at least your SLR camera. Check everything before you leave home for proper operation and again when you arrive at your destination. If you are going to a site in the southern hemisphere, remember that your telescope's drive motor will be useless to you because it turns the wrong way. You will need a special motor that will drive the scope counterclockwise for a trip south of the equator. If you are using AC power, make sure that your equipment is up to foreign specifications, such as 220 volt European power supplies, which can fry a 110 volt U.S. made motor.

As important as it is to prepare your equipment, it is important to prepare yourself as well. Make sure you eat a good meal before heading out, sleep well the night before, be well organized and in all seriousness make sure you go before you leave. There will likely not be a readily available outhouse in the Outback. Also take the time to be very familiar with exactly what will happen exactly when. There are four major events that you must know the precise times of. The point where the Moon first touches the Sun is called "first contact" and this is where the partial phase of the eclipse begins. Over the next hour, the Moon will progressively hide more and more of the Sun's disk. Here is where eclipse watching can become very dangerous. The Sun's *surface brightness* is what causes eye damage when you look at it for even a fleeting glance but the reason you look away by reflex is because of the *total light* present. The total light from the Sun rapidly decreases as more and more of it is hidden, but the surface brightness remains unchanged and thus its potential to injure your eyes is even greater because your natural reflex to look away is inhibited.

As totality nears, the Sun wanes to a thinner and thinner crescent. Take the time to notice what is going on around you. As the sky begins to darken, animals will begin to behave as though night is setting in. Birds will stop flying and ground-based animals will retreat to their shelters. If you are standing near a tree, the pin-holes in the leaves will cast hundreds of tiny little crescent projections of the Sun on the ground. The temperature will also begin to fall. If conditions were mild before first contact, you may need a jacket by the time totality nears. In the sky, stars and the bright planets may begin to appear. This is an excellent chance to spend a few minutes with normally elusive Mercury if it is a safe distance from the Sun. Venus will also appear in all its glorious brilliance along with many other bright stars.

The second event that you must know the precise moment of is "second contact," where the Moon's advancing limb touches the opposite limb of the Sun and totality begins. You may even be able to see the shadow of the Moon advancing towards you from the west, a dark and foreboding apparition hovering in the sky. In the final minute before second contact, any breeze present will likely cease and a series of *shadow bands* will run across the ground. These are an eerie harbinger of what will come next. In the final seconds before second contact, the hairline crescent of the Sun will break up into a series of bright segments as lunar mountains block the Sun, but lunar valleys allow light to pass. These segments are known as *Bailey's Beads* and will persist for a few seconds before the Sun is completely extinguished. As the Sun is covered and darkness falls, the pinkish red chromosphere and the pearly white corona will appear in the sky around the silhouette of the Moon. The total light present here is less than that of a full Moon and thus is safe to view without a filter. Make sure you have a precise plan for how you will spend these precious few minutes or seconds for which you have spent years in planning. Take note of the size of the corona. This tells us a great deal about how much dust is present in space close to the Sun. A large corona suggests the presence of a large amount of dust in the upper atmosphere of the Sun while a lack of dust will cause the upper F corona to be suppressed.

The next and perhaps most critical event to know is the time of "third contact," the instant that total eclipse ends. In this instant, the Sun peaks back out from behind the Moon. You must know exactly when third contact will occur because in that instant the Sun becomes dangerous again and your equipment must be protected before that happens. Don't lose track of time and be late. Your telescope, camera, or worst of all your eyes will regret it deeply. As the Sun begins to reappear at third contact, all the events that took place prior to second contact will occur in reverse. Watch again as Bailey's Beads come and go on the opposite limb and the post eclipse shadow bands race by again. As the partial phase of the eclipse ends, watch the world come back to life around you. Birds will fly again and animals will come out of hiding. The final event is "fourth contact." This is the last instant where any part of the Moon hides any part of the Sun and the eclipse ends.

You have now completed that once in a lifetime journey to see a total solar eclipse. Or so you think. Having viewed one eclipse just about guarantees one thing. You will just *have* to go back somewhere and see another. When you get home, getting that film developed will certainly be your first task. Seeing your travel agent will likely be the second.

Observing Project 6C – The Analemma and the Equation of Time

Have you ever looked at a globe of Earth and noticed the odd figure eight shape that is placed usually somewhere over the Pacific Ocean and wondered what it was? The figure eight pattern is called the *analemma*. The analemma is a representation of where the Sun appears in the sky at exactly the same day on each day of the year. Two important motions that are fundamental to astronomy and to our lives on Earth are responsible for the motion. These motions are Earth's axial tilt and Earth's elliptical orbit around the Sun.

Earth's axial tilt causes the Sun to alternately travel north, then south in the sky reaching extremes of 23.5 degrees north of the equator in June and the same distance south of the equator in December. So if the Sun's position on the sky were to be plotted each day starting in December, it would move a bit farther north each day until June 21, then it would stop and begin moving south again. So this explains the up/down pattern on your globe. Now why the left/right motion?

If Earth moved in a perfectly circular orbit around the Sun, then it's orbital speed would be constant at all times. Then each day at local noon, the Sun would cross the local meridian. But Earth's orbit around the sun is slightly elliptical, so while Earth's rotational speed is constant its orbital speed is not. In January when Earth is at perihelion, it is traveling at its fastest around the sun. After 24 hours, Earth rotates through 361 degrees completing one solar day.[18] But because Earth is moving slightly faster than normal it takes slightly more than 24 hours for the Sun to catch up. Now it will only take about eight seconds for Earth to rotate the additional distance necessary to bring the Sun to the meridian. What is important to understand is that difference is *cumulative* and builds up each day until Earth's orbital speed slows down enough so that the actual solar day is an even 24 hours again. During that time the Sun is drifting slowly eastward in the sky each day at local noon. By early spring the cumulative delay in the Sun's crossing the meridian will have built up to approximately eight minutes. The difference between local noon and the time when the Sun crosses the meridian is called the *equation of time*. After April 2, the effect is reversed because Earth is slowing down in orbit as

---

[18] A solar day is the time from noon to noon and is exactly 24 hours. The time it takes Earth to complete one full rotation (360 degrees) is slightly less than that, 23 hours and 56 minutes. The difference is caused by the additional one degree of rotation needed to make up the change in point of view caused by Earth's motion around the Sun.

it drifts farther from the Sun. This will cause the Sun to drift westward in the sky until October. This turns the straight north–south line of the Sun's seasonal motion into the figure eight illustrated on the globe.

One of the most difficult photographic achievements in astronomy is a photograph of the full analemma. This will require the use of a camera mounted on a stationary pier that will not move over the course of the entire year. The camera must be able to open the shutter repeatedly without having to advance the film. Probably the right camera for this task would be an older model Hasselblad. Such cameras have not been built for many years, so just the difficulty in finding the right equipment makes imaging the analemma very difficult. The next thing you then need is a lot of luck. You want to be able to image the Sun at regular intervals, approximately once each five days. The right time to begin your project would be at the equinox, when the Sun is at the crossing point of the figure eight and fairly early in the morning. This provides smooth air and maximizes the probability of clear skies. What you must account for next is that the two loops are not the same size on film. When the Sun is in the south, the low altitude will cause the lower loop of the analemma to appear much larger than the upper one so you must position your camera so that the upper loop is targeted into the upper right of the frame. If you are successful, when you are done you will have achieved something that few astronomers at any level have ever accomplished, an image of the Sun's annual journey across the sky.

There is a simpler way of documenting the analemma if you don't want to depend on luck and spend a lot of money on photographic equipment. Find a spot where the Sun shines all year around without ever being hidden behind a house, a tree or anything else where you can drive a rod in the ground. The rod should be about three feet tall and be constructed in such a way that it will not move over the course of the year. Each few days at exactly the same time (noon at local standard time works best and one hour later during daylight savings time) note where the end of the shadow cast by the rod falls and mark that spot. Do this for a year and when the year is over, your marks on the ground will have traced the analemma.

The analemma also tells us something important about our Earth. The figure eight shape is very narrow and the length of it only spans some 47 degrees of sky. This tells us that Earth's orbit is very close to circular and the change of seasons is not unusually extreme. Imagine what the analemma might look like if Earth's orbit were as eccentric as Pluto's were and its axis were tilted to the extremes that Uranus' was. The extremes of season and temperature between perihelion and aphelion would make life on this planet very difficult to enjoy at best.

Observing Project 6D – The Invisible Sun

In visible light, the Sun shows us dramatic examples of its activity level. Sunspots show signs of powerful action in the solar subsurface but beyond the photosphere there is little to see in visible light. Massive solar flares and prominences generally hide from view when they are on the solar limb and cannot be seen unless a total eclipse is in progress. At least that is if you are observing in visible light. At this point you're probably asking, "What else would I be looking at?" Is there something more perhaps to see?

Actually to see more of the Sun, you need to see less. The Sun produces energy across the entire spectrum and when we look at it through a telescope, we are seeing the entire visible band from deepest red to deepest blue or what is called "white light." Red dominates the solar chromosphere because hydrogen emits energy principally at the red end of the spectrum. When we limit our view to a very narrow part of the spectrum and eliminate everything else, amazing detail jumps into view. What we are most interested in is the light of what is called *hydrogen-alpha*. Hydrogen is the simplest element in the universe. A hydrogen atom consists of one electron orbiting one proton. That electron can orbit in one of several different orbits, which are numbered outward from the nucleus of the atom. When an electron gains energy, it jumps to a higher orbit and when it does so, it creates what is called an *absorption line* in its spectrum. When it loses energy or when a proton gains energy the electron drops to a lower orbit and creates an *emission line* in its spectrum. An electron that drops from the fourth orbital level to the second emits a special kind of light called *hydrogen-beta*. This kind of light is emitted by many deep space phenomena such as emission nebulae like the famous Orion nebula. Hydrogen-beta is a higher energy light so is visible more readily than is hydrogen-alpha. The Orion Nebula does emit hydrogen-alpha but is too faint to see in most cases. But this is not the case with the Sun. The Sun emits hydrogen-alpha that is very bright just as it does at every other wavelength. Hydrogen-alpha is produced when the electron drops from the third level orbit to the second. This causes an emission line to appear in the Sun's spectrum about 2 angstroms wide centered on 6,562.8 angstroms[19]. A hydrogen-alpha filter limits our view of the Sun to only that light emitted by hydrogen-alpha at 6,562.8 angstroms. These filters basically come in two types for your telescope, front mounted and rear mounted. The rear mounted filter is much more complex but offers much narrower bandpass, as little as 0.1 angstroms deviation from 6,562.8 angstroms. It will usually require an electrical power source to heat an internal oven. The front mounted filter is mechanically very simple requiring no electrical input, but is not as precise. The light passed through can be off by as much as 0.7 angstroms. This is important because if the bandpass gets any wider, then the detail of hydrogen-alpha will disappear and be overwhelmed by white light. What type you choose will be determined primarily by budget. Hydrogen-alpha filters start out at nearly $500.

All H-alpha filters systems consist of three main elements, an energy rejection filter (ERF), a telecentric lens system and the H-alpha filter itself. The energy rejection filter removes unwanted light from the ultraviolet and infrared bands. The telecentric lens system straightens light prior to passing the filter. It typically consists of a 2x Barlow, an optically neutral spacer and convergent doublet. Light emerges from the telecentric system traveling straight and enlarged by a factor of three. In effect an f/10 telescope becomes an f/30. The final element is the actual H-alpha filter. When light strikes the filter, it first passes an antireflective element that enhances contrast and transmission. Light then passes through a narrow-band filter, which further restricts passage of light to very little beyond the H-alpha band. The heart of the filter is a *Fabry–Perot etalon*. The etalon consists of two panes of parallel glass or quartz coated with a reflecting material that passes only a very narrow band of

---

[19] A nanometer is equal to exactly one-billionth of a meter. An angstrom is equal to one-tenth of a nanometer or about the size of an atom.

light. The two filter elements are separated by a spacer that maintains a very precise distance between the two elements. The spacing between the two etalon elements is critical because it very precisely determines what the final bandpass of the filter will be. The two elements are spaced by somewhere between 100 and 200 nanometers. Since such a simple etalon will pass multiple secondary wavelengths of light, a broadband filter backs up the etalon and eliminates the secondary wavelengths and a final antireflective glass element forms the end of the filter.

With the H-alpha filter installed, the Sun comes to life in dramatic new ways. Though sunspots are best viewed in ordinary white light, the active regions surrounding them become more apparent including surface features such as spicules that point radially away from sunspots, or fibrils and loops that occur when the magnetic field are stronger close to the surface. Solar flares are caused by extremely severe magnetic field stresses and are transient events that can last between a few minutes and several hours. Normally the matter blown out of the photosphere will settle back onto the surface in the form of a mist raining out of the corona. If sufficiently energetic, the rising solar material could be blown clear at escape velocity and leave the Sun in the form of a coronal mass ejection. This in turn can have serious impact on the magnetic and near-space environment here on Earth.

Observing Project 6E – Solar Astronomy for a Rainy Day

So you think the clouds got the best of your plans to view that huge new sunspot group? Fear not for SOHO is on the job. The Solar and Heliospheric Observatory has been one of the most productive space astronomy missions ever launched.

Figure 6.3. A LASCO C3 image from the SOHO spacecraft. NASA image.

2005/10/30 18:42

Despite running into several problems, including a loss of communication and orientation that nearly caused the loss of the mission, SOHO on a continuous basis beams back pictures of the Sun through its many different cameras.

SOHO carries 12 instruments on board to study the Sun's entire structure. Throughout each day, the SOHO web site updates eight images of the Sun. Four images are taken from Extreme Ultraviolet Imaging Telescope (EIT) at varying ultraviolet wavelengths. The various images show the movements of gas at several different levels of the Sun. The hotter the gasses, the higher you are looking in the solar atmosphere. At 304 angstroms, the brightest gas is about 60,000 to 80,000 degrees Kelvin. The 284-angstrom images are the hottest gasses at over 200 million degrees!

There are two images taken by SOHO's Large Angle and Spectrometric Coronagraph (LASCO). The images taken by the wide-angle C3 imager are the ones that have produced the most unexpected scientific rewards when it began picking up Sun grazing comets! Astonished astronomers have just found as of this writing their 1,000th comet using SOHO. SOHO also uses a narrow-angle instrument called the C2, which images the inner solar corona out to about five million miles of the Sun's surface.

The last two images are taken by the Michelson Doppler Imager (MDI), which images the Sun across the entire continuum and is close to what the Sun looks like in visible light. The magnetogram highlights areas of powerful magnetic activity.

Our Sun is the dynamo that drives all life here on Earth. Light and heat radiate from it to illuminate and warm our world. Yet in the grand scheme of things our Sun is quite normal, a very average star in terms of energy output, luminosity, size and temperature. It is the very model of an ordinary hydrogen fusing star, rock steady in its output and unchanged for billions of years. We take it so for granted and yet it means everything to us. It affects our lives in subtle ways as well. By blinding our communications satellites in a fit of temper, it can affect services that are a given part of our daily lives. By learning more about it and how to predict and understand its more violent tendencies, we can better learn to prepare our technology and ourselves for those times when the Sun is not so even-tempered.

CHAPTER SEVEN

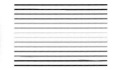

Mercury, Venus, and the Inner Solar System

Figure 7.1. Venus at greatest brilliancy on May 14, 2004. Image by author using a Celestron Super C8 Plus and a low-resolution video eyepiece.

Every object in the heavens conveniently passes opposite the Sun in the sky and makes it very easy for us to see them except for two. These two enigmatic objects make it very difficult for observers to see and enjoy because they rarely if ever wander far from the Sun. Instead they stray east of the Sun coming into view for a few short weeks or months, hugging the horizon murk, then they dart back towards the Sun disappearing in front of it, then emerge west of the Sun into the predawn skies, playing the same tantalizing games of hide and seek before sunrise before circling behind the Sun gradually sinking out of view. The planets Mercury and Venus have therefore hidden their secrets from us more effectively than any other objects in the solar system. Because they always hide low in the twilight murk, the seeing through a telescope is always poor, leaving the tiny disks of the planets swarming in the turbulent atmosphere close to the horizon. Yet with work and patience, Venus and Mercury slowly yield their secrets and visual treats to the patient amateur astronomer. Both planets require more careful planning to observe than do the other planets or nighttime objects. When the outer planets are at their best and brightest, they are visible all night, so you can set up outside about anytime you want. The inner planets may only offer you an opportunity that is just a few minutes long. One must plan wisely and know what you're going to do long before you actually observe. One must also realize that the rules for observing Venus and those for observing Mercury turn out to be very different from each other.

Venus, Goddess of Love

Venus is the far easier planet to observe. Its brilliant white color led the ancient Greeks to name it in honor of their goddess of love. Venus' brilliance comes from the reflection of light from those thick white clouds, which cause the planet to reflect some 80% of the sunlight, which falls on it. It comes closer to Earth than any other body in the solar system except the Moon. It is very nearly identical to Earth in size, measuring 12,103 kilometers in diameter. It is remarkably similar to Earth in density and mass. It was thought for centuries that Venus was a twin of Earth. But for the entire age of the telescope, its surface has been a mystery to us because it is shrouded in clouds. Spectrographic analysis gave us the first sign that Venus is very different from Earth. Venus' atmosphere turns out to be about 96% carbon dioxide and exerts pressures on the surface equal to 90 times that on the surface of Earth. This combination of pressure, cover and carbon dioxide drives temperatures on the surface to levels around 500 degrees Celsius (932 degrees Fahrenheit). That temperature is sufficient to melt lead and other soft metals. That combination of temperature and pressure makes even robotic exploration of the surface incredibly difficult. Though four Russian built spacecraft have made it to the surface intact and returned pictures, all expired after less than two hours. American spacecraft have observed the planet from orbit. The Pioneer Venus 1 orbiter made the first global radar map of the surface in the late 1970s. In the early 1990s, the Magellan probe improved dramatically on the resolution produced by Pioneer Venus 1. In fact, the Magellan maps are so good they are an order of magnitude better than any produced for any application on Earth, at least until the Shuttle Radar Topography Mission of 2000. It was a common joke among astronomers during the decade between those two missions that if you were going to get lost in the solar system do it on Venus. It has better maps. So some four hundred years after Galileo first observed Venus through his telescope, we know how the surface is shaped and contoured, but it still remains a mystery what the surface is made of and what helped shape it. Many of Venus' surface features appear to be sculpted by volcanism. Magellan radar images show indications of lava flows on the surface in the somewhat recent past. What the source of volcanism is however is different from Earth. Venus does not appear to have any surface plates like Earth has. The crust is all one piece. There are surface features called "rift valleys" that resemble surface plates pulling apart from one another. The East African rift and the Dead Sea are both formed by this process on Earth, by the Eurasian and African plates moving apart. Large valleys on Venus of this same geological type must have been formed by more local activity. It was thought for many years that Venus had no magnetic field but it does appear that Venus does have a very weak field. The planet has an average density of 5.2 grams per cubic centimeter, suggesting an Earth-like iron core. The weakness of the field may be related to the slowness of the planet's rotation.

Venus rotates on its axis like all the other planets do, but Venus does two very unusual things. First as viewed from above, Venus rotates retrograde, that is to say that it rotates clockwise rather than the counterclockwise direction that all the other planets do. It also takes longer to complete a rotation than it takes to go around the Sun. A sidereal day on Venus is 243 Earth days long, but the planet requires only 225 Earth days to make one orbit around the Sun. Why this is so is poorly understood. Though there is little empirical evidence to back up any theory, the leading theories available suggest that Venus was formed from two large masses that impacted in such a way so as to almost completely null the rotation rate of the joint mass. Imagine what an analemma looks like on Venus! While the planet rotates slowly, its atmosphere moves quickly at altitude. As strange as the planet's rotation speed is, the planet's orbit is neat and orderly. Venus' orbit is the most nearly circular of all the planets in the solar system.

Weather at Venus also behaves nothing like on Earth. Ultraviolet imagery of the Venusian cloud tops by the Pioneer Venus orbiter shows us that the cloud tops rotate about the planet at speeds of well over 300 kilometers per hour. At that speed, the entire upper atmosphere rotates as a whole around the planet every four days. But winds on the surface are very different. The Venera 9 and Venera 10 landers found during the brief time that they functioned on the surface that winds at surface level are nearly calm and the atmosphere is completely stagnant. The atmosphere is almost completely lacking in water vapor. The dominant feature of Venusian weather is the amazing temperatures. The runaway temperatures are the result of an out-of-control greenhouse effect. Venus appears to have no more in the way of available greenhouse gases than Earth does, but the planet's closer proximity to the Sun appears to be the key difference. Most of Earth's carbon dioxide is bound up in the rocks. Such rocks, like limestone, prevent the gas from flooding the atmosphere. But on Venus the rocks are heated more than on Earth, forcing them to release carbon dioxide to the atmosphere. As the volume of carbon dioxide increases in the atmosphere, the greenhouse effect accelerates to the point that any water that might have been on the surface boiled off into the upper atmosphere. Once there, ultraviolet radiation would break apart the water molecules into hydrogen and oxygen, which then could easily escape the planet's gravity and depart into interplanetary space.

The age of the telescope led us to believe that Venus was a world very much like Earth. The age of space flight taught us that nothing could be farther from the truth. Venus is a world much like hell. Crushing pressures, scorching temperatures and corrosive clouds make it anything but a romantic place.

Venus in our Sky and our Telescopes

Because Venus orbits inside the orbit of Earth, the planet always stays relatively close to the Sun, but at times Venus can become easily observable. After the planet passes behind the Sun, a position called *superior conjunction* the planet will slowly begin to appear in the evening sky. At this time, the Sun is traveling about one degree per day across the sky eastward against the background of the stars, or 30 degrees per month. Venus at this point is traveling 38 degrees eastward per month, so the planet only very slowly pulls away from the Sun. By a month after superior conjunction the planet has pulled only seven or eight degrees ahead of the Sun. At that pace, it usually takes two to three months for Venus to pull far enough away from the Sun to become easily visible. How long it takes to actually begin to see the planet as it rises above the horizon depends not only on angular distance from the Sun but also upon the angle of the ecliptic relative to the horizon, as we discussed during the chapter one observing projects. If it is springtime in the north, the planet will become visible much more rapidly than it would during autumn. If you can train a telescope on the planet, it shows you a tiny disk barely ten arc seconds across, nearly fully illuminated. As the months go on, it will become obvious that Venus is becoming more and more gibbous in shape, exactly as a waning Moon does. As Venus circles the Sun, it slowly begins to present more and more of its dark side towards Earth. As the planet begins to grow larger in telescopes, it also grows brighter. From magnitude −3.3 at superior conjunction the planet will surpass magnitude −4 by the end of the fifth month of the apparition. The planet's phase continues to wane as is draws closer to Earth. By five months after superior conjunction, the planet has grown to about 17 arc seconds in diameter and has waned to about two-thirds illuminated. Now Venus begins "rounding the corner" in its orbit, traveling less tangentially to our line of sight and more directly towards us. This causes the planet's eastward motion to slow gradually. By the time the planet is eight months past superior conjunction, Venus has swelled to 24 arc seconds in size and it reaches a position in space where it is now at the corner of a right angle with Earth and Sun. At this time, Venus is at its maximum angular distance from the Sun. This value ranges from between 46 and 48 degrees depending whether Venus is at perihelion or aphelion at the time. With the right angle geometry, Venus at this point shows us exactly a half illuminated disk, a condition called *dichotomy*. Now Venus' rate of eastward motion against the stars slows to less than that of the Sun and the Sun begins to overtake it in the sky. The half-phase rapidly begins to wane to a crescent as the planet moves ever more rapidly towards the Sun. Eventually the planet's eastward motion ceases and it begins to move westward on the sky, an effect called *retrograde motion*. The planet continues to swell, wane and brighten until about 38 days prior to passing between Earth and the Sun, a point called *inferior conjunction*. At this point Venus is about 38% illuminated and spans about 38 arc seconds. At this time the planet achieves its maximum brightness or what astronomers refer to as *greatest brilliancy*. At magnitude −4.8, Venus shines so brightly in the evening sky that it can cast shadows on the ground. The planet's brightness is dependent entirely on how much surface area is visible to us. Until now, the planet has been growing in size faster than its waning phase can diminish the total surface area. So at this time, Venus shows us the largest amount of surface area expressed in square arc seconds and thus achieves its maximum brightness. This is the time of *greatest illuminated extent* and this is when an inferior planet is at its brightest. But from this point forward, the planet's waning phase begins to diminish the total surface area faster than its rapidly growing size can make up for. Over the final month of the apparition Venus begins to fade and fall from the sky, closing on the Sun at faster than a degree per day. The planet's crescent phase fades to a hairline even as it lengthens to as much as a full arc minute. The planet's brightness fades from its peak to about −3.6. Before long, the planet is completely lost from view. How long you can keep it in sight will again depend upon the angle of the ecliptic relative to the horizon. Venus's orbital inclination can also be a factor. Although the planet's orbit is inclined only a few degrees from Earth's, the difference is enough that if Venus is on the north side of its orbit, it can add quite a bit of altitude to the planet's position and delay its disappearance by a few days. But in the end, disappear it will as it overtakes Earth and passes between the Sun and us.

As Venus moves rapidly to the west of the Sun, it will begin to reappear in the morning sky as fast as it fell from the evening sky and thus become what the ancients called the "morning star." Venus runs through a morning apparition in exactly the reverse order that it ran through an evening apparition. The planet appears within a few days or weeks, depending upon the angle the ecliptic makes with the morning horizon. Late summer and early autumn offer the most favorable conditions. Just over a month after inferior conjunction, the planet achieves greatest brilliancy and a month later the planet's westward motion stops and *direct motion* resumes. About ten weeks after inferior conjunction, the planet reaches greatest elongation west of the Sun and then again its eastward motion against the stars causes it to again begin to overtake the Sun. Over the next eight months, the planet slowly closes the gap between it and the Sun, waxing from dichotomy at greatest elongation to a gibbous phase that grows more and more full as the planet shrinks in size back to a minimum of about nine or ten arc seconds by the time the planet reaches superior conjunction again about twenty months after the last time it passed behind the Sun.

The Resonance of Venus

In Chapter 5, we discussed the remarkable resonance between the Moon's synodic and Draconian periods, which produce eclipses in continuing series. Gravitational interactions between Earth and Venus cause the orbits of the two planets to resonate in such a way that for every eight times Earth orbits the Sun, Venus completes fourteen. So every eight years Venus, Earth and the Sun return to almost exactly the same alignment in space. This creates a remarkable regularity in Venus's apparitions in both the morning and the evening skies. During each eight-year period, Venus will overtake Earth five times, producing five sets of apparitions that repeat themselves over and over each eight years. For example, on June 8, 2004 Venus reached inferior conjunction. Almost exactly eight years and five inferior conjunctions later, on June 6, 2012, the planets will again line up in just about the exact same position. The conjunction will occur about forty hours earlier than it did eight years before. Each of the five apparitions is distinctly different from each other. In the apparition that ended in June 2004 Venus emerged from behind the Sun in August of the previous year and needed about four months to appear in the evening sky before the combination of increasing elongation and rising ecliptic brought the planet more rapidly into view. This apparition eventually becomes the most favorable of the five in the cycle because the planet reaches greatest elongation about ten days after the vernal equinox. About this time, Venus stands some 40 degrees high at sunset and sets as much as four hours after the sun. The morning apparition that follows is almost as good. The only difference is that Venus is south of the ecliptic at greatest elongation so its altitude is not quite as high. When Venus next returns to the evening sky for the second apparition of the cycle, it enters the evening sky quickly, but stalls out because it is pulling away from the Sun during summertime so the ecliptic lies progressively flatter with respect to the horizon even as Venus pulls away from the Sun, so the planet never climbs more than 20 degrees above the horizon. After inferior conjunction, in January 2006 the planet will have an equally poor apparition in the morning sky. After six more apparitions, Venus will return to the evening sky in August, 2011 and repeat almost exactly the evening apparition of 2003–2004, reaching inferior conjunction again on June 6, 2012. When Venus reaches inferior conjunction in March, as it will in 2008, it is near the northernmost point of its orbit and from our point of view appears to pass about eight degrees north of the Sun at inferior conjunction. This creates an opportunity each eight years to view the planet on conjunction day both in the evening sky at sunset and the morning sky at sunrise.

Another fascinating aspect of Venus–Earth resonance is that the planet's rotation is timed such that every fifth midnight on Venus, the planet presents exactly the same face to Earth. That means that at the end of each resonance cycle at a given inferior conjunction the planet will always present the same face towards Earth. If we could see surface features both through clouds and in the dark, Venus shows us exactly the same face at its inferior conjunctions in June 1980, 1988, 1996, 2004 and 2012. Venus will do so again at the inferior conjunctions of June 2020, 2028, 2036 and so on. The term "resonance" takes something of a beating here though because a resonance requires some type of gravitational interaction to occur and for their respective masses; Venus and Earth are just too far apart to have any effect on each other's rotational periods. It would appear that the rotational resonance of Venus is one of the more remarkable coincidences in science.

Transits of Venus

Venus's orbit is inclined slightly with respect to Earth's, by 3.394 degrees. But there are two points in the orbit where Venus will cross the ecliptic traveling north (ascending node) or south (descending node). If Venus reaches inferior conjunction within about forty hours of crossing a node, its orbital path will carry it across the face of the Sun creating what is called a *transit*. Transits of Venus are exquisitely rare events. They can only occur if Venus passes inferior conjunction and crosses the node of its orbit within about two days of each other. So if inferior conjunction occurs within a four-day wide window in early June or early December, then a transit will occur. Since the time of inferior conjunction backs up by about two days with each eight-year cycle of apparitions, transits usually occur in pairs spaced eight years apart. A transit occurred with Venus just past the descending node of its orbit on June 8, 2004 and another will occur just prior to Venus reaching the descending node on June 6, 2012. These transit pairs are spaced by 122 years. The last transits prior to 2004 were in 1874 and 1882, respectively. Transit pairs take place on opposing sides of the Sun at each cycle. The 1874 and 1882 events were at the ascending node and took place in December. The next pair on December 11, 2117 and December 6, 2125 will also be ascending node events. If however a transit occurs right on the node, the rate at which inferior conjunction will regress on the calendar is greater than the radius of the transit window of opportunity can contain and in such a case there will only be a single transit in the cycle. The time between the first ascending node transit of one cycle and the first ascending node transit of the next such pair is called the *Venus Cycle* and lasts 243 years. The two pairs of transits are not precisely spaced. The time between the first ascending node transit of one pair and first descending node transit of the next pair is always 129.5 years while the time from the first descending node transit to the first ascending node transit of the following pair is always 113.5 years.

The same resonant motion of the two planets that we have been discussing is what causes transits to occur in such a regular cycle. Since the planet repeats the same five apparitions basically over and over again, there are five points along the orbits of both planets where inferior conjunction occurs. These five points form a pentagon shape in space. The fourteen-to-eight resonance of the two planets is not quite that precise. The time between the June 2004 and 2012 inferior conjunctions is actually 7.997 years, not eight years. As a result of this, the point of each inferior conjunction on the pentagon slowly drifts westward around the Sun with each cycle. Each 243 years, one of the pentagon points will drift past each one of Venus' nodes enabling two transit sets to occur. One pair will be at the descending node in June and the next pair at the ascending node in December. Over the course of thousands of years, the passages drift away from the previous locations. For example each 243 years, the path followed by the 2004 and 2012 transits will slowly drift northward until the 2012 path drifts off the Sun entirely and the 2004 path will be nearly at the center of the Sun's disk. When this occurs, the pentagonal point for the next passage will drift too far away from the node after the eight-year cycle and no second transit can occur. This last occurred in the year AD 60 and will next occur in 3956.

The Messenger of the Gods

Venus' brilliant beacon enables it to be easily viewed even during times when the planet is buried low in the twilight of morning or evening and often rises well above the murky horizon to allow for easy viewing. During at least four of its five evening and morning apparitions, Venus will at some point rise or set at least three hours before or after the Sun. Mercury enjoys none of these advantages and in fact for northern hemisphere observers the combination of favorable circumstances that often come together to make Venus easy to see even at modest distances from the Sun never at any time come together for Mercury.

Despite the difficulties of viewing it, Mercury was known even in antiquity. The ancient Greeks saw Mercury alternately in the evening sky and in the morning sky but they did not comprehend that they were seeing the same object and so gave it two different names. Because of the speed with which the planet moves in the evening sky, they named it for the messenger of the Greek gods, Hermes. When the planet appeared in the morning sky, it was called "Apollo" the Greek god of the Sun. The Romans did know it was a single object and called the planet "Mercury" for the Roman god of commerce, travel and thievery (of all things!). After the invention of the telescope, it was eventually discovered that like Venus, Mercury displays Moon-like phases, but the disk of the planet was too small and it was too difficult to get steady views of the planet to learn what the surface of the planet was really like. We knew little about the planet physically during the pre-space age era of astronomy. We know the planet is small, just 4,880 km in diameter. We also knew that the planet flies around the Sun at more than twice Earth's average speed, nearly 30 miles every second. At that speed Mercury scoots around the Sun once every 88 Earth days. It was also believed until the mid-1960s that Mercury's rotation period was synchronous with the Sun, as the Moon's is with Earth. Studies using Doppler radar in 1965 determined that Mercury has a rotation period of 59 days. This creates yet another example of a resonance between Mercury's rotation period and its revolution period. For every two revolutions Mercury makes around the Sun (88 × 2 = 176 days), the planet revolves three times (59 × 3 = 177 days). Because of this, over two revolutions around the Sun, no place on Mercury sees the Sun rise or set more than once, sort of. Unlike Venus, Mercury's orbit is very eccentric carrying it from a minimum of 46 million kilometers from the Sun at perihelion to a maximum distance of 70 million kilometers. Mercury's orbital speed varies dramatically as it speeds around the Sun. When the planet is at perihelion, it moves so fast in its orbit that if you are experiencing sunset on the planet's trailing limb, the planet's orbital velocity is so great that the Sun will actually peak back over the horizon creating a double sunset, an event unique in all the solar system. Mercury's orbital period also created dilemmas for physicists of the time because its precise orbital period defied attempts to predict it. Though the discrepancies were minute, they were nonetheless troubling. It was thought for a time that a planet inside Mercury's orbit was responsible for the disturbances in the planet's orbit, just as Neptune was found as a result of the effect it has on Uranus. Scientists had even named the theoretical unnamed planet "Vulcan" for the Greek god of the forge. Vulcan was never found despite an extensive search. The real answer turned out to be a result of the effects of general relativity. The ability of Einstein's theory to correctly predict the orbit of Mercury was one of the first great triumphs for Einstein in the battle for acceptance of the theory of general relativity.

Only one spacecraft has ever visited Mercury. In 1973, the United States sent the Mariner 10 spacecraft to conduct three flybys of Mercury. The probe first flew past Venus, using its gravity to slow the spacecraft and cause it to fall deeper into the inner solar system into a solar orbit that would carry it close to Mercury three times. Each time Mariner 10 trained its TV cameras on the tiny scorched world. Mercury turns out to be a world that is as heavily battered as the Moon is. During its three flybys Mariner 10 did tell us some surprising things about Mercury. Though it was similar in size to the Moon and looked a lot like the Moon, it's internal composition seems to be more like Earth. The planet has a mean density of 5.5 grams per cubic centimeter, about the same as Earth. This tells us that the planet's core is largely iron. The core itself is the largest in relation to its planet's size in the solar system. Mercury is almost all core, with a thin silicate mantle and crust. Much to the surprise of astronomers studying the Mariner 10 data, Mercury has a substantial magnetic field. While it is not nearly so strong as that of Earth (about 1%), it is more than strong enough to turn away the solar wind. Mercury also boasts an enormous impact basin which scientists named "Caloris Basin." The impact the created Caloris was so powerful that it created ripples in the terrain clear on the opposite side of the planet. Unfortunately at each of Mariner 10's three encounters with Mercury, the planet showed the same side to the probe and its cameras so to this day, we have mapped barely one half of the planet's surface. In 2004 the United States launched the first mission to Mercury in more than three decades. Dubbed MESSENGER, the probe will begin orbiting the planet after having made one flyby of Earth, two of Venus and three of Mercury before braking into orbit around the planet in 2011. MESSENGER will conduct the most extensive survey of Mercury conducted to date and reveal the planet's hidden side for the first time.

Since Mercury is so close to the Sun, it receives about nine times as much solar energy as Earth does. Daytime high temperatures at Mercury are nearly as hot as they are at Venus, over 500 degrees Celsius. Since the planet has no appreciable atmosphere to retain heat, when the long Mercury night sets in, the temperature falls more than 700 degrees Celsius. This makes Mercury not only one of the hottest places in the solar system, but one of the coldest too. Mercury's axis also has just about no tilt so the planet's poles never view the Sun at a very steep angle and certain crater bottoms, like on the Moon are never exposed to sunlight. Scientists speculate that these permanently shadowed craters might also contain deposits of water ice from primeval comets. One of MESSENGER's most important assignments will be to use a spectrometer to check for the presence of hydrogen emission lines at Mercury's poles as Lunar Prospector did in the late 1990s. Also because the planet's solar period is so slow, areas near the terminator are slow to heat while the Sun is close to the horizon. When the Sun is about 10 degrees above the horizon, the temperatures are approximately 25–40 degrees Celsius, a warm to hot summer day on Earth. If man were to ever visit Mercury he would likely choose a sunrise location for his first landing.

Though we have known of Mercury since the days of antiquity, it remains the planet we know the least about. Its position deep in the solar system makes it very difficult to observe by telescope and also very difficult to reach by spacecraft because of the amount of fuel needed to decelerate a spacecraft into the realm of Mercury. But slowly as man turns his energy, telescopes and ingenuity to the problems, the speedy, stealthy planet named for the ancient Roman god of thieves is slowly yielding to us its secrets.

Mercury in the Sky and in our Telescopes

Because Mercury orbits inside the orbit of Earth, many of the same basic rules that apply to the movement of Venus though the sky also apply to the movements of Mercury. Mercury just does everything much faster than Venus does. From behind the Sun, the planet first appears in the evening sky pulling rapidly away from the Sun. Because the planet's orbit is considerably more elongated than is that of Venus, Mercury's angular distance from the Sun at greatest elongation can range anywhere from 18 to 28 degrees. The planet then rapidly closes back in on the Sun, within a matter of a few weeks and passes through inferior conjunction. Within a few days if circumstances are favorable, the planet will rapidly appear in the morning sky, pulling out to its greatest angular distance in just about three weeks. The planet will then circle around behind the Sun coming back to superior conjunction about 120 days after previously being there. That means that Mercury overtakes and passes Earth three times every year. By comparison, Venus needs about 540 days to go from one superior conjunction to the next. Mercury will run through the same range of phases as Venus does. There are however many differences in the way Mercury behaves from Venus.

The first is the unfortunate coincidence for northern hemisphere observers is that when Mercury reaches greatest elongation at aphelion; it does so when the viewing geometry is at its most unfavorable. This occurs in March for morning apparitions and September for evening apparitions. When Mercury reaches greatest elongation in March for evening apparitions or September for morning apparitions, it is always near perihelion. So this gives rise to a startling contradiction. An evening apparition of Mercury is much more favorable in March, when the planet strays only 18 degrees from the Sun. That 18 degrees is almost vertically above the Sun and Mercury is in the sky for as much as an hour and a half after sunset. When the planet reaches greatest elongation in the evening sky in September, it will stretch to 28 degrees away from the Sun, but that 28 degrees is almost straight along the horizon. The planet never rises out of the horizon haze and sets within 40 minutes after the Sun.

While Venus is brilliantly reflective, Mercury is a dullard. The planet has no atmosphere, no bright reflective surface features such as oceans and much of the surface is made of dark materials that are likely volcanic in nature. While Venus reflects 80% of the light that strikes it back to space, Mercury is more like the Moon and reflects back only about 7% of the light that strikes it. Mercury is also less than half the size of Venus and so never has as much surface area to show as Venus does. Venus at its smallest apparent size is nearly as big as Mercury can ever appear at its largest.

In a telescope, Mercury runs through phases the same way Venus does. Mercury is much smaller than Venus is. At superior conjunction, the planet's disk is slightly less than four arc seconds across and by the time the planet is lost near inferior conjunction, its disk is between 10 and 12 arc seconds across. So as it draws nearer to us, Mercury's disk increases in size by a factor of two and half from the time it first appears to the time it disappears. By comparison, Venus disk increases by a factor of about six over the same period of time. Because of this, Venus reaches its point of greatest illuminated extent when it is a thick crescent. Mercury is at its greatest illuminated extent within a couple of weeks of superior conjunction. To illustrate, a fully illuminated Mercury four arc seconds across shows more illuminated area than a one-third illuminated Mercury eight seconds across. This gives Mercury a property unique in all the universe. By comparison, a fully illuminated Venus is nine arc seconds across. When it is one-third illuminated, it is 45 seconds across. This explains the mystery of Mercury's changing brightness that we illustrated in Chapter 1. Mercury is the only object in all the heavens that grows *fainter* as it gets *closer*! After passing superior conjunction, Mercury will shine at anywhere between magnitude −2.0 and −1.5 depending upon its distance from us. As the planet emerges into the evening sky and begins to become visible, this is the best time to see it. By the time the planet reaches greatest elongation, it has faded from anywhere between about −0.5 to +0.5. After greatest elongation, the planet will be visible for only a few more days because as it falls back into the solar glare it fades dramatically, by more than a tenth of a magnitude per day until it can no longer be picked out of the twilight glow.

Venus' phases are very regular. The planet reaches dichotomy within a very short range of time centered on greatest elongation. Mercury can be very irregular. The planet is always half-illuminated when it is at the bend in a right angle between the Sun, itself and Earth. But this is not always the point of greatest elongation. If its orbit were a perfect circle, as Venus' nearly is, then it would be so. But the eccentricity of Mercury's orbit causes some unusual things to happen. If Mercury is out at aphelion well before reaching that right-angle position then greatest elongation will come early just because it is several million miles farther from the Sun then than it is at the geometric position where greatest elongation should appear. Thus when Mercury is at greatest elongation in such instances it may be as much as 60% illuminated and thus brighter. When the reverse is true, with the planet at perihelion about two or three weeks before greatest elongation, the planet will pass through dichotomy and continue to pull away from the Sun in our sky because it really is pulling farther away from the Sun. So in this case, greatest elongation may occur a week after passing the right angle position. The planet may be barely 40% illuminated at greatest elongation as a result.

Because Mercury's synodic period is so short, it presents us with plenty of opportunities to see it each year. The planet will typically reach greatest

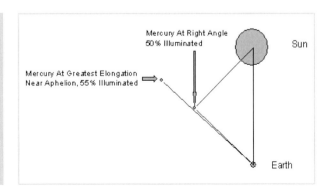

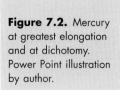

Figure 7.2. Mercury at greatest elongation and at dichotomy. Power Point illustration by author.

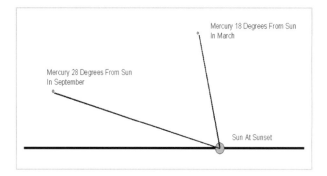

Mercury 18 Degrees From Sun
In March

Mercury 28 Degrees From Sun
In September

Sun At Sunset

Figure 7.3. Mercury at sunset at its best and its worst. Power Point illustration by author.

elongation six times each year, three times to the east of the Sun (evening sky) and three times west of the Sun (morning sky). Typically in a given year in the evening sky, Mercury will give us one good apparition in the late winter or early spring, one very poor apparition in the late summer or early autumn and then one apparition that will occur around one of the solstices which will be somewhere in between the two. In the morning sky, Mercury will have a good apparition in late summer or early fall, a very poor one in late winter or early spring and then one apparition around the winter or summer solstice that will also be mid-level in viewing quality. At its best, Mercury may be in the sky for as long as 90–100 minutes after sunset. At an elongation that occurs around the solstice, Mercury will be in the sky for around 60–80 minutes after sunset and at its worst, even though it may be 28 degrees from the Sun, will not stay in the sky for more than 35–50 minutes after sunset. Four of the six apparitions give us a good chance to see Mercury. Remember that the best time for viewing Mercury, unlike Venus, is before greatest elongation when it is at its brightest, not later when the planet's disk may be larger but it will be fainter against a bright sky.

There is one other time for viewing Mercury that many people do not think of and that is during the day. Mercury can shine as bright as the brightest stars in the sky, so it is actually very easy to find in daylight. The problem is of course scanning for it in an unfiltered telescope in close proximity to the Sun. There are two easy and safe ways to view Mercury in daylight. The first and simpler way is to sight the planet in the morning sky before sunrise and continue to follow it away from the horizon as morning progresses. During the first two hours after sunrise, Mercury can climb as much as 35–40 degrees above the horizon during a favorable apparition. When Mercury is trailing the Sun during an evening apparition, install your solar filter and sight on the Sun. Then noting the difference between the right ascension of the Sun and Mercury, turn off the drive on your telescope for a period of time exactly equal to the difference in right ascension. If Mercury happens to be at exactly the same declination as the Sun, when the time is up and you turn the drive back on, the planet will be exactly in the center of your field of view after you remove the solar filter. If it is not, then you may have to use the setting circles of your telescope to move it north or south by a distance equal to the difference in declination difference between the two bodies.

Transits of Mercury

Just as Venus does, Mercury on rare occasion passes across the face of the Sun when Mercury reaches inferior conjunction within a few days of passing the ascending or descending node of its orbit. Like with transits of Venus, there is a regularity to transits of Mercury. Successive transits of the planet are separated by periods of 3.5, 7, 9.5, 10 or 13 years. This is because of a harmonic that brings Mercury and Earth back to the same position each 16,802 Earth days or 46 Earth years with Mercury arriving at inferior conjunction 0.37 days later at the end of each cycle. There are several mathematical harmonics within this more precise group that are not as exact which cause the uneven spacing of transits due to the elliptical nature of Mercury's orbit. Because of this, transits of Mercury can be neatly organized into groups much like solar and lunar eclipses fall into saros series. For example the transit of May 2003 is part of a family that includes transits that occurred in May 1957 and May 2049. The November 2006 transit is part of a family that includes transits in November 1960 and November 2052. There are not nearly so many series in progress as there are with eclipses and they do not last as long either. Currently there are six series of transits in progress. Transits also are very different depending upon which side of the Sun they occur.

Mercury passes the descending node of its orbit at a position that corresponds to May on Earth and passes the ascending node on the opposite side of the Sun at a position corresponding to November. Thus transits can only occur within a pair of windows currently centered on about May 10 and November 10, respectively. When Mercury transits the Sun at the descending node in May, it is also only 30 degrees of orbital travel past aphelion and thus travels relatively slowly in its orbit. Being far from the Sun also means it is closer to Earth so the planet appears larger than in November events, about 12 seconds across. Each time a new member of a transit series takes place, the track shifts about 200 arc seconds further south because Mercury arrives at inferior conjunction about eight hours later than it did at the event 46 years earlier. After approximately 10 events, the track shifts off the Sun and the series ends after a run of approximately 400–450 Earth years. November events are different. November events take place with Mercury about 30 degrees of orbital travel past perihelion, so the planet is traveling at near its maximum velocity in space and scoots across the Sun at twice the speed that it does in a May event. If Mercury were to travel directly across the center of the Sun in both a May and a November event, the November event would last only half as long. Also Mercury's speed means that the planet's track across the Sun shifts only 100 arc seconds northward across the Sun with each successive event in a series. As a result of this, a November transit series will have about twice as many members and last twice as long as a May series. Overall therefore there are twice as many transits of Mercury that take place in November as take place in May. Mercury is also much farther from Earth during November events so the planet is somewhat smaller, only about ten arc seconds across during November transits.

Observing a transit of Mercury is a bit tougher than Venus in that optical aid is likely needed to pick out Mercury's tiny disk. Venus can be more than one full minute across when at inferior conjunction but Mercury is only ten to twelve arc seconds across when it transits the Sun. You certainly have more opportunities to view Mercury transits in that they will occur about 13 or 14 times per century, so you will have opportunities to view a few in a lifetime, unlike those of Venus which are once in a lifetime events.

Interlopers in the Inner Solar System

The inner solar system is at first glance a pretty clean place, having long since been swept clean of most interplanetary debris by the gravity of the Sun or that of the planets in the area. Every now and then there will be an interloper that will drop down into the inner solar system after a gravitational interaction with Jupiter or Saturn or out beyond the planets in the Kuiper Belt or Oort Cloud. It has been the secret fear of astronomers for many years that a large asteroid or comet will drop into the inner solar system on an impact trajectory with Earth. The potential for such a cataclysm is sufficiently terrifying that the U.S. government has funded two studies to scan the skies for such objects and inspired Hollywood to produce two movies based on such a disaster called *Deep Impact* and *Armageddon*. Such an impact is believed responsible for the extinction of the dinosaurs and many other reptilian forms of life on Earth many years ago.

The two government funded robotic programs are designed to scan the entire northern sky at regular intervals for undiscovered asteroids. These robotic programs called LINEAR (Lincoln Near Earth Asteroid Research) and NEAT (Near Earth Asteroid Tracking) operate every night at desert locations. LINEAR uses two one-meter telescopes based at White Sands Space Harbor in New Mexico. One telescope is dedicated to the LINEAR program; the other is primarily used by the Air Force for space surveillance. NEAT uses two 48-inch telescopes. One is based on Maui and the other is based at Mount Palomar. NASA's Jet Propulsion Laboratory manages the NEAT program. LINEAR is operated by the Lincoln Laboratory of the Massachusetts Institute of Technology. The hope is that any asteroid moving into the inner solar system that might become a threat will be found in time to allow the government to devise a plan to cope with it prior to the disaster striking. Each survey is responsible for finding thousands of these space rocks. Learning about the interior of these bodies was the principal objective of the 2005 Deep Impact mission (which NASA claims was not named for the aforementioned movie) to excavate a crater in the surface of Comet Tempel 1. Through such studies, we hope to learn whether comets and asteroids are solid rocky or metallic objects or primarily loosely associated clumps of ice and dirt. The answers would be critically important to scientists and Defense Department researchers who might one day be forced to deal with answering the question of how to destroy or deflect an asteroid or comet on a collision course with Earth.

The objects of greatest potential concern are asteroids with orbits that are known to cross inside that of Earth, which astronomers classify as *Apollos*. The two robotic telescopes also survey asteroids from the *Amor* group, which approach Earth from outside our orbit and *Atens,* which approach our orbit from the inside. Amors and Atens do not actually cross our orbit in space but can approach close enough to Earth to possibly some day present a danger. These asteroids are normally brought into the inner solar system from the main asteroid belt between Mars and Jupiter after a gravitational interaction with Jupiter. As Jupiter tugs at the asteroid from behind, it loses energy and falls deeper into the inner solar system on its next orbit. Fortunately most of these asteroids are in orbits that are inclined several degrees from that of Earth so even though the asteroids might come closer to the Sun than Earth does, they do not actually cross the plane of Earth's orbit and thus are not a collision threat. But our planet does bear the scars of the occasional hunk of rock that has found its way across our path. Meteor Crater in Arizona is an enormous one-mile across impact scar formed within the last few million years by an object about the size of a railroad car.

NEAT and LINEAR not only sweep up asteroids but also voluminous numbers of comets. Each has found more than 100 comets to date. Most are far too faint to be viewed in amateur telescopes but the two robotic telescopes have found some of the brighter ones in the last decade as well including two naked-eye comets which put on nice displays in 2004. Many comets coming from deep space from the Oort Cloud come not just inside the orbit of Earth but well inside the orbit of Mercury as well. Such comets, appropriately called "Sun-grazers," can become the most spectacular comets of all. The great comets of 1882 and 1876 as well as 1965's Comet Ikeya–Seki all came within one million miles of the Sun and all grew so bright after perihelion that they could be easily viewed in daylight. It had been thought that sun-grazing comets were rare phenomena but the SOHO spacecraft has taught us that they are in fact rather common. Since its launch, SOHO's C2 coronagraph has found more than 100 comets, which were too faint to be detected by conventional means until they flared to life in close to the solar photosphere.

Now having studied the various movements, tricks and surprises of the inner solar system, let's put our new knowledge to the test as Mercury, Venus and the occasional surprise guest spin through space between Earth and the Sun. While Venus moves with great regularity repeating the same eight-year cycle over and over again, Mercury darts in and out in apparitions that vary from one month to another in viewing quality. Then there are the surprise visitors, asteroids that come from both inside and outside of our orbit and from the deepest realms of the solar system, the frozen snowballs called comets that terrorized the ancients and thrilled amateur astronomers for centuries.

Observing Projects 7 – The Inner Solar System

Observing Project 7A – Charting the Movements of Venus

This is a non-telescopic project that can keep you busy for many years. It simply involves charting the position of Venus in the evening sky over the course of several apparitions. Here we will see two things. First, how the path of Venus in the sky and favorability of a given apparition will vary based upon the angle the ecliptic makes with the morning or evening horizon depending upon time of year. The other will show you over a period of ten years or so how apparitions repeat themselves over and over again in a regular cycle. During favorable apparitions, the time of greatest elongation occurs in late March when the ecliptic stands nearly vertical over the setting Sun, so Venus stands nearly 40 degrees above the horizon at sunset. Conversely at the unfavorable 1997 and 2005 apparitions the time of greatest elongation is in November when the ecliptic is flat relative to the horizon and thus the planet never at any time climbs more than 20 degrees above the horizon and sets just after twilight ends. The viewing in March 1996 and 2004 is further enhanced by the fact that Venus is about as far north of the ecliptic as is possible, giving the planet a bit of extra altitude. At the 1997 and 2005 elongations the planet is situated as far south in the sky as it can ever get so it lies even closer to the horizon than the angle of the ecliptic would suggest.

In addition to its movements, you can also try your hand at judging Venus' changing brightness. Venus will change in brightness over the course of an apparition from a minimum of about −3.3 to a maximum of about −4.7 before fading to just below magnitude −4.0 just before it disappears into the solar glare. This is something that will prove extremely challenging because there is so little to compare Venus to that is comparable in brightness. Jupiter can come within two magnitudes when it is at opposition, but when Venus goes past it each year Jupiter is near its minimum brightness. It may not be a great yardstick but it is as close to one as you are going to find.

Observing Project 7B – Dichotomy of Venus and Mercury

In this project, we will use our telescopes to judge the time at which Venus and Mercury reach *dichotomy* or precisely half-illuminated. For Venus, this will come almost exactly on the day of greatest elongation when Venus is at the corner of a right angle involving itself, Earth and the Sun. For Mercury though, this can come several days before or after the day of greatest elongation. The principle reason is due to the elongated nature of Mercury's orbit as opposed to the nearly circular orbit of Venus.

An inferior planet always shows its one-half phase when it reached that position at the corner of that right angle. If the orbit of the planet is perfectly circular or nearly so, then that will also be the time of greatest elongation. Venus' orbit is the closest to a perfect circle in the entire solar system so it is very likely that dichotomy and greatest elongation will always occur close to the same time. For Mercury, however, its orbit is one of the most eccentric of any major body in the entire solar system. Only that of Pluto is more elongated. This can cause Mercury to reach greatest elongation early simply because it may be several million miles farther from the Sun and thus reach greatest elongation while still gibbous. Or if it is near perihelion, the planet may reach half-illuminated on perihelion day, then continue to climb away from the Sun as it pulls away in its orbit. This will happen very quickly because Mercury's orbital speed is double at perihelion what it is at aphelion. A few days difference therefore can mean several million kilometers difference in Mercury's distance from the Sun.

Study Mercury and Venus carefully as they swing around the Sun towards their half-illuminated phase. Know when the date of greatest elongation is for the planet and make a careful estimate of how much of the planet's surface is illuminated on that date. For Venus, it is a good bet that the planet will be half-lit on the day of greatest elongation, or fairly close to it. For Mercury, the only time that dichotomy and greatest elongation will occur on the same day is if the planet is also at perihelion or aphelion on or about that exact date. Remember also for Mercury that the planet's brightness will also vary dramatically on the day of greatest elongation depending upon what its phase is. If the planet is less than half-illuminated it will likely be fainter than zero magnitude while if it is gibbous, then it will likely be brighter than zero magnitude. Mercury's distance from Earth will also be a factor in determining how bright the planet will appear at any given time. Watch Mercury carefully from apparition to apparition and see how its behavior changes each time it comes around the Sun.

Observing Project 7C – Catching Interlopers in the Darkness

The NEAT and LINEAR robotic telescopes have charted thousands of asteroids orbiting in the vicinity of Earth. In this project, we will use a simple 35-mm camera to demonstrate how NEAT and LINEAR work to expose potential trespassers in the near-Earth environment. You will also need your clock driven telescope and the longest telephoto lens in your bag.

NEAT and LINEAR use computer databases to compare the images that they take on a rapid-fire basis against the known sky. Anything that does not belong there is instantly identified and astronomers then can use sequential images of the object to determine its path through the sky. We will use the telescope's clock drive and a camera with a telephoto to try and duplicate what NEAT and LINEAR do to much fainter magnitudes.

Before going out, you will need to do some careful study and planning. NEAT and LINEAR have all the information they need in their computerized brains but you do not. You will want to use a star chart that shows the sky down to at least eighth magnitude (tenth would be better, but more expensive) and pick an area of the sky that will be no more than 30 degrees high in the west at the end of twilight. It should be about 5×10 degrees wide, which is about the width of the field of view in your camera with a long telephoto lens. This area of sky is roughly where the orbit of Venus is, looking behind our path around the Sun. Get to know this little patch of sky very well. When something enters this area of sky that does not belong there, you will know it right away. Use your chart to locate a reasonably bright star that you will be easily able to find in the sky once you are out in the field. You will use this star to center your telescope and camera.

Align the telescope as you normally would and use a piggyback bracket (likely supplied with your telescope) to bolt the camera to the top of the telescope so the scope will drive the camera along with it. Set up the camera and lens at the shortest possible focal length. In normal photography you would not want to do this because the short focal length creates a very flat field, but for the purposes of astrophotography the sky should always be considered to be flat. Once set up, aim the telescope at a bright star and use that star to carefully align the camera so that the star is also in the center of the field of view of your camera. You are using a bright star here because the focusing screen of your camera is likely not very clear so you must ensure that you can see what you are aiming at. You likely will not have that luxury later when you move the scope to the target area.

Turn the telescope towards your target area and center on the target star you selected earlier. Look through the camera and make sure that there are no obstructions such as trees or buildings blocking your view or immediately below. Once you are ready, select the "B" setting on the camera and attach the cable release. As you do so, note the *exact* time. This is important because if any streaks appear in your film, you will want to compare that track against a satellite-tracking program to make sure you have not imaged the ISS or an iridium flare. When you are ready, open and lock the shutter. Take an image that is anywhere from five to fifteen minutes in length. The maximum length of your exposure will be determined by how accurately you polar aligned the scope and how much light pollution is present. You will have to experiment. The longer the exposure, the easier it will get to find an asteroid. There are two reasons for this. First the longer you expose the film, the fainter the objects that will appear. Even without a telescope in a dark sky site, you will be able to image stellar objects as faint as eleventh magnitude. Secondly, remember that the objects you seek are moving targets. The longer the shutter is open, the longer a trail a moving asteroid might make on your frame. A fast-moving Apollo or Aten asteroid will leave a noticeable streak in a fairly short period of time. This will clearly stand out against your pinpoint stars.

Over time, you can experiment with using different lengths of exposures and aiming at different parts of the sky. Asteroids that have fallen into the inner solar system from the main asteroid belt tend to remain pretty close to the plane in which all the planets orbit relatively closely so while you may experiment with aiming closer to or farther away from the Sun, you should not aim too far away from the ecliptic. Now that you know where to aim and how to plan and shoot your pictures, let's talk about the last two elements you will need for success. First is patience. You will not likely strike gold the first night or with the first image. Or the second. Or the tenth. It takes time, work, planning and patience. And it takes one more thing. It takes some luck. The luck the William Herschel had when he found Uranus. The luck Clyde Tombaugh had the night Pluto blinked at him from a pair of alternately flashing slides. Or the luck that Yuji Hyakutake, Alan Hale and Thomas Bopp had when these dedicated amateurs discovered the great comets Hyakutake and Hale–Bopp in the mid-1990s. Hyakutake scanned the skies for decades in search of his first comet before bagging two in less than a year. There was a lot of luck involved. But that luck was engendered by a lot of hard work. Keep taking those pictures. Some day you might be rewarded.

Observing Project 7D – The Tricks of Venus' Atmosphere

The ashen light is one of the great mysteries of the inner solar system. Giovanni Riccioli first reported it in 1643. Looking through his telescope at the crescent

Venus he reported seeing a strange and unexplainable luminescence on Venus night side similar to Earthshine on a crescent Moon. Sir William Herschel also reportedly viewed the ashen light. The sharp-eyed E.E. Barnard never saw it but much more recent contemporary accounts made by prominent scientists reveal simultaneous viewing of the ashen light from two separate locations.

So is the ashen light for real and if so then what causes it? Astronomers using the 10-meter Keck I telescope on Mauna Kea have recently detected a greenish luminescence from the dark side of Venus. Spectrographic analysis of the glow is consistent with the 5,580-angstrom emission line of molecular oxygen (O_2). Astronomers have hypothesized that ultraviolet radiation at the upper levels of the Venusian atmosphere might break down carbon dioxide into its constituent atoms, then high-level winds in the Venusian atmosphere would rapidly carry those atoms into darkness where the atoms would recombine into molecular oxygen, emitting green light as it does so. The problem with using this as the source of the ashen light is that molecular oxygen emissions are very weak and likely undetectable with amateur telescopes, especially in such close quarters with the −4 magnitude crescent Venus.

Could lighting flashes be the cause? If this is so and if they occur in sufficient quantities then it is possible that it may excite Venus' upper atmosphere to glow. But many scientists think this unlikely as well. In 1998 and 1999, the Cassini spacecraft made two close flybys of Venus to gain a gravitational boost from the planet to enable it to reach Saturn. Cassini buzzed just a few hundred miles above the cloud tops of Venus each time. If lighting strikes were common in the Venusian atmosphere, it would have created distortions in the spacecrafts low-frequency radio transmissions, much the same as lighting on Earth distorts AM radio. No such distortions were heard.

Many believe the ashen light is simply an optical illusion, a retinal after-effect created by the dazzling brilliance of Venus itself. But others believe it is real. On rare occasion an opportunity presents itself to search for the ashen light when the Moon intervenes and passes in front of a crescent Venus. When a waxing crescent Moon passes in front of a crescent Venus, then the Moon will first block the bright crescent. The Moon will then require as much as sixty to ninety seconds depending upon how big Venus itself appears to occult the entire planet. During this time, the ashen light may present itself. But such lunar occulations are rare and because they must occur close to sunset, only a very narrow area will get to see the event in a dark sky.

Through our telescopes, Venus does not ordinarily reveal much beyond its blinding white brilliance. But in the late 1970s, NASA's approaching Pioneer Venus orbiter imaged the planet's cloud tops in ultraviolet light and the planet revealed itself in surprising ways. The uniformly white cloud tops gave way to patterns and swirls as they raced around the planet. Though you cannot see ultraviolet light even if it could penetrate to Earth's surface, you might be able to pick out some detail in the Venusian surface with a violet colored filter, which screens out all visible light but that at the very end of the spectrum. Many amateurs have had success with this technique beginning to see hints of detail in the Venusian clouds. Seeing patterns and features in the clouds of Venus will require patience and a steady atmosphere for those precious few seconds you need to see, but you can see them.

Try to see these effects for yourself. For the ashen light, do not use any more magnification than is absolutely necessary or the high power may squelch the faint light you're trying to see. Do not use any filters either because the first light to be lost will be the ashen light. During four of its five evening apparitions, Venus is high enough in a dark sky while crescent that it sets at least two and a half hours after the Sun. The ashen light has escaped the notice of some of the world's most renowned astronomers. How about you? Many of the great astronomers of antiquity have never seen the movement of Venus' cloud tops either. But they also did not have the advantages of modern equipment. With the help of a quality telescope and a piece of purple colored glass, here's your chance to see what few others have and that the great astronomers of two and three centuries ago could only have dreamed of seeing.

Observing Project 7E – Transits of Venus and Mercury

Among the rarest sights in astronomy are transits of the inner planets across the face of the Sun. We've already discussed at length what makes these events so rare. Until 2004, no person living on the face of Earth had ever viewed a transit of Venus, the last one having occurred in 1882. The next opportunities to view transits of the inner planets will occur when Mercury next transits the Sun on November 8, 2006 and Venus on June 6, 2012.

Transits are even more rare than solar eclipses so if you're planning to venture abroad to see either of these events, think carefully about weather. For both of the upcoming events in 2006 and 2012, the eastern third of the United States will be able to view the beginning of the event and have a couple of hours of good visibility. Europe and Africa will be able to see the entire event in both cases. June can be rainy on the U.S. east coast but not nearly so much so as earlier in the spring. November is climatologically the driest time of the year in the eastern half of the U.S. for the 2006 transit of Mercury.

Read back in Chapter 6 our notes about planning to travel to a solar eclipse. You may wind up in sub-Saharan Africa and that's a lousy place to discover you forgot your solar filter. If you elect to observe from home, make sure that your western horizon is clear for each event. Make sure your telescope is carefully aligned and aimed at the Sun with solar filter in place about a half-hour before the time of first contact.

Transits, like eclipses, have four distinct contact points as each edge of the planet touches the appropriate edge of the Sun. For the transit of Mercury, what will be striking is the speed with which the planet scoots across the Sun. Mercury's 10 arc second disk will require less than two minutes to completely enter the surface of the Sun, the time between first contact and second contact. Since it is at perihelion, Mercury is moving at its maximum orbital speed. In its transit of 2012, Venus moves at a much more leisurely pace requiring a full eighteen minutes to heave its 59 arc second disk across the limb of the Sun. At second contact (the point at which you can clearly see the photosphere between the planet's limb and the Sun's limb) watch for what is called the "teardrop effect." Many observers have reported seeing

Mercury or Venus stretched out slightly, appearing almost to cling the edge of the Sun before finally fully emerging into the Sun's disk. In 2012, Venus will spend a total of over six hours fully immersed on the Sun's disk so up to 75% of Earth's surface will be exposed to some part of the transit. This transit is actually somewhat short because the planet will never be more than 5 arc minutes from the edge of the Sun. By comparison the Christmas transit of 3,818 will last nearly eight hours and Venus will pass within 48 arc seconds of the Sun's center. The Mercury event will be shorter, lasting 3 hours, 54 minutes. Mercury will travel about half way from the limb of the Sun to the center.

Once it became evident that the Sun was at the center of the solar system and not Earth, astronomers of the eighteenth century realized that transits could be used to help determine critical distances such as the distance from Earth to the Sun. In 1761, the British explorer James Cook sailed his vessel *Endeavour* to the South Pacific to observe the transit of Venus that year and helped make the first reasonable estimates of the distance from Earth to Sun. His contributions to science are recognized today by NASA, which named its sixth Space Shuttle for Cook's first ship. Later on Cook commanded another ship named *Discovery*, with which he is credited for discovering Hawaii and the Aleutian Islands. NASA's fourth space shuttle is named for Cook's second ship.

The inner planets provide intriguing tests of our observing skills. Because they hide low in the turbulent air near the horizon, they test our patience by making us wait for still air. They test our ability to plan an observing period because the opportunity to view them on a given night is very short. Even at its best, Mercury might only be observable for a few minutes each night between the time it first appears and the time it becomes too low to see anymore. In the telescope, they behave in ways that sometimes contradict simple assumptions. Mercury tests our ability to resolve tiny details in poor seeing conditions while Venus challenges us to find the subtle details hidden in its brilliance. There are no great craters or lava plains to see here as there are on the Moon. Venus and Mercury also do not have enormous sunspots or prominences to jump out and grab your attention. Here we must work patiently and use all the resources of our eyes and our equipment to tease out the secrets of the inner planets and the interplanetary transients that sometimes join them. But with time, effort and practice, you will discover these hot, small rocky worlds can be truly amazing.

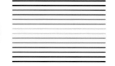

CHAPTER EIGHT

The Enigmas
of Mars, the
Red Planet

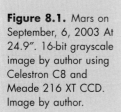

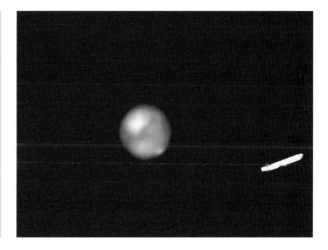

Figure 8.1. Mars on September, 6, 2003 At 24.9″. 16-bit grayscale image by author using Celestron C8 and Meade 216 XT CCD. Image by author.

There is no world in the solar system that has captured the fancy of the human imagination more than Mars. The mysterious red planet comes to glow brightly in our evening sky each twenty-six months commanding our attention and sparking our natural urge to explore. In each two-year window available since 1997, the United States has always dispatched at least one spacecraft and usually two to explore the solar system's fourth planet. Mercury hides too close to the Sun and is buried too deep in the glare of twilight limiting views of its surface. Venus hides behind choking clouds of carbon dioxide and sulfuric acid. Mars at its best stands high in a dark sky and bare for our telescopes to see, tantalizing us with the mysteries of its surface. Mars can often be an extremely difficult target to see. For the majority of the time during each twenty-six month apparition, the planet resides within 90 degrees of the Sun. The planet's dim glow is fainter than magnitude +1 and its disk is no larger than that of tiny Mercury. But as the planet nears the day of opposition, more and more telescopes begin to turn to it. As the planet nears Earth, it swells tremendously in brightness and in size. At its very best, Mars can grow brighter than Jupiter, a brilliant orange beacon in the night sky and its disk can become larger than that of Saturn. Then for a precious few weeks, it tells to our eyes amazing tales of a tiny world endowed with awesome and enormous landforms that was once very different than what we see today.

The God of War

The ancient Romans named Mars for their god of war because of what they perceived to be its blood-like color in the sky. Mars is actually more of an orange-yellow in the sky at its brightest and some people who have less than perfect color vision may have trouble seeing a pronounced color in Mars at all. Its bright biennial appearances spread great fear and terror through the ancient world as a harbinger of war or disaster.

The invention of the telescope elevated our fascination with Mars to an entirely different level. Astronomers viewing the red planet with telescopes that were not as high in quality as those that amateurs enjoy today saw a planet with pronounced markings across its surface that many believed were green in color. This suggested the presence of vegetation growing on the surface. During the twentieth century the evolving techniques of spectroscopy taught us that the planet's thin atmosphere was made mostly of carbon dioxide, perfect for growing green leafy things! Through each generation, astronomers and common folk looked to Mars with growing fascination and excitement. The question of whether or not we are alone in the universe is one of the most profound in all of science and philosophy. Mars seemed to scientists as late as the 1960s to be a place where life would find a promising foothold to grow. Stories centered on Martian civilizations grew in popularity in popular culture. H.G. Wells classic *War Of The Worlds* was one of the first such great stories. Wells' story was an allegory to the rise of German militarism after the 1870 unification of Germany and the fear created by the horrifying ease and brutality with which it defeated France in the Franco-Prussian War[20]. He was further inspired by the 1894 observations of the Italian astronomer Giovanni Schiaparelli who thought he saw channels on Mars which he chronicled in his logs using the Italian word *canali*, which in turn was mistranslated into the word "canals," which are artificial in origin. Later on, the American astronomer Percival Lowell also reported seeing channels on the Martian surface. Wells used what was known of Mars at the time to portray it as an old and exhausted world with dwindling supplies of oxygen that was becoming captured by iron in the surface (which

[20] The 1870 Franco-Prussian War was the first conflict in human history to claim more than one million civilian casualties, all French.

gives the planet its red color). In 1938, Orson Wells frightened an entire nation by reading *War Of The Worlds* on the radio convincing thousands of people who missed the disclaimer at the opening of the broadcast that it was a work of fiction, especially in New Jersey where the Martian ship supposedly crashed in the Pine Barrens. The classic story was "reimagined" in a 2005 motion picture. Other well-known works of science fiction about the red planet include Ray Bradbury's 1970s classic *The Martian Chronicles,* which became a network television miniseries and the Kim Stanley Robinson novel *Red Mars.* Robinson's work inspired two best-selling sequels *Green Mars* and *Blue Mars.*

As the Space Age dawned, many yearned for the opportunity for a close up look at Mars. Many probes were dispatched to Mars from both the United States and Soviet Union. Most met with disaster. Rockets like Atlas today are renowned for their amazing reliability but in 1964 were considered notoriously temperamental. On November 28, 1964 an Atlas rocket with an Agena upper stage (also a very failure prone vehicle) launched the American probe Mariner 4. After a cruise of more than seven months, Mariner 4 reached Mars and became the first spacecraft to return usable science and images from another planet. Mariner 4 returned 21 low-contrast grayscale images of Mars at distances ranging from 14,000 to 10,000 kilometers. Some of the images showed impact craters but they did not tell us a great deal about what the global planetary environment was really like. Two more flyby missions followed in the summer of 1969 shortly after the Apollo 11 landing on the Moon. Mariner 6 and Mariner 7 after successful Atlas–Centaur launches passed Mars on July 30 and August 4 of that year, respectively. In 1971, Mars came exceptionally close to Earth and this fact made it possible to send much heavier and more capable spacecraft to Mars. The United States thus planned a two space-craft mission to Mars consisting of Mariner 8, which was to photograph up to 70% of the planet's surface and Mariner 9, which was to study the changes in the Martian atmosphere and surface over time. Both Mariners carried rocket engines that would enable them to brake into orbit around Mars, rather than just fly past it. The project ran into immediate trouble when a launch accident claimed Mariner 8 on May 7, 1971 after the Centaur upper stage failed barely a minute into its burn. Mariner 9 would have to carry out Mariner 8's photographic objectives as well as its own environmental studies. After a quick investigation found Mariner 9's Centaur was not suspect, the spacecraft was launched to Mars on May 30, 1971.

When Mariner 9 arrived on November 14, 1971 after a cruise of five and a half months, it successfully fired its rocket engine and became the first probe to orbit another planet. Unfortunately Mars had little to show us for an enormous global dust storm enveloped the entire planet in a bright yellowish haze. After three months in orbit waiting, the planet's atmosphere settled down and Mariner 9 went to work. While Mariner 9 completely revolutionized our knowledge of Mars, it was also one of the greatest disappointments in the history of science. As thousands of crystal clear (by 1971 standards) images came back, the reality of Mars set in. There were no beds of vegetation, no canals, no Martians. The spacecraft instead saw only Moon-like craters, enormous canyons and extinct volcanoes of mind-blowing size. Mariner 9 from its perch in space circled Mars twice per day in a high orbit that eventually enabled it to map the entire planet. In all, Mariner 9 returned a total of 54 million bits of data, more than 27 times the previous three flyby missions combined.

Though Mariner 9 showed a tortured surface, it also showed some promising signs too. The planet's northern polar ice cap was revealed to be water ice and the atmosphere's pressure was accurately measured. Mariner 9 was able to image clouds in the sky over Mars and its instruments were also able to estimate surface temperatures that during southern hemisphere summer could range as high as 27°C (80°F). NASA's planners were already hard at work on a bold new mission that would take twin landers to the surface to search for life in a mission that would come to be known as Project Viking.

Each Viking was a massive spacecraft consisting of an orbiter and a Volkswagen sized lander. The total project cost over one billion (1975) dollars. Two spacecraft pairs were built for the mission because of the likelihood that one of the two probes would not survive its launch on its Titan 3C rocket. The Titan 3C[21] was America's most powerful launcher in service at the time but like the Atlas in the 1960s was considered only marginally reliable. The Atlas–Centaur had evolved into a far more reliable vehicle but it was not powerful enough to push the Viking stack into a transfer orbit to Mars. To the joy of the researchers, both launches were perfect. Viking 1 went first on August 20, 1975 and Viking 2 followed it on September 9, 1975. After a ten-month cruise, Viking 1 arrived on June 16, 1976 and began scouting its planned landing area in support of a landing on the nation's bicentennial on July 4, 1976. But the originally planned landing area turned out to be unsafe and mission planners delayed landing by more than two weeks while an alternative landing site was scouted. On July 20, 1976, the seventh anniversary of Apollo 11's Moon landing, Viking 1 descended through the atmosphere on a parachute to about 5,000 feet above Chryse Planitia before completing a rocket powered landing. Viking 2 entered orbit on August 7, 1976 and dispatched its lander on September 3, 1976 to a landing at Utopia Planitia. Viking 2 transmitted information from the surface for a bit more than $3\frac{1}{2}$ years before expiring on April 11, 1980. Viking 1 returned data until November 13, 1982. The two orbiters were not as long lived with Viking 2 giving out after just under two years and 706 orbits. Viking 1 survived in orbit until August 1980 and 1,400 orbits.

The Viking landers were specifically designed to probe the question of whether or not the Martian surface harbored microbiological life of any kind. Though the experiments were thought to be foolproof, they in fact returned inconclusive results. Many scientists believe that Mars today is a self-sterilizing environment with the combination of the oxidation of the soil and the interaction of ultraviolet radiation combining to make it impossible for organic compounds to form in the soil. We also learned though that Mars was once a very different place from what we see today.

The United States stopped Mars exploration after Viking, putting its next large investment into the Voyager program to the outer planets and a Space Shuttle program that was swallowing up more and more of a shrinking space budget. The Russians made an effort to go to Mars with twin 1988 orbiter/landers called Phobos 1 and Phobos 2. Both vehicles failed in a massive embarrassment to the Soviet government. Phobos 1 was lost enroute when controllers uploaded erroneous software, which caused the spacecraft's solar panels to point away from the Sun, and

---

[21] The Titan 3C is a Titan 2 ICBM flanked by two large solid-fuel boosters with a two-engine Centaur upper stage with stretched fuel tanks.

depleting the batteries. Phobos 2 reached Mars and began a survey of the planet. The last phase of the mission involved descending Phobos 2 to only 80 kilometers above the surface and releasing two landing vehicles. Contact was lost with Phobos 2 during the descent and never reacquired. The Russians have had very little success with Mars missions. Of the 19 that they have dispatched, only two flyby missions were completely successful. Two others released landers that were unsuccessful while a parent craft flew past the planet. One lander did apparently reach the surface but failed almost instantly. The other fourteen were complete failures, many never getting out of Earth orbit. Four of those were lost during the launch.

In 1993, the United States finally returned to Mars with a mission designed to study the geology and climate of Mars. The Mars Observer weighed more than 5,000 pounds, to that date the largest deep space probe ever built and at a cost of one billion dollars the most expensive to that time. A Titan 3 successfully launched the monster probe and it flew a flawless ten-month flight to Mars. But as the spacecraft pressurized its liquid fueled engine for orbital entry, contact with the probe was suddenly lost. A ruptured fuel line was blamed for the mishap. The loss of the probe led NASA under its new administrator Daniel Goldin to shift exploration efforts to a new paradigm, which he called "faster, better, cheaper." The first two missions of that program were highly successful. In 1997, Mars Global Surveyor entered orbit and began high-resolution explorations of the surface. MGS's mission was to replace the geology science lost on Mars Observer. On July 4 of that summer, Mars Pathfinder made a direct entry and bounced to a landing using an innovative airbag system. Pathfinder then released the tiny Sojourner rover to explore the surface and mesmerized millions of Americans. Goldin's approach seemed to be working, using small spacecraft launched by lighter Delta 2 rockets rather than the more expensive Titan 3 and 4 vehicles. But in 1999 came dual setbacks. First the Mars Climate Orbiter, which was to duplicate lost climate experiments from Mars Observer was lost when it dipped too close to the surface on approach due to a confusion among controllers between metric and English units. Then a poorly designed switch caused the engine of the Mars Polar Lander to shut down too early leaving the probe about fifty feet above the surface with no power. Under pressure from President Bill Clinton, NASA reevaluated the way in which it was supporting its Mars programs. It cancelled a 2001 lander, flying only an orbiter that year, and then with strong support from the new Bush Administration following up with its greatest Mars success, the 2003 Mars Exploration Rovers. The two rovers were named by schoolchildren "Spirit" and "Opportunity." They have trundled about the Martian surface for over a year and a half now beyond their 90-day design life. As of this writing, the probes have been on the Martian surface for one full Martian year. Their mission was to find out if liquid water every existed on the surface and both rovers have found ample evidence that each of their landing sites, on opposite sides of the planet from each other that water once flowed in large quantities across Mars. Opportunity scored first, demonstrating that many of the rocks at its landing site contained a mineral called *hematite*. Hematite forms principally in the presence of large quantities of water, though it could form volcanically. Spirit took longer, but while climbing a rise called "Husband Hill[22]," Spirit found rocks with

---

[22] Named for the commander of the final mission of Space Shuttle *Columbia,* Rick Husband. It is the largest of seven hills near the landing site, each named for one of the *Columbia* astronauts.

extremely high salt contents, almost proof positive that its landing site in an ancient crater called Gusev also was once awash with water. Mars may not have any life today and may in fact be sterile as Viking indicated, but at one point it may well have. Future missions may well answer this question. The Mars Reconnaissance Orbiter launched in August 2005 is designed to study the Martian surface to an entirely new level of detail paving the way for future landings. In 2007 a small fixed lander called Phoenix will attempt to carry out many of the objectives of the Mars Polar Lander and in 2009, a huge lander called will arrive with an advanced biology laboratory. Also in 2009 we will see the launch of the Mars Telecommunications Orbiter, the first communications satellite intended for another planet. These are all progenitors to President George W. Bush's proposed Mars initiative to take Americans to Mars by about 2025. Whether or not future presidents have the courage and will to carry out Bush's vision remains to be seen, but we can go and explore the mysteries of Mars from our backyards without billions from the politicians.

A Season of Mars

Observing Mars is almost always about the timing. During the twenty-six months between Martian conjunctions with the Sun, there are maybe only about four or five months where it is nearly impossible to see. During all the rest of the time, the red planet is visible and reasonably easy to observe, but for most of that time, Mars offers very little to see. An apparition of Mars begins at the time it passes behind the Sun into *conjunction*. We don't use the terms "superior" or "inferior" conjunction with Mars because it can only pass behind the Sun, never in front of it as Mercury and Venus do. After passing behind the Sun, Mars will slowly emerge into morning twilight. Because it orbits farther out than Earth does, Mars travels more slowly around the Sun that Earth does so Mars will appear to fall behind the Sun on the sky rather than pull out ahead of it like the inferior planets. Mars will appear to travel somewhere between 18 and 25 degrees eastward per month depending upon whether the planet is near perihelion or aphelion. The distance to Mars will also affect the view through the telescope. If Mars is emerging from conjunction in late August and is at or near aphelion, then the planet will be at its smallest and faintest. Mars will be barely three seconds across and fainter than magnitude +2. Even though the angle of the ecliptic with the morning horizon is favorable in late summer and early fall, the planet is so faint that it will be difficult to pick out of the predawn twilight for at least two to three months. For this first eight to nine months after solar conjunction, Mars does not typically offer you much to see in the telescope. The planet gains size only very slowly during this time and observers who are not prepared for what they are going to see tend to be very disappointed. During this early period in the apparition there is just nothing to see because Mars is so tiny in telescopes. Know what to expect because if you are waiting around until three in the morning to see Mars, you should not expect the planet to show you anything more than a tiny, featureless reddish dot. What you *can* expect to see as the planet slowly gains size is that the planet begins to show a slight phase in your telescope. Because Mars' orbit is not that far outside of Earth's, the viewing geometry allows us a peak behind Mars' terminator at its night side. As Mars' approaches a position about ninety degrees from the Sun, it only appears about 87% to 91% illuminated depending upon its distance from us. This will occur about ten months after solar conjunction and the exact positioning of Mars 90 degrees from the Sun is called *quadrature*. By now, Mars is also about 10 arc seconds across and nearing zero magnitude. This is considered the minimum size for the planet to begin showing some of its surface detail. Depending upon when we encounter Mars during our trip around the Sun, Mars may show us either its north pole or its south pole. The oppositions of 2001, 2003, and 2005 occurred in late southern spring or southern summer. Subsequent oppositions will more favor the northern hemisphere. The first prominent feature you will have a chance to see is the polar cap of the exposed hemisphere. Because it is somewhere between early spring and late summer at whatever pole you are looking at, it will be shrinking, if not already gone, you want to get a look at it as soon as you can see it. As the season progresses, the cap will rapidly shrink away. Use high power and watch that cap very carefully.

Watch Mars's path against the stars as it passes quadrature. You will notice that the planet's steady eastward march against the zodiac will begin to slow and will eventually stop altogether. The planet itself has not stopped moving but the perspective caused by our beginning to overtake it in space causes the planet to appear to stop and then begin to move backwards against the stars or towards the west. Like with the inner planets, this is also a form of retrograde motion. In this case it is an illusion caused by the fact that Earth's speed overtaking Mars causes the planet to appear to move backwards against the zodiac.

Mars will be at its best when it is directly opposite the Sun in the sky and this point is oddly enough called *opposition*. If two planets are in a circular orbit, then this point where the inner planet overtakes the outer planet is the point where they are closest to each other. Neither Earth nor Mars has a circular orbit and the orbit of Mars is in fact more elliptical than any other in the solar system other than Mercury or Pluto. Because of this, the time of opposition and closest approach for

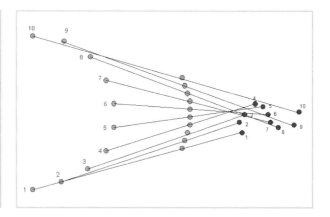

Figure 8.2. The illusion of retrograde motion. Power Point illustration by author.

Mars may differ by a few days. Viewing is still at its best right around opposition because Mars will rise at the same time as the Sun sets and sets as the Sun rises in the morning, thus it is in the sky all night long. During the weeks leading up to opposition, Mars will brighten and grow explosively in size. How much it does so will depend on whether the planet is near perihelion or aphelion in its orbit. Within the past twenty-five years, we've viewed oppositions of Mars at both extremes. On February 25, 1980 Mars came to opposition within five hours of reaching aphelion. That year Mars reached only 13.8 arc seconds in diameter, about the smallest it can ever be at opposition and brightened only to magnitude −1.0, also a minimum value for opposition[23]. The period of time where Mars was larger than 10 arc seconds, that prime viewing time, lasted for less than sixty days. Conversely on August 25, 2003 Mars reached opposition within 40 hours of reaching perihelion. On that night, Mars peaked at 25.1 arc seconds in size and blazed in the late summer sky at magnitude −2.8. Mars was larger than 20 arc seconds for a longer period of time in 2003 than it was larger than 10 arc seconds in 1980. Mars was larger than 10 arc seconds for nearly six months. This was the closest that Mars has come to Earth in over 50,000 years. Favorable oppositions of Mars do occur more frequently than that though. In 1986 and 1988 the planet came to opposition within a month of perihelion and displayed disks of 22.8″ and 23.2″, respectively and was brighter than magnitude −2.5. The oppositions occurring two years before and after the 2003 event both had Mars sporting a 20.2″ disk and brightening to magnitude −2.2. With Mars, though, the period of time where the planet shows a disk large enough to show you large surface features easily in a telescope is very short and so you have to dedicate yourself to getting out there and spend the time observing Mars when the planet is favorably placed in the sky and also close to Earth. The window of opportunity with Mars is always short. For the outer planets for example, they will spend about 45% of their time more than 90 degrees from the Sun. Because Mars is closer to us, it spends as little as 20% of its time more than 90 degrees from the Sun and even less time larger than that precious 10 arc second threshold.

After opposition, the planet begins to fall behind Earth in its orbit and will rapidly begin to shrink and fade. Within about six weeks, another sign that the best part of Mars biennial show is over will occur when the planet's westward motion against the stars slows, stops and then reverses itself again and the planet resumes *direct* or eastward motion against the stars. Mars' phase will shift back towards a pronounced gibbous by the time the planet reaches quadrature in the evening sky. By this time the planet will have shrunk back below 10 arc seconds and typically have faded back to below zero magnitude. During the final eight to ten months of the apparition, Mars will slip lower and lower in the western evening sky, fading to first and then second magnitude before disappearing in the twilight glow for good about two years after you first saw it emerge into the morning sky. Finally Mars will pass behind the Sun into conjunction and then begin the cycle anew. Because the average speeds of Earth and Mars are relatively close to each other,

---

[23] As a historical footnote, this opposition was one of the most watched in history because Mars was also in a close conjunction with Jupiter, which came to opposition that same night.

the apparitions of Mars, or the planet's *synodic period* (the time from one conjunction to the next), is the longest of any of the planets in the solar system. So opportunities to view Mars at its best occur with less frequency than that for any other planet and they last a shorter time than any other planet except for Mercury. So when the red planet burns bright at opposition, make it a priority for viewing because a good chance to get a good look at Mars does not last for very long. What will last for long is your wait for another chance if you don't spend time with it.

Observing Mars' Surface, Atmosphere and Moons

Even at its best, Mars is a comparatively tiny target. And for northern hemisphere observers, when Mars is at its very best, it stays low in the south at its most favorable oppositions. Patience is critical in observing Mars because those few moments when the planet is in steady air only come along a few times in a night and only last for a fraction of a second. But for those who wait, the detail Mars can show during those few precious seconds can be absolutely phenomenal. Then if you are patient, persistent and paying attention to the fine detail, some of the solar system's most massive formations come into view. One feature of Mars that does jump out is its color, especially in a telescope. Mars' bright orange-red hue will easily trip the cones of your eyes even when not at it's brightest.

With Mars at its best, watch the planet's changing markings as the night wears on. Mars rotates on its axis once each 24 hours, 37 minutes, or just over a half-hour longer than Earth takes. So Mars' surface features parade across the face of the disk during the night. Compare the features as they slide by to a map of the Martian surface and see if you can identify approximately where the historic explorations of Viking, Pathfinder and the MER's took place. The surface of Mars you will eventually realize presents an interesting dichotomy. You will not be able to see anywhere near the kind of detail that you did in the Moon, but you can learn to see fine detail in the Martian geography. The south is primarily bright highlands similar to those of the Moon, except for the high iron content of the rocks. The north is primarily low-lying plain. There is a sharp dividing line between the two, which will become evident as you watch the planet rotate. The average elevation sharply falls off by several thousand meters. There is no good theory to explain how this happened. In addition the highland hemisphere is very heavily cratered while the lowland hemisphere is very smooth. For every crater in the north larger than 28 kilometers there are 15 in the south. This too is one of Mars's great mysteries.

Like the Moon and Mercury, Mars has no plate tectonic activity as evidenced by the lack of folded mountain formations that so abound on Earth. This means that local hot spots in the mantle can sit under the same area of crust for millennia and build massive volcanoes. One of the great challenges that amateurs tried to undertake during the close approach of Mars in 2003 was to try and sight the massive Tharsis Montes and Olympus Mons. These are massive shield volcanoes that dwarf anything on Earth. Olympus Mons is the largest mountain in the solar system, as big as Texas and towering some 80,000 feet above the plain that surrounds it. The forces that formed the Tharsis Montes also created an enormous crack in the surface known as Valles Marineris, the solar system's largest known canyon. It stretches for 4,000 kilometers around the planet and is a much as seven kilometers deep in places.

Maps of Mars and freeware computer programs abound on the Internet that will show you the planet's geography and what side of the planet will be visible. You can watch the planet's parade of high and lowlands go by as regular as clockwork each day. As you look at the Martian surface, try if you can to imagine the scale of some of the features at which you are looking. Mars is home to the largest volcanoes in the solar system, the deepest canyons and somewhere on or below that surface there might be large quantities of frozen water. The European probe Mars Express has found what scientists believe is an enormous subsurface ocean of frozen water; a critical discovery that future probes will follow up on.

Speaking of water, watch the behavior of the exposed polar cap. The European Mars Express probe has been watching the polar caps closely since arriving at Mars in December 2003. The caps each consist of a permanent component that consists of alternating layers of ice and dirt. The mechanism that creates the layering is still unknown but a likely culprit would seem to be the unstable nature of Mars' axial tilt. The planet's axis is currently inclined just over 25 degrees with respect to the plane of its orbit, slightly more than Earth's is. Earth's axial inclination is stable however. Though Earth's axis precesses in a circle over time, the angle of inclination does not change with respect to its orbit. Mars wobbles like a top about to fall over, its axial inclination approaching angles as great as 60 degrees over many millennia. This causes dramatic changes in climate over many millions of years and leads to the kind of layering that we see in the cap. Each layer represents a different epoch of climatological and geological change. During the winter, a overlying cap of carbon dioxide ice overlies the permanent cap. During the summer months the carbon dioxide cap completely sublimes back into the atmosphere. At the Viking landing sites, this process of building the polar caps and vaporizing them again in the summer was found to vary atmospheric pressure by as much as 25% over the course of a Martian year. As the cap sublimates away with the onset of spring and summer, it may begin to break up into segments around the outer pole. Using high magnification you can view this effect in medium telescopes with high magnification. As the caps melt and atmospheric pressure rises, winds will begin to pick up in the higher latitudes. This can be a prime time to watch out for dust storms on the planet.

Mars' atmosphere also offers the patient and skilled observer some interesting details. The atmosphere is excruciatingly thin with a mean surface pressure of only about 7 millibars (sea level pressure on Earth is 1013.2 millibars, on Venus it is 90,000 millibars). Locally though pressures can vary dramatically. At the peak of Olympus Mons, the pressure is barely 1 millibar while in the deepest basins pressure can build as high as 9 millibars. The makeup of the atmosphere is almost entirely carbon dioxide gas that was not absorbed into the rocks during the planet's formation. Small amounts of nitrogen, argon, water vapor and oxygen are also present. The atmosphere can on occasion present water vapor clouds in thin white wisps in the upper atmosphere. Much more dramatic are when the winds kick up huge dust storms that begin in a plain and within a few days completely envelop the entire planet. Such storms are most likely when Mars is near the Sun and the sublimating ice cap pumps large amounts of gas into the atmosphere. Massive global storms were visible during the planet's close oppositions in 1971 and 2001.

There is more to Mars than just Mars. Two tiny moons also accompany the planet, discovered by the American astronomer Asaph Hall. Phobos is the larger of the two moons. Hall discovered it on August 18, 1877. It is irregular in shape measuring about 27 by 22 by 18 kilometers in diameter. The moon orbits Mars at a mean distance of only 9,378 kilometers from the center of Mars. This puts the moon lower than the synchronous orbit altitude so Phobos revolves around Mars faster than Mars rotates. In fact the moon is so low that not every location on the side of the planet facing Phobos can see it. Where it can be seen, Phobos moves across the sky in the opposite direction of everything else, rising in the west and setting in the east three times each day as the satellite scoots around the planet three times each day. Phobos is also condemned to destruction. Because Phobos orbits below the synchronous orbit altitude[24], tidal forces are dragging it down towards Mars at a rate of about 1.8 meters per century. Within fifty million years it will be gone (barely a blink of an eye astronomically speaking), either dragged down to the surface or more likely broken up into a ring.

Though it is the smaller moon, Deimos was actually discovered first, six days before Phobos. Deimos orbits above the synchronous orbit altitude, so tidal forces are actually driving it slowly away from Mars, just as our own Moon is drifting slowly away. Deimos is also irregular in shape and measures about 15 by 12 by 11 kilometers. The little rock orbits Mars every thirty hours at a distance of 23,459 kilometers from Mars' center and is only visible to an observer on the surface as a very bright star moving across the background of the stars. Both are small rocky bodies similar to the rocky asteroids that orbit in the belt just beyond the orbit of Mars. There is no completely satisfactory theory that explains the presence of the moons because the gravity of Mars should not be strong enough to capture and hold them. But there they are, faithfully circling Mars. You can see them both in telescopes of 8 inches and larger under dark skies and favorable conditions. Phobos is the brighter of the two and can become as bright as magnitude 10 during the closest oppositions. But Phobos orbits so close to Mars that it can be difficult to see in the glare of the planet. Deimos circles out farther from the planet and thus is not as likely to be washed by Mars' brilliance, but is two magnitudes fainter putting it near the limit of an 8-inch scope under dark skies.

Mars offers many treats for the observer. Its surface is readily visible, it has an atmosphere that can show clouds and storms to the patient observer and it presents a constantly changing face to the observer with some of the most astonishing terrain features in the solar system. For the skilled observer willing to work, Mars's two moons can be teased out of the planet's glare. Lets go to Mars and have a look.

---

[24] If a satellite orbits a planet so low that it revolves faster than the planet rotates, then the satellite loses energy to the planet's rotation and falls. If the satellite is high enough that it revolves slower than the planet rotates, then the satellite gains energy from the planet and drifts higher while the planet's rotation slows.

Observing Projects VIII – Observing the Dynamics of Mars

Observing Project 8A – The Rotation of Mars

Mars rotates on its axis just like Earth does and in almost the same amount of time. The planet's sidereal period is only 41 minutes longer than Earth's is. This is one of the easiest things to gauge about the planet because its surface markings are consistent and easy to see. You can study the rotation of Mars in one of two ways.

If you want to spend a few hours at the telescope you can watch the planet's features slide across its disk. Pick an object near the planet's central meridian and time how long it takes for it to reach the planet's limb. In just a bit over six hours your point should roll out of sight and into Martian night.

A simpler way to do the same thing would be to watch the planet on a successive series of nights. Since Mars takes about forty minutes longer to complete a rotation than Earth does, each night a feature on the planet's central meridian will retreat a bit further onto the planet's morning hemisphere each day. Over the course of about a month, you can track the planet through an entire reverse rotation in this way. Pick a prominent feature like Hellas or Sinus Sabaeus and track it from night to night. This drill will also help you begin to grow familiar with the geography of Mars.

When working with Mars use all the power the seeing will tolerate. Also use filters. To enhance surface features you should use red and orange filters. These block light from haze and clouds and increase contrast among surface features. Study your maps and know the planet's features and geography and you can recreate one of the most important early discoveries about Mars in the telescopic era.

Observing Project 8B – Cheshire Polar Caps and Other Atmospheric Phenomena

Mars has a tenuous atmosphere of carbon dioxide that provides less than 1/100th the surface pressure of Earth's atmosphere but that does not mean that Mars' atmosphere is not capable of providing some interesting displays. Even with only 7 millibars of pressure to work with, a good gale force wind can wreak havoc on a surface covered with a lot of loose sand. And on Mars, it often does. Watch for yellow discolorations particularly rising from the northern plains or from large impact basins like Hellas. The polar cap may also disappear into a haze of yellow and red at times. Also watch for the planet's limbs to occasionally become fuzzy and dark instead of sharp and bright. All of these events are signs of dust rising in the atmosphere riding the Martian winds. When Mars is near opposition, particularly when the planet is near perihelion, is when seasoned Mars watchers keep a close watch on the planet's weather. The example of a hide and seek polar cap is just one manifestation of the planet's weather patterns caused by dust rising into the atmosphere. Several locations on Mars are very suspect to dust storm formation, especially the wide-open plains areas.

Polar hoods also tend to form during local winter. These are clouds of ice crystals that form over the poles when large amounts of carbon dioxide sublimate into ice in the upper atmosphere over the polar regions where temperatures during the eleven-month long polar night reach temperatures nearing −170 degrees Celsius. The hoods may grow large enough, particularly during southern winter, which occurs with the planet at aphelion and temperatures in illuminated areas can remain at −50 degrees even at high noon.

Thin wispy cirrus clouds may also form anywhere on the planet, though they are difficult to see. The clouds may be formed either by water vapor (which can be nearly 1% of the Martian atmosphere) or sublimated carbon dioxide in the winter hemisphere. Try using a blue filter to reduce glare and also enhance the white colored clouds against the red surface and see if any of the planet's weather patterns come into view for you. A violet filter will enhance atmospheric features even more but will cause surface features to all but disappear. These high-altitude cirrus clouds are an extreme challenge for amateur size telescopes but the level of satisfaction you will get for bagging them can be just as extreme.

Observing Project 8C – Looking for Deimos and Phobos

Mars' two tiny moons, Deimos and Phobos, circle the planet with blistering speed and never stay very far from it. When Phobos is at its greatest elongation from Mars, it is never more than three planetary radii from the planet. Since the moon circles Mars in just over seven hours, the window of opportunity to see it while it is away from the planet is only about an hour long. It is critically important therefore to know when the moon is at that point of greatest elongation and have a plan. Use an astronomy software program to plot Phobos position so you will know what it looks like relative to the planet. Make certain that you protect your night vision carefully and then execute your observing plan on time. At opposition, Phobos can range anywhere from about magnitude +10.2 when Mars is closest to Earth and down to magnitude +12 is Mars is near aphelion at the time of opposition. Once you have found Phobos, watch it for a little while to verify that you have not found a background star. The moon from greatest elongation should begin to move quickly back towards Mars. This motion should be discernible in as little as thirty minutes.

Deimos is not so subject to washout as Phobos is because it is not nearly so close to Mars at greatest elongation. Still it is just as important to find that time of greatest elongation to go looking for Deimos. Something else that is important that one does not normally worry about with solar system observing is a dark sky. Even with Mars at a perihelic opposition, Deimos will barely be brighter than magnitude +11.9. At less favorable oppositions, Deimos is beyond even the reach of medium size telescopes never becoming brighter than magnitude +13.7. I have viewed Phobos successfully under moderate light pollution with my 8-inch Celestron at favorable oppositions in 1988, 2001 and 2003. Deimos is impossible to see under any light-pollution conditions because it is already near the limit of the 8-inch scope so a dark sky is absolutely critical to success. As with Phobos, verify that Deimos shifts its position over time. The motion is not nearly so rapid as that of Phobos but over the course of an hour the motion of Deimos relative to Mars should be apparent.

If the glare of Mars is a bit too much here is a dirty little trick you can try. Try running a $\frac{1}{8}$-inch wide strip of aluminum foil across the field stop of your eyepiece (don't let the tape touch any optics!) and put the eyepiece in your telescope. The foil will form what is called an *occulting bar*. Use the bar to block Mars from your sight and this will eliminate the glare from the planet. Now see if Phobos, Deimos or both pop into view. You may have to experiment with the width of your occulting bar and it will have to vary in width depending upon magnification. Still, getting rid of that glare from Mars itself may prove to be the difference in seeing or not seeing the moons.

Observing Deimos and Phobos is a challenge that will stretch both your eyes, your sky and your equipment to their limits and may even require a bit of outside interference. They represent two of the great mysteries of the solar system in that no one can explain how a planet of relatively weak gravity managed to capture two asteroids into prograde circular orbits. For the amateur willing to do the work, they can be coaxed out into view. Do your studying and go find them.

Observing Projects 8D – Surveying the Martian Geography: Quadrant 1 Centered on 45 Degrees

Mars is a world much like the Moon in geological diversity and though we can't see nearly the detail in Mars that we can in the Moon, it offers us a lot of interesting things to see in what is there. To begin working this project, the first thing you will need is a global topographic map of Mars. You can find this on several Internet sites or if your astronomy software has a sufficiently detailed Mars model, it will suffice.

The first thing you always notice right away about Mars is the highland versus lowland dichotomy. As we discussed earlier, Mars has two distinctly different surfaces separated by thousands of meters of average elevation. Highland areas dominate the south, while the lowlands dominate the north. This strange split personality defies any likelihood that mere chance caused this. But the alternative is that something catastrophic occurred early in the planet's history.

A map of Mars uses a latitude system that is identical to that used on Earth. The longitude system used however has one important difference. On Earth, longitude is counted in both the easterly and westerly directions from a designated prime meridian that runs through the Royal Greenwich Observatory in England. On Mars, there is a designated prime meridian but longitude is only counted upwards in a westerly direction for a full 360 degrees around the planet. In this project we will survey the planet in four quadrants, first from 0 to 90 degrees, then in 90-degree increments traveling westward around the planet.

In our first project, we will view the planet starting with the 45 degree meridian as the central meridian and working 45 degrees on either side. In this way, we are never viewing close to the Martian limb where foreshortening makes it difficult to see objects with clarity. The most prominent feature visible in your telescope here is the large dark area just south of the equator. This area, known as Mare Erytharaeum, is a highland area. The large bright area to its north is Chryse Planitia. This is a broad, flat low-lying plain, which holds a special place in Martian history. It is where the first successful visitor from Earth arrived. On July 20, 1976 the Viking 1 lander set down in the northern region of Chryse near a smaller dark area that is at about 30 degrees latitude called Nilokeras. Not far to the east of Viking 1, about a thousand kilometers, is where in 1997 the Mars Pathfinder bounced to a stop on the spot now called Sagan Memorial Station. A large broad dark area to the north and east of Nilokeras, which extends well towards the north Martian pole, is called Mare Acidalium. Even though this area is similar in color to Erytharaeum, Acidalium is a lowland area, some of the lowest on Mars. South of the Erytharaeum area is a broad open highland area called Argyre. This plain is a bit lighter in color than Erytharaeum. Mare Oceanidium is located south of Argyre and is much more bright in color than Argyre. As we look further towards the south pole of Mars, the terrain becomes much more mountainous for a distance before giving way to another broad flat plain called Mare Australe. This broad flat highland will eventually give way to the south polar cap.

The most prominent geological feature within this first quadrant is the famous Valles Marineris. This massive canyon is the longest such structure in the solar system. The Valles Marineris extends westward away from Mare Erytharaeum towards the 90-degree meridian about 10 degrees south of the Martian equator towards the Tharsis region, which we explore next. With Mars at its best and the skies clear and calm and with the planet favorably placed, you might be able to trace the hairline of Valles Marineris across the Martian disk.

Observing Projects 8E – Surveying the Martian Geography: Quadrant 2 Centered on 135 Degrees

In this project, we'll continue highlighting surface features on Mars, this time moving 90 degrees further to the west. Here is where we find Mars' most dramatic surface features. Here we are looking at Mars with the 135-degree meridian centered and we are scanning the surface 45 degrees either side of the central meridian.

At about 30 degrees south latitude is a large dark area called Mare Sirenum. This is a rugged and heavily cratered highland zone. This is the most prominent feature in this quadrant that is easily visible. On the extreme northwest edge of Mare Sirenum is a 130-kilometer wide ancient crater called Gusev. Here scientists studying images from Mars Global Surveyor believed that they have viewed the telltale signs of flowing water on the ancient surface and for this reason the crater was targeted as the first landing site for the Mars Exploration Rover Program. The rover Spirit bounced to a landing here on the night of January 4, 2004. South of Mare Sirenum the terrain brightens again rapidly as we come to a higher region called Phaethontis. As you scan further south towards the polar region, the terrain gives way to the flat area that we saw in quadrant one, Mare Australe, which extends about one-third of the way around the planet from where we saw it originate.

Scanning back to the equator is where we find Mars' crowning glory of features. The Tharsis Montes are three of the largest volcanic structures in the solar system and they would be the largest if it were not for what looms to the northwest. The three Tharsis volcanoes each tower twice the height of Mount Everest and each would completely cover the state of New York. These are called *shield volcanoes*. They are built up from below by magma pushing up weak spots in the Martian curst. On Earth, the ability of such mountains to build is limited by the presence of tectonic motion in the crust, which over millions of years will carry the building volcano away from the hot spot in the mantle beneath it. But on Mars there are no plate tectonics so the Tharsis range was able to build for as long as the volcanic pressures were present because the building mountain never moves. The mountains however barely subtend one arc second when Mars is at a median opposition diameter. That is big enough to be viewed in an 8-inch or larger telescope but conditions need to be perfect. Yet there is something bigger still to see.

Just to the northwest of the Tharsis is the solar system's largest mountain, Olympus Mons. Olympus Mons is thrust up on a base which itself towers more than a kilometer above the surrounding plains. The total height of the caldera above the surrounding plains is more than 80,000 feet and the volcano is large enough to cover the entire state of Texas! Olympus Mons was formed in the same manner as the Tharsis Montes. The uplifting of these four mountains stretched and split the Martian crust and created the Valles Marineris over to the east. This was a blow to those who thought that the massive canyon system might have been formed by flowing water.

Several large plains sit to the north of Olympus Mons. The large dark area is called Arcadia. Arcadia appears to be flat and smooth like much of the rest of the Martian lowlands. Arcadia is shaped like a downward pointing arrowhead coming out of the north polar region. The bright area to the east of Arcadia is a plain called Ceraunius. The bright area to the west of the arrowhead is called Diacria. All of these areas are part of the Martian northern lowlands. North of this area, low flat plains continue until reaching the polar cap area.

After viewing the first half of the Martian surface, one interesting thing becomes apparent. The color of the terrain seems to have little bearing upon its elevation. Dark markings appear at both highlands and lowland areas of the planet. The difference in relative reflectivity of the rock and dust of that given area is what causes the dark markings. Though Mars is a small world, half the size of Earth, that has not stopped it from building geological features that are Olympian in size.

Observing Projects 8F – Surveying the Martian Geography: Quadrant 3 Centered on 225 Degrees

If you've spent a long time in the field at night and you made Mars your first stop at about 8 PM local time, consider coming back for another look at 2 AM for Mars is now showing you a very different face. If the 135-degree meridian was centered then, now you're looking at the 225-degree meridian. This line of longitude was on the limb six hours before and now it's just about dead center.

The southern hemisphere of the planet at 225-degrees longitude is dominated by the large moderately dark area stretching from east to west called Mare Climmerium. Cimmerium is dominated by heavily cratered highlands.

Climmerium is a broad albedo feature that spans from the equator southward to about −30 degrees latitude and spanning the entire area between 200 degrees and about 270 degrees longitude. South of this area is the bright highland Eridania, and then a somewhat darker highland called Mare Chronium. South of Chronium, the terrain begins to merge into the south polar cap. The terrain south of Mare Chronium is some of the roughest on the planet. It is both very mountainous and very heavily cratered. The terrain remains rough until reaching the south polar cap.

Moving to the north about on the equator or just north of it on the central meridian is a bright area called Cyclopia. There is a small dark albedo feature where the terrain is heavily depressed called Cevberus. Further to the north the terrain brightens and descends into lowland plains. The plains area immediately to the north of Cevberus is Elysium Planitia. Just about on the central meridian and just north of 45 degrees north latitude is the broad plain Utopia Planitia where in September 1976, the Viking 2 lander set down. On the border between Elysium and Utopia, the terrain then again rapidly begins to darken. This is about the only place in the northern polar region of Mars where the terrain darkens markedly over a wide area. The terrain in this area remains dark, but smooth and relatively flat.

Observing Project 8G – Surveying the Martian Geography: Quadrant 4 Centered on 315 Degrees

This region of Mars is perhaps the planet's most fascinating. There is one bright area which dominates the view in the south and two large dark areas in the tropics and north which dominate the view in that part of the surface.

The large bright area in the southern hemisphere is called Hellas and it is one of the largest impact basins in the solar system. Mars does not do anything small! Hellas measures more than 2,100 kilometers across, is more than 6 kilometers deep and the force that excavated it has flattened and rewritten all the terrain

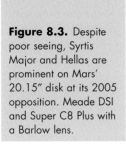

Figure 8.3. Despite poor seeing, Syrtis Major and Hellas are prominent on Mars' 20.15" disk at its 2005 opposition. Meade DSI and Super C8 Plus with a Barlow lens.

surrounding it, including clearing a large circular area in the dark albedo feature to the north of it. A ring of materials raised up by the impact rises nearly a full 2 kilometers above the floor of the basin and extends out about 4,000 kilometers from the basin center. Hellas is the king of Mars' hemisphere of craters. Mars' southern highlands are much more heavily cratered than are the northern lowlands. Including Hellas, Mars hemisphere of craters contains 3,068 craters that are larger than 20 miles (28 kilometers) in diameter while the opposite hemisphere has only 237! The search for ancient life not withstanding this remains one of the greatest mysteries of Mars. Was the planet's southern hemisphere exposed to something the north was not? Or was the northern hemisphere reshaped by some internal mechanism that healed its wounds? Hellas is the greatest of the planet's wounds and the planet's brightest feature, polar caps not withstanding.

Hellas excavated a circular area out of the long dark albedo feature to its north called Mare Tyrrhenum. This feature is nearly as dark as Hellas is bright, giving a beautiful contrast between the huge impact basin and the dark highlands on the equator. To the north of Tyrrhenum is an even darker area, perhaps the darkest on all of Mars, called Syrtis Major. This is a major peninsula of highland area extending well into the lowlands of the northern hemisphere. Above the northern edge of Syrtis Major, the terrain falls off sharply in elevation and brightens in color. At about 45 degrees north latitude is a narrow albedo feature called Umbra, then above that the terrain brightens for a short distance before merging into the western edge of Utopia Planitia.

From the center of Mare Tyrrhenum, a very dark finger-like peninsula extends for about 2,000 kilometers to the west. This area is called Sinus Sabaeus and it trails behind the other two features like a tail as the planet rotates. The most notable thing about Sinus Sabaeus is that right about in the middle of the peninsula is a small dark area called Meridiani Planum. Here scientists studying the area with the Mars Global Surveyor spacecraft found the spectral signature of a mineral called "hematite." Although hematite can form in volcanic environments, it also forms prodigiously in water. Suspecting that the Meridiani Planum area may have once been awash in water, NASA planners targeted this spot as the secondary landing site for the Mars Exploration Rover program. Assuming that the first lander was successful in reaching Gusev crater, the second rover, called Opportunity, would be targeted into Meridiani. After Spirit was successful in early January, Opportunity was allowed to bounce to a landing on Meridiani on January 25, 2004. Scientists were soon rewarded when Opportunity rolled to a stop in a crater where an outcropping of bedrock was exposed. It was not long before hematite was found at the site and that the hematite was formed as a result of the rocks of the region having been awash in water.

Each two years, the red planet Mars rises high and bright in our sky to tantalize us with its brilliant orange-red hue and parade of light and dark markings between polar caps. It is home to the largest known mountains, biggest impact scars and deepest canyons in the solar system. Geology on a grand scale tantalized the imagination at the eyepiece and challenges even the most skilled observers to unlock its secrets. The promise of water and along with it, the possibility of life on another world at least at some time in its distant past serves to unlock the power of our dreams and ability to explore. Each two years an army of telescopes, big and small and new fleets of spacecraft swarm upon Mars seeking to unlock the answers.

CHAPTER NINE

Comets and Asteroids, the Cosmic Leftovers of Creation

Figure 9.1. Comet Hale–Bopp. 35-mm SLR camera, 210-mm telephoto, ISO 400, 5 minute exposure. Photo by author.

They are the leftovers from the creation of the solar system, the matter, dirt, dust and detritus that did not find its way into one of the nine major bodies of the solar system. They have collectively gathered for the most part in neat bands between the orbits of Mars and Jupiter. At first there was but one. Then there were four. Then they were found by the dozens, and then hundreds and today we know them by the tens of thousands. They were too small to be planets so a new name was created. Astronomers called them *asteroids*. Meanwhile out in the frozen depths beyond Pluto thousands more frozen chunks of rock and ice circle in the distant darkness. They sit dormant until an interaction with another frozen chunk of rock causes it to fall into the inner solar system. As it does so it heats up and volatile materials trapped within begin to erupt to the surface blowing off gas and dust. The hunk of rock is now surrounded by a cloud of highly reflective materials which begin to trail off behind the falling nucleus. In ancient times the appearance of a *comet* in our skies inspired terror as they were considered a harbinger of disaster to come. Today we admire them in awe on the rare occasions when one of the bright ones flare in our sky.

There is far more to the solar system than the nine major planets and the Sun. There are literally tens of thousands of other bodies floating around out there around the Sun. Knowing about them is critical to us for one day; one of them may have our name on it. Some 65 million years ago, an asteroid with our name on it made sure that you would never have a Tyrannosaurus Rex for your next-door neighbor. So before continuing our grand tour of the grand bodies of the solar system, lets take a look at the solar system's smallest bodies.

The Asteroid Belt

On New Year's night in 1801, the Italian astronomer Giuseppe Piazzi found a small object wandering across the sky. Piazzi at first thought that he had discovered a new comet. After study however it was found that the object was moving in a nearly circular orbit around the Sun. Piazzi named the little body "Ceres" for the Sicilian goddess of grain[25]. Ceres turned out to be way too small to be a planet, measuring 933 kilometers in size. Even at that size, Ceres contains about 25% of all the mass in the asteroid belt and is by far the largest object in the belt. It orbits the Sun at a mean distance of 414 million kilometers and orbits the Sun each 4.6 years. That places it about 162 million kilometers outside the orbit of Mars and about 305 million kilometers inside the orbit of Jupiter. Ceres' orbit lies not far out of the plane of the solar system inclined only about 10 degrees from the ecliptic. It was not long before Ceres had company. The next year a second body was discovered by Heinrich Olbers and he named it Pallas. Olbers was attempting to recover Ceres and refine our knowledge of its orbit when he happened on Pallas. Pallas' has a mean distance from the Sun about equal to that of Ceres but travels in an orbit that is somewhat more eccentric than that of Ceres. While Ceres orbit has an eccentricity of about .07, Pallas is about .23. This is far more than any other planet. Pallas orbit is inclined substantially as well, more than 34 degrees. Pallas is considerably smaller than is Ceres measuring about 525 kilometers in diameter. Olbers named Pallas for the ancient Roman god of wisdom. Juno was the third asteroid to be discovered by Karl Harding in 1804. Juno is barely half the size of Pallas. In 1839, Juno's orbit was determined to change appreciably. Today, Juno's orbit is actually slightly more eccentric than is that of Pallas and is inclined 13 degrees to the ecliptic plane. Juno orbits the Sun each 4.35 years. In 1807, Olbers found his second

---

[25] By tradition asteroids are named by their discoverers. Early asteroids were named for ancient Greek or Roman deities as the planets were. Many asteroids are now named for contemporary heroes (3,350 Scobee) or family members or even pets. Today, an asteroid is always referenced by a catalog number issued in order of its discovery and a proper name.

asteroid and named it "Vesta" for the Roman goddess of virtue. Vesta measures 285 miles in diameter (460 kilometers) and orbits the Sun each 3.63 years, the shortest period of the four largest asteroids. Vesta is very different from the first three asteroids. Though it is approximately the same size as Pallas, its surface is much brighter than any other asteroid and belongs to a special class of asteroids different from the others. Vesta's surface is so reflective that it is the only asteroid that ever reaches naked-eye visibility. Today we know of several asteroids that share similar spectral properties with Vesta and occupy similar orbits. Some call them "Vestoids." Scientifically they are given a special classification based on composition and albedo. We'll discuss different asteroid classes in just a bit. These first four asteroids came to be generally known as the "Big Four" among the asteroid belt. By the time the century had ended, several hundred asteroids had been found. Twenty-six of those were found to be larger than 200 kilometers. As we learned more about the asteroids, we discovered that the idea that they were possible leftovers from a shattered planet began to lose appeal. All the known asteroids added together have less total mass than the Moon. Scientists actually do not have a great deal of interest in the large ones because we've found them all and know their orbits very precisely. It's the small ones that worry astronomers because we have probably found only a fraction of 1% of the estimated one million asteroids in the one-kilometer size range.

Asteroids are organized into groups by means of their orbital characteristics. Most asteroids are found in the area between Mars and Jupiter between 1.7 and 3.0 astronomical units from the Sun. These asteroids are called *main belt* asteroids. The main belt is divided into several different sub-groupings each named for a principal asteroid within that group. Such groups include *Hungarias, Floras, Phocaea, Koronis, Eos, Themis, Cybeles* and *Hildas*. Small relatively empty zones called *Kirkwood Gaps* separate each of these groups from each other. The Kirkwood Gaps are kept relatively free of asteroids as a result of gravitational interaction with Jupiter. An object within a Kirkwood Gap would likely have an orbital period that would equal a simple fraction of Jupiter's (one-quarter, one-third, one-half). An object in such an orbit would resonate with Jupiter causing energy to be added to the asteroid and causing it to move into a higher orbit.

In Chapter 7 we talked a bit about near-Earth asteroids (NEA). NEAs are divided into three classes. *Atens* orbit inside of Earth's orbit with an aphelion that is greater than Earth's perihelion. *Apollos* orbit outside of Earth's orbit with a perihelion distance that is less than Earth's aphelion distance. *Amours* circle the Sun just outside Earth's orbit and approach Earth's aphelion distance but never actually cross Earth's orbit. This makes them less threatening than Apollos and Atens, which actually pass through the same space that Earth does.

Astronomers have also charted two groups of asteroids, which are called *Trojans*. The two Trojan groups are in gravitational balance between the Sun and Jupiter at the Lagrangian points 60 degrees ahead and 60 degrees behind Jupiter and travel together along with Jupiter in this odd gravitational dance. Astronomers also speculate that there may be some Trojan-type asteroids bound to the Lagrangian points of Venus and Earth. The asteroid 5621 Eureka is bound to one of Mars' Lagrangian points.

Asteroids are classified (different from specifying orbital categories) by what they are made of and by how much light they reflect back into space (albedo). Most asteroids are what are called *C-type*. These asteroids are rocky in nature, made mostly of carbonaceous materials and are similar to the Sun except for a lack of hydrogen, helium and other volatiles. These make up about 75% of all known asteroids, but because they tend to be so dark and hard to see, they may actually be underrepresented in the total asteroid population. About 1 in 7 asteroids are classified as *S-type*. These asteroids are made of metallic nickel-iron and iron and magnesium-silicates. These asteroids are much brighter than the C-types are. The last major classification is the *M-type* asteroids. These asteroids are the brightest being nearly pure nickel-iron. They make up only about 7% of all known asteroids. Over the years several different systems for differentiating asteroids have emerged and new categories have emerged in old systems. But for simplicity, it is probably easiest to simply discuss asteroids as "rocky" and "metallic." Ceres, Pallas and Juno are rocky. Vesta is heavily metallic.

We did not know much more about asteroids until the opportunity came along to explore them with spacecraft. Asteroids were not considered to be objects of sufficient interest by NASA or other space agencies to send dedicated missions to them, but NASA planners saw an opportunity when redesigning the trajectory of the Galileo probe to Jupiter. Galileo was to be launched directly to Jupiter by the Space Shuttle *Challenger* in May 1986 with a two engine Centaur upper stage. But *Challenger* was destroyed on the mission immediately preceding its scheduled Galileo flight and in the aftermath of the accident, NASA decided against flying the hydrogen-fueled Centaur in the shuttle cargo bay. A less powerful two-stage solid fuel rocket called an Inertial Upper Stage would instead launch Galileo. The IUS did not have the power to propel the massive Galileo probe directly to Jupiter, so a series of slingshots would have to be devised. Galileo would reach Jupiter by gaining speed after launch from the Space Shuttle *Atlantis* by making a flyby of Venus, then Earth. Along the way, planners realized that Galileo would fly close by two asteroids. First it would fly by the asteroid 243 Ida. When it did so, astronomers turned Galileo's cameras on it and got a completely unexpected shock. Ida had a satellite! A tiny companion that astronomers named "Dactyl" was circling Ida. Later, Galileo also flew past the asteroid 951 Gaspra. In later years the advanced technology demonstrator spacecraft, Deep Space 1 flew past the asteroid 9969 Braille before traveling on to its planned destination, Comet Borrelly. The Stardust spacecraft made a flyby of the asteroid 5535 Annefrank before going on to its dust-collecting mission at Comet Wild 2. The only dedicated mission to an asteroid to date was the Near Earth Asteroid Rendezvous mission in the late 1990s. NEAR nearly ended in disaster. It conducted a successful flyby of the asteroid 253 Mathilde before flying to 433 Eros and entering orbit. But when it was time to enter orbit around Eros, a software error caused the engine to fail to fire. The trajectory gnomes jumped in and saved the day and after flying an orbit around the Sun, NEAR caught Eros again one year late and slipped into orbit. After circling the rock for a year, NEAR gently settled on the surface of Eros and transmitted data from the surface for a short time. What these missions have taught us is that C-type asteroids are not always the solid rocky bodies we thought they were but rather amount to little more than lumps of detritus and rock weakly held together by gravity.

An exciting future mission is NASA's DAWN mission scheduled to launch in 2007. This will be the first mission to travel to any of the Big Four. DAWN carries a re-startable ion engine that will allow it to first enter orbit 4 Vesta in June 2010. After spending a year studying 4 Vesta, DAWN will then blast out of orbit and fly to 1 Ceres, entering orbit there in 2014.

Observing Asteroids

The asteroids orbit mainly outside the orbit of Earth and so they behave much the same way as Mars does. The best time to view an asteroid then is when it is at opposition to the Sun. For each of the big four, this occurs each sixteen to twenty months. Since the asteroids are very faint, you will always need optical aid to find them, except for Vesta, which can reach naked-eye visibility at very favorable oppositions. Each of the Big Four can be seen with binoculars.

Ceres is not only the largest asteroid, but has the orbit most like that of the planets. It stays fairly close to the ecliptic and has the lowest eccentricity. Pallas and Juno stray a considerable distance from the ecliptic with orbits inclined 34 and 13 degrees, respectively, so it is possible to find them wandering through constellations not normally associated with planetary visitors. Vesta's orbit is more like that of Ceres, inclined only about 8 degrees but somewhat closer to the Sun than is Ceres.

It is interesting to track the movement of the asteroids as the come around to opposition. Because they pass often so far north or south of Earth's orbital plane, they will often trace interesting tracks in the sky. Ceres and Vesta tend to behave more like Mars usually tracing flat oval-shaped tracks. Asteroids with higher eccentricity orbits tend to make patterns that are more exaggerated in shape. Asteroids near the upper or lower culmination of their orbits will make broad ovals that are almost circular in shape. Asteroids that come to opposition near their node will make broad "Z" shape patterns.

Identifying asteroids in your telescope requires some discipline. No asteroid grows large enough in angular size to be visible as a disk in a telescope so what you are looking for is the extra star in a given area. There are then three ways to identify an asteroid. The first is to know a star field well enough that any extra star

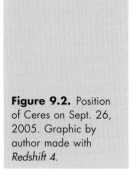

Figure 9.2. Position of Ceres on Sept. 26, 2005. Graphic by author made with *Redshift 4.*

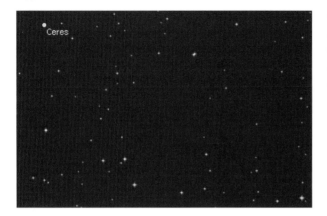

Figure 9.3. Position of Ceres on Oct. 2, 2005. Graphic by author made with *Redshift 4*.

in the area will be apparent to you. You can also study a given an area over a few days and watch for the star that moves. In either case, you need a good star chart to pick out the extra star in a given area. The third option is to use long-exposure photography. Over the course of time, an asteroid on the move will leave a short streak on your film as we illustrated in Observing Project 7C. If the asteroid is slow moving, it may not leave such a streak but if you take images over successive nights you can chronicle the movement of any faint background object.

An amateur activity that has great potential for scientific value is timing asteroid occulations of background stars. Each year, *Sky & Telescope* and other major astronomy publications list all predicted asteroid occulations of background stars for the coming year. Timing and track locations are very important to astronomers. By timing occultation duration, we get a much more precise idea about how large that pinpoint of light really is. By combining that information with a spectrographic analysis, we can determine how massive that point of light is. Gaining precise ground track locations will tell us more precisely about the orbit of that asteroid. Gaining this precise data will teach us a great deal about the rock that one day in the future might have Earth in its sights. In this chapter's Observing Projects, we'll talk about how to time asteroid occulations.

Hairy Stars

In the not too distant past the appearance of comets in the sky was a reason for great terror and fear. With their bright and stellar like nuclei and long fanned tails, some called them "hairy stars." Comets were believed to be harbingers of doom and destruction. Even as recently as 1910, thousands of people reacted in terror to the news that the tail of Halley's Comet (properly referenced 1P/Halley) would sweep across Earth possibly poisoning the atmosphere of Earth with deadly gases. Salesman pushing gas masks made a fortune during the spring of 1910. But the comet passed with not a single noxious odor. Today comets are viewed with wonder and awe when one of the beautiful and rare bright ones flies past Earth.

The British astronomer Edmund Halley was among the first to begin to demystify comets. In the year 1705 after careful study of the orbits of bright comets that appeared in 1531, 1607 and 1682 he realized that they followed paths that were virtually identical. Halley postulated from this that they must be the same object and boldly predicted that the comet would return in 1758. Halley died in 1742, but on Christmas night 1758 a German farmer and amateur astronomer named Johann Georg Palitzsch recovered the comet. Astronomers in recognition of Halley's achievement named the comet for him[26]. Astronomers of the time then backtracked through history and identified twenty-three previous appearances of the comet proving that it had been viewed by people at every return dating all the way back to the year 240 BC when Chinese observers wrote of a "broom star" and reported its motion from the eastern sky to the northern sky. Recreations of the 240 BC apparition bear out this observation. The ancient Babylonians saw the comet at its return of 164 BC. Sadly the comet's most observed return, its most recent was also the poorest in recorded history. 1P/Halley brightened to only magnitude +2.6 in March 1986 and its tail spanned only about 15 degrees in the early morning sky that year.

Though the return of 1986 was poor, amateur astronomers will get a treat at Halley's next return in 2061. A close fly-by of Jupiter in September 2060 will raise the comet's perihelion slightly and shorten the comet's orbital period by 1.3 years. The comet will then pass within .48 AU of Earth and only .05 AU from Venus (that's only about eight million kilometers!). Halley actually came closer to Earth in 1986, but did so while far from the Sun. The pass of 2061 will occur with the comet much closer to the Sun.

So what determines whether or not a comet will become bright in our skies? The first thing that one must always remember about predicting cometary brightness is that comets are notoriously unpredictable. Several factors must be considered when astronomers try to tell us whether or not a comet will become spectacular. First is whether or not the comet will pass close to the Sun. As a comet plummets into the inner solar system, the Sun heats up volatiles within the solid cometary nucleus, which burst through its icy covering throwing ice, dust and gas into space. These begin to reflect the Sun's light in an enormous area around the nucleus, which is called the coma or "head" of the comet. As the comet nears the Sun, the solar wind begins to push the dust and gas downwind from the comet forming a tail. The tail can have two distinct components. A whitish-yellow dust tail is the most prominent feature, but careful examination will also reveal the presence of a bluish gas tail. Both tails are obvious in the picture of Comet 1995 O1 Hale–Bopp. The gas tail tends to stream directly downwind from the Sun while the dust tail may fan out along the comet's direction of motion. This spreading of the tails is also evident in the image at the opening of this chapter. If the comet does not pass close to the Sun, then it will not be excited to an adequate degree or may not begin to discharge matter at all.

---

[26] By tradition, new comets are named for their discoverers, rather than by them. A new comet is also given a designation number which begins with their year of discovery, a letter which denotes what two-week segment of the year they were discovered in, followed by a number issued in chronological order of discovery. Comets with known orbits are designated differently. They are given a sequential "P/" numbers such as 1P/Halley or 2P/Encke.

The comet must also pass close to Earth. Most comets are extremely small with a nucleus that measures less than one kilometer across. In order for such a tiny object to become bright, it must pass close enough to Earth for us to get a good look at it. If it does not come close to Earth, then it will appear many magnitudes fainter. A good example is that of Comet 2P/Encke, which has the shortest period of all of the known returning comets. Encke returns each 3.3 years. At two of its apparitions it passes fairly close to Earth and reaches minimal naked-eye brightness, but at each third apparition it is far from Earth and never brightens above magnitude +10.

Size also matters. The bigger a comet is, the more surface area it has available to blast dust and gas into space. It will also have more reservoirs of such material to exhaust into the void. So larger comets tend to have the opportunity to become brighter. But size is not the only key feature of the comet itself that determines how bright it might become. Astronomers now know that comets that are passing the inner solar system for the first time are often covered in a dark material that prevents them from outgassing at a substantial rate.

This was the unfortunate situation in 1974 when astronomers prematurely predicted that the massive Comet 1973 E1 Kohoutek would become one of the most brilliant in history. But as Kohoutek neared the Sun in 1974, its virgin dark coating prevented it from reaching the predicted brilliance and many never saw it, dealing the entire astronomical profession an embarrassing blow. Kohoutek did actually become relatively bright, reaching magnitude +3.0 in the morning sky. As the comet reached perihelion close to the Sun, viewers on Earth lost sight of it but astronauts on Skylab could see it against a dark sky and reported that it might have become as bright as magnitude −3.0! Earth-based viewers got a brief look at the comet at zero magnitude in early January 1974 but the comet rapidly faded as it moved away from the Sun, disappearing completely from naked-eye sight by the end of the month. Astronomers were understandably gun shy when in 1975, Comet 1975 V1 West was discovered and showed promise of becoming brilliant in the morning sky before dawn after passing perihelion less than ten million miles from the Sun on February 25, 1976. Unlike Kohoutek, Comet West delivered, becoming bright enough to be viewed in daylight for several days shortly after perihelion and sporting as many as five distinct tails. Comet West is considered one of the premiere comets of the twentieth century.

So we have now learned that comets can become bright in some combination of four ways. They can pass very close to the Sun, they can pass very close to Earth, they can be enormously large or they can simply be enormously energetic. Comet 1P/Halley is fairly large (about seven kilometers along its longest axis), passes well within the orbit of Venus, usually passes within 0.5 AU of Earth at the same time and is a prodigious producer of dust and gas. So Halley combines all four elements. In 1994, amateur astronomers Alan Hale and Thomas Bopp independently found a new comet of about magnitude +10. That in itself is not surprising, but what was shocking was where the comet was when it was found. Comet 1995 O1 Hale–Bopp was still well beyond the orbit of *Saturn*! When it was found it was still nearly two years away from perihelion! The geometric circumstances of the coming apparition were to be very poor. Hale–Bopp would not come within 1 AU of the Sun, nor would it come within 1 AU of Earth. But Hale–Bopp was enormous. Estimates of its nucleus placed it at over 70 kilometers in size. The comet brightened to magnitude −1 and displayed a beautiful fanned 20 degree long tail. Hale–Bopp is an example of a comet that is *intrinsically bright*. It did not matter how far away it was, Hale–Bopp was going to put on a magnificent show. And boy did it ever! Hale–Bopp was visible to the unaided eye under dark skies for more than fifteen months and astronomers are still tracking the outgassing nucleus nine years later! Had Earth encountered the comet just five months later, it would have become brighter than Venus!

While the astronomical world was abuzz over the approach of Comet Hale–Bopp, the dedicated Japanese amateur Yuji Hyakutake was scanning the skies in January 1996 when he happened upon a faint fuzz ball in the sky. Comet 1996 B2 Hyakutake was the amateur's second comet in less than a year after more than twenty years of searching. The first comet was a nondescript tenth magnitude object that never developed, remaining well outside the orbit of Mars. But Hyakutake's second comet would prove to be one of the grandest in history in its own right, upstaging the approaching Hale–Bopp. It has been twenty years since the grand display of Comet West and now we would be treated to two cometary spectacles in less than a year. Roaring out of the southern morning sky, Hyakutake brightened rapidly and reached zero magnitude as it passed almost directly over Earth's north pole at a distance of less than 13 million kilometers heading inbound towards the Sun. As it did so, some observers reported an ion tail of astonishing length. No account claimed less than 70 degrees and some observers claimed the tail stretched across more than 100 degrees of sky! Hyakutake was not a big comet, barely a kilometer across nor was it intrinsically bright, but it came very close to us. Hyakutake went on to curve within 5 million miles of the Sun before heading out of the inner solar system and putting on a less grand show for southern hemisphere observers.

Comet Hyakutake became brilliant because it passed near Earth, Hale–Bopp never came anywhere near Earth but became bright simply because of its brute size. There are many variables that contribute to cometary brightness, including whether or not they have been here before. Comet Kohoutek was a first-time visitor, but careful calculations show that Comet Hale–Bopp had been through the inner solar system before. That is, fifty thousand years before! Comet Halley is neither as big as Hale–Bopp, nor does it come as close to Earth as Hyakutake did. But a good combination of size, proximity and other factors combine to make it the Old Faithful of comets.

Though we were certainly spoiled by the appearance of two bright comets within a twelve-month span in 1996 and 1997, these are very rare events. The last comet to exceed magnitude zero before Hyakutake was Comet West twenty years prior and West was the first since Comet Ikeya–Seki blazed close to the Sun in 1965 becoming the brightest comet in recorded history. Ikeya–Seki was found a few weeks before perihelion and after solar conjunction blazed brighter than the full moon. Most comets remain extremely faint and are usually a challenge for even the largest amateur telescopes.

Since the mid-1980s comets have been prime targets for spacecraft exploration. An international armada greeted Comet Halley at its return in 1986. Five dedicated missions flew to Halley including the Russian probes Vega 1 and Vega 2, which studied the comet at medium range. Long-range investigations were carried out by the Japanese probes Suisei and Sakigake. The most prized observations were made by Europe's Giotto probe, which flew only 300 miles from Halley's nucleus and actually survived to the surprise of its controllers. Giotto returned the first images of the nucleus of a comet. Images returned showed a dark, potato-shaped rock spewing jets of gas and dust into space around it. Budgetary issues caused the United States to sit out Comet Halley but it since has launched several more missions to other comets since. These include the Deep Space 1 flyby of Comet 19P/Borrelly and the Stardust mission to obtain samples of material from the coma of Comet 81P/Wild 2. The European Rosetta mission is enroute to reach Comet 67P/Churyumov–Gerasimenko and enter orbit around it in the year 2011.

What Comets Leave Behind

When a comet passes through the inner solar system, it ejects large amounts of material into its coma and tail. The comet settles back into its inactive state as it departs the inner solar system, but what becomes of all the dust and other material ejected from the comet?

All that dust become spread out along the comet's orbit over time, continuing to circle the Sun along the same path as the comet does. Several of these streams cross near or through Earth's orbital plane, so once or twice (if both the inbound and outbound legs of the comet's orbit cross Earth's orbit) each year Earth passes through the stream of dust. As the particles enter Earth's atmosphere, they flare brightly in our sky as a *meteor*. Each of these streams varies in character according to the angle at which it encounters Earth, the speed of the particles and their size. A regular display of such meteors, appearing to come from the same place in the sky is called a *meteor shower*. The most prominent of the annual showers is the Perseid shower in August. A meteor shower is generally named for the constellation in the sky from which the meteors appear radiate from and that exact point in the sky is called the *radiant*.

Each of the known meteor showers is tied to a particular periodic comet. For example the Eta Aquarid meteor shower in May and the October Orionid showers are tied to 1P/Halley. These are relatively minor showers producing about 20 meteors per hour each. The Perseid shower is created by debris from Comet 109P/Swift–Tuttle. The Perseids are considered to be the most persistent and reliable annual shower reliably producing a zenithal hourly rate of about 100. The November Leonid shower is created by Comet 55P/Tempel–Tuttle. The Leonid shower is normally fairly weak producing about 15–20 meteors per hour. Each thirty-three years[27] the parent comet returns to the inner solar system and when it does, it drops a large volume of fresh material behind it. Then for a couple of years, the Leonids historically turn into a storm, producing not just a few dozen meteors per hour but many thousands. In the early part of the 2000s the Leonids

---

[27] During the next return a gravitational interaction with Jupiter will disturb the orbit of Comet 55P/Tempel–Tuttle so that the next two major meteor streams will miss Earth and thus storms such as those viewed earlier in the decade will not recur for at least the next two returns of the comet. The next chance for meteor storms from the Leonids will not occur until after 2099.

produced meteors at rates of over 6,000 per hour or almost two every second! Meteor showers are very easy to observe and enjoy. All you need is a dark sky, including a lack of a bright Moon in the sky and a lawn chair. Then just sit back and enjoy. Another major shower that has been gaining in prominence in recent years is the December Geminid meteor shower. The Geminids have been increasing in strength over the past few years so that they now rival the Perseids in intensity. Unlike the Perseids, the Geminids are known for creating bright fireballs or *bolides*. The parent body of the Geminids is a body that actually was originally identified as an asteroid and is designated 3,200 Phaethon. Phaethon has never been observed ejecting gas or dust or exhibiting any other comet-like behavior but it follows a comet-like orbit. Phaethon is now believed to be an extinct comet and is the source body of the Geminids.

For the most part meteor watching is an endeavor for insomniacs. The reason why is that as Earth passes through the meteor stream, the leading side of the planet is its morning side. The evening side trails as we circle the Sun so to see the meteors we must wait until Earth's rotation carries us into a position where we are facing into our direction of orbital travel. So the best time to view most of the major showers is in the morning hours before sunrise. The exception to the rule is the Quadrantid[28] shower, which takes place in January. These meteors overtake Earth from behind in space and so are best viewed in the evening hours after twilight has ended. The Quadrantids produce some 100 meteors per hour but they tend to be faint and slow moving because their relative speed as they enter the atmosphere is reduced by Earth's motion through space.

Let's now study some of the small bodies of the solar system a bit more closely and discover what we can learn and what wonder there is to see.

Observing Projects IX – Asteroids and Comets and Meteors, Oh My!

Observing Project 9A – Tracking Asteroids

With a little bit of effort, you can learn the location of just about any asteroid in the sky for the purposes of planning an observing session. Popular astronomy magazines will publish articles containing finder charts when asteroids of note are favorably placed for viewing. You can also find the location of most asteroids using just about any planetarium program. After that finding your target is up to you.

The problem with locating an asteroid as opposed to some other faint object like Uranus or Neptune is that they look like any other star. Uranus or Neptune clearly do not look like stars in a telescope. Picking an asteroid out of the background requires that you know what point of light in the picture is the one that does not

---

[28] This shower is named for the now defunct constellation Quadrans, which is now in the northeastern area of Bootes.

belong there. Imaging the area where you believe the asteroid is located with your telescope will show clearly that one of the points of light in the field will move over the course of a few days. If you do not have a camera, then make drawings of the area over the course of a few days. Use low to moderate power in your telescope and chart the motion of your asteroid in a field of view that is about one to one and a half degrees wide.

It is also interesting to note the wide variety of patterns that an asteroid makes as it nears and passes opposition. Because the planets all orbit in close to the same plane as Earth, the retrograde patterns that they make can be very subtle. The planets tend to make narrow loops or "Z" shapes that have very narrow declination spreads. Some asteroids have orbital inclinations that are well offset from that of Earth. Of the Big Four, Ceres and Vesta orbit close to the plane of the ecliptic while Juno and Pallas have orbits that are more highly inclined. Pallas in particular because it is in an orbit inclined 34 degrees to the ecliptic can make patterns in the sky that are much more exaggerated than Ceres might make. You can plot the motions of both asteroids against the sky over several months around the time of opposition and see the difference for yourself.

Observing Project 9B – Timing Asteroid Occultations

Here is a chance for you to take part in some serious science. Timing asteroid occultations allows us to determine the size and shape of asteroids to an order of magnitude that would not otherwise be possible. For example if the orbital velocity of an asteroid is well known, then by timing how long it hides a star, we can directly infer how large it is.

Now let's assume an asteroid is round, like a planet. An occultation will take place of a given star along a track that we'll say is 200 kilometers wide. If an observer at the center of the track sees the star disappear for sixty seconds, then observers at either edge of the track will only see the star vanish for a second or two. As you get closer to the track center, the duration of the occultation will continue to increase. Now what if the asteroid is shaped like a pyramid, with the base at the south and the point oriented towards celestial north. Then an observer at the north end of the track would only see the star vanish for a second or so. At the track center, the duration would be perhaps a minute or so and then as you proceed towards the south limit of the track, the duration would continue to increase. This allows us to infer the shape of the asteroid. If we can get two or three timings from various points within the track, then we can get a rough idea of the shape of the asteroid. To get a solid model, then we need very precise timings from many observers stationed along the track.

You can obtain very precise occultation information from organizations such as the International Occultation Timing Association. Once you know when an occultation will occur, you need three things. First is the willingness to move and do so on short notice. Unlike with the planets or comets or the Big Four, we don't necessarily know the orbits of asteroids with exact precision. So as an occultation draws near, the track of the event you thought you were sitting right in the center of can shift by a considerable distance as more refined predictions are made close to the event. If you are serious about helping to time occultations, you need to be ready to get up and move with only a few hours notice.

The second key thing you need is the ability to very precisely determine your position. Fortunately technology has made this very easy to do. For less than $200 now, you can purchase a Global Positioning System receiver that will allow you to plot your position on Earth's surface, including elevation to an accuracy of only 30 meters (100 feet). This allows astronomers calculating an asteroid's profile to know exactly where you are with respect to the asteroid's track over the ground.

The third thing you must have is a very precise knowledge of the time. The best way to do this is to monitor short-wave time signals such as those provided by radio stations such as WWV. This station broadcasts time signals on 5 MHz, 10 MHz and 15 MHz with 10,000 watts of power and also broadcasts on 2.5 MHz and 20 MHz with 2,500 watts of power. WWV and its sister station WWVH are both operated by the National Institute Of Standards and Technology. They are based in Fort Collins, CO and Kauai, HI, respectively, and will broadcast voice signals at the top of each minute. WWVH will have a voice signal fifteen seconds before the beginning of the next minute and WWV will do so 7.5 seconds prior. With time and practice, you will learn to tell time on WWV or WWVH by means of the tones broadcast. A specific pitched tone tells you what minute it is within an hour. Even if you don't know how to interpret the time signals, all you need to do is record the signals in conjunction with a video of the occultation. For occultations of bright objects, you can use a simple camcorder to do this. You can receive the short wave signals with an inexpensive receiver that is available at many consumer electronics shops. You can also mount a camcorder to a telescope with a special mount for imaging fainter events.

Timing asteroid occultations requires hard work and the occasional bit of very good luck. But for those willing to do the work this is an opportunity to take part in some exciting research that sometimes delivers unexpected results. In one recent case, an observer watched as a star was occulted by an asteroid and then reappeared. A few seconds later the star vanished again, this time hidden by a previously unknown satellite of that asteroid! What surprises await you?

Observing Project 9C – How to Discover Your very own Comet

Few if anybody outside his or her own local observing circles knew of Alan Hale prior to 1995. Some might have thought that at the mention of the name that you were talking about the deceased actor who portrayed the "Skipper" on Gilligan's Island (I made that mistake). The same would be true for names like Thomas Bopp or Yuji Hyakutake. They became famous for the bright comets they discovered that would forever carry their names. How did they do it?

Persistence is one key tool you must possess. Yuji Hyakutake scanned the skies for more than twenty years before finding his very first comet. He then found two in less than twelve months, the second being one of the most spectacular of the twentieth century. Others like Karou Ikeya find comets in droves. In his mid and late twenties he became famous in Japan after finding five of them, the last of which was the famous sungrazer, Comet Ikeya–Seki, which he found at age 31. Ikeya built a business around his fame, designing and manufacturing high-quality telescope mirrors. Ikeya never stopped searching for comets, but had to wait thirty-seven years before finding his next one, Comet 2002 C1 Ikeya–Zhang[29]. Ikeya–Zhang eventually became a modest naked-eye comet. Other dedicated amateurs like Carolyn Shoemaker (wife of the famed geologist) and William Bradfield have discovered many comets. Bradfield, who is a seventy-seven year old retired rocket scientist working for the Australian government, has bagged eighteen comets since his first find thirty-three years ago. His 2004 comet though was his first after an eight-year slump.

Bradfield does have the advantage of searching the southern hemisphere. Most bright comets that venture close to the Sun seem to come from one of five families, which are named for the astronomers who have identified them. All five groups have orbits that bring them close to the Sun from well south of the ecliptic plane. Comet Ikeya–Seki was a member of the Kreutz group, a series of comets which all move in nearly identical orbits. All members of this group are likely derived from one parent body. In the case of the Kreutz group, the progenitor is a brilliant comet, which passed within some 500,000 miles of the Sun in the year 372 BC and was observed to split in half. The great comet of 1882 and Ikeya–Seki were both comets of this group as were some 85% of all the comets discovered by SOHO. Most Kreutz comets do not survive their passage within less than 500,000 miles of the Sun's surface but those that do like the Comet of 1882 and Ikeya–Seki went on to become legend. Southern hemisphere observers have the advantage of being able to view Kreutz group sungrazers first as well as those from the Kracht I and Kracht II groups, the Marsden group and the Meyer group.

Pick what you believe is a likely patch of sky, say about 1–3 square degrees and regularly image that patch of space. Like with our asteroid search projects, get to know this little area of space like it was the back of your hand. When something unusual passes through it, it should immediately grab your attention. When comets are close to Earth, nearing the perihelion of a highly elliptical orbit they pick up an enormous amount of speed and thus will leave long streaks on film if left exposed. In fact, in order to get a clear exposure of a comet, it is necessary to track on the comet because it is moving so rapidly in its orbit. So clean exposures of comets taken through a telescope usually show star trails.

So if you want to discover a comet, remember the fundamentals. Be patient and persistent, know your target area of the sky very well and search areas that seem more favorable for cometary discoveries. A disproportionate share of comets are discovered in locations south of the ecliptic. Don't be discouraged if you go weeks, months or even years without finding anything. The greatest comet hunters in our science go many years between finds, but the one thing they have in common is that they never quit. Yuji Hyakutake searched the heavens in a fruitless hunt for comets until he was an old man, but never gave up on his dream. His two decades of persistence paid off with greater rewards than he ever could have dreamed and

---

[29] Ikeya–Zhang is now known to be a short-period comet, orbiting the Sun in just under 200 years and so has been given the designation Comet 153P/Ikeya–Zhang.

now one of the most beautiful comets in history bears his name. Even the greatest of the comet seekers sometimes go many years and even decades between making discoveries, such as Japan's Karou Ikeya. Maybe someday your persistence will pay off too.

Observing Project 9D – Meteor Showers

Enjoying the beauty of a meteor shower is largely dependent on two things. First is a dark, moonless night. Second is an adequate supply of caffeine because observing a meteor shower will keep you up into the wee hours of the night. The reason for this is that Earth revolves around the Sun in a counterclockwise direction and also rotates counterclockwise. So as Earth plows through the dust stream of a comet, our morning side is facing the stream while Earth's evening side is hidden. The only major shower that defies this logic are the Quadrantids of January, which overtake Earth from behind.

We need to breathe a bit of reality into the concept of hourly meteor rates and what you can expect when you are told a shower will produce 100 meteors per hour. That value is what is called a "zenithal hourly rate." That means the shower will produce 100 meteors per hour of sixth magnitude or brighter when the radiant of the shower is at the zenith, or directly overhead. So you need to consider two things. First the shower radiant rarely if ever passes directly overhead and second the sky must be dark enough to allow you to see down to sixth magnitude. If neither condition exists, the shower will not produce anywhere near that kind of rate.

Of interest to the observer are two things. What kind of color and brightness do the meteors produce and are they fast moving or slow moving? The Geminids for example are very fast moving meteors and the shower tends to produce many fireballs. The Perseids produce a steady rate of meteors that tend to be more uniform in brightness. The Quadrantids are characterized by their almost stately movement across the sky. Their slowness is caused by the fact that they are overtaking Earth from behind.

Meteor showers are easy and fun to observe and since you don't need a telescope they are something that you can share easily with friends. So break out the lawn chairs, fire up the grill or pop the popcorn and treat your friends to the best show they likely never knew was there.

CHAPTER TEN

Jupiter and Saturn, Kings of Worlds

Figure 10.1. Mighty Jupiter with two of its moons, Io (left) and Europa (right). The planet is 45″ in diameter. The image is made with a Celestron Super C8 Plus and Meade DSI. Image by author.

They are the mighty sentries of the solar system. Their powerful gravity shapes and shepherds the paths of the small bodies of the solar system, accelerating asteroids into organized bands beyond the orbit of Mars, keeping the area inside the orbit of Mars relatively free of thousands of free floating rocks that could possibly visit a cataclysm upon Earth. They grace our skies with their brilliant beacons. They fill our telescopes with expansive disks filled with bands and belts and swirling storms or delicate and beautiful rings. Each is escorted by dozens of moons, many of which are bright enough to be easily viewed in amateur telescopes. They entertain us with their nightly dance, occulting and eclipsing and transiting each other and their parent bodies. They are Jupiter and Saturn. They are the kings of the solar system's planets. Jupiter alone contains twice as much mass than all the other planets put together. Saturn is the sight that propels many of us into a lifelong love of astronomy. None of us will ever forget our first viewing of the planet's magnificent ring system. For me, that was in late 1979 when the ring system was inclined barely two degrees to our line of sight. The thin bright line through Saturn's disk was awe inspiring to me. Jupiter and Saturn offer us enormous beauty that is easy to see and also detail that is subtle and challenging. Let's look in depth now at mighty Jupiter and beautiful Saturn.

The King of the Gods, the Failure of a Star

The ancient Greeks named Jupiter for their ancient king of the gods because of the brilliance of its yellow-white beacon that rules the night sky from high overhead when at opposition. Though not as bright as Venus, Jupiter comes high in the sky at night to rule over the heavens. Even at its faintest it is still about as bright as Sirius and can approach magnitude −2.8 at its closest oppositions. Only Mars can shine in a dark sky at magnitudes even approaching that of Jupiter, and as we have already demonstrated, Mars can only become that bright about once in each fifteen-year period. Jupiter does it every year, soaring high in the sky, dominating all the stars in the heavens and outshines every other stellar object in the heavens by at least one magnitude.

As we gaze out beyond the asteroid belt and contemplate Jupiter, it becomes obvious right away that this is a very different kind of world than any of the other bodies that we've studied thus far. First of all, it is enormous. Of all the rocky bodies of the solar system we've toured thus far, the largest is the one you're standing on, Earth. Jupiter is so immense that it would swallow more than 1,300 Earths within its volume. Jupiter's diameter of 142,984 kilometers is by far the largest body in the solar system other than the Sun. Despite having a volume of 1,300 Earths, Jupiter's mass is only equal to about 318 Earth's suggesting that the planet's density is very low. On average, a cubic centimeter of Jovian matter weighs only 1.33 grams, compared to 5.5 for Earth and Mercury. Jupiter is almost entirely gas. The planet by volume consists of approximately 90% hydrogen and 10% helium. Since helium is much heavier than hydrogen the helium actually makes up about 25% of the planet's total mass. In terms of its chemical makeup, Jupiter much more closely resembles the Sun than any of the rocky planets we've viewed thus far. Jupiter has a highly complex structure with a small rocky core at the center surrounded by a vast layer of liquid metallic hydrogen, an exotic form of hydrogen that can only exist under the greatest extremes of temperature and pressure. Since liquid metallic hydrogen is a great conductor of electricity, its presence is the basis for Jupiter's massive magnetic field.

So if Jupiter is so much like the Sun, why doesn't Jupiter shine like the Sun does? The answer is that Jupiter does not have sufficient mass to build temperatures in its interior high enough to begin the process of nuclear fusion. For hydrogen fusion to commence, internal temperatures must reach at least 10,000,000 Kelvin. Jupiter in order to produce those temperatures in its core would need to be about eighty times more massive than it actually is. So while gravitational energy produces a great deal of energy and heat in the Jovian core, it is nowhere near sufficient to trigger nuclear fusion in its core. This process of gravitational compression is called the *Kelvin–Helmholtz mechanism.* Jupiter began accreting matter in much the same manner as the Sun did but there was not enough hydrogen gas left in the primordial solar system to create a second star and instead the mass of accreted hydrogen became the solar system's king of the planets. But Jupiter was a failure as a star. Still, Jupiter's mass allows it to actually radiate more heat from its core than it receives from the Sun. Gravitational energy released by the Kelvin–Helmholtz process raises temperatures at Jupiter's core to around 20,000 Kelvin.

The heat from within Jupiter drives the planet's very dynamic weather patterns. Even the smallest of telescopes will show the planet's active atmosphere. In the image at the opening of this chapter, you can clearly see the planet's clouds divided into belts and zones. Larger telescopes begin to show that the borderline between the belts are not smooth but interrupted by swirls and turbulence. Jupiter's atmosphere is also riddled with storms that are extremely violent in nature. The clouds are wracked with lightning that generates enormous amounts of radio noise and have been viewed on the planet's night side. The storms are also immense in scale. Some are large enough to swallow Earth whole.

When Galileo turned his first primitive telescope on Jupiter, it was not the planet itself though that rocked his world and in fact the entire civilized world. Galileo discovered that Jupiter had companions, four of them. At first Galileo thought that the objects were background stars but after several nights he realized that they were moving around Jupiter. The discovery was astounding because it was the first evidence that not everything in the universe circled Earth. This was an enormous boost to the Copernican theory that the Sun was the center of the solar system and not Earth. Galileo so stated in his 1610 paper *Sidereus Nuncius* and the motion of the "moons" of Jupiter were his proof. In 1615, Galileo was summoned by Pope Paul V to appear before the Inquisition. Here none other than Robert Cardinal Bellarmine, considered the most knowledgeable theologian ever produced by the Catholic Church, interrogated him. Though the match was of two of the finest minds of the seventeenth century, the playing field was hardly level. Bellarmine wielded an order from Pope Paul V ordering him to renounce the Copernican theory or be confined to the dungeons of the Inquisition. This was for all intents and purposes, a death threat. Faced with a choice between life and death, Galileo capitulated.

Pope Paul V allowed Galileo to return home and continue his research, he was simply forbidden to publish any result. The situation changed again in 1623 with the ascension of Maffeo Cardinal Barberini to the papacy as Pope Urban VIII. Since Urban was a fellow Florentine, Galileo thought he'd found a sympathetic audience. But when it became known that Galileo had not yet given up on the Copernican system, he was summoned back to Rome where Urban himself tried to sway Galileo and he was forced to endure the cowardly attacks of theologians who knew that Galileo could not fight back. Finally after eight years of effort, Galileo was permitted by Urban to publish a rebuttal under the condition that it carry a derogatory preface. Galileo's book was a huge popular success and Urban's ploy failed. Even worse, the *Dialogo* was written in Italian, the language of the masses rather than in Latin, which was the language of the elite. For Pope Urban VIII, humiliated and hated, the matter was now not theological, but personal. Galileo's allies were silenced and in June 1633 Galileo facing being burned at the stake, was forced on his knees to publicly recant. Galileo spent the last nine years of his life under house arrest until his death in 1642. His works were expunged and historical descriptions of him were changed. Adjectives like "renowned" were replaced with "notorious." It was nearly 200 years before the Catholic Church came to accept the Sun-centered solar system and not until the early 1990s that Pope John Paul II vindicated Galileo and reminded us all that "two realms of knowledge ought not be viewed as opposition." Faith and science each have their place.

As Jupiter and its moons circled on around the Sun each twelve years, more details about the giant planet began to emerge. More moons too. By the time exploration of the planet began, twelve moons had been found. Jupiter's rotation speed had been accurately timed at only 9 hours 50 minutes. Jupiter in fact rotates so rapidly that centrifugal force causes the planet's globe to bulge at the equator so severely that the planet's polar diameter is only about 90% of its equatorial diameter. Jupiter was a world that cried to be explored and the United States answered the call in the early 1970s. The first attempt to probe Jupiter was the Pioneer 10 mission. An Atlas–Centaur launched the probe on March 2, 1972. Carrying an extra rocket stage, Pioneer 10 was accelerated to over 50,000 kph. Pioneer 10 crossed the orbit of the Moon in only eleven hours and the orbit of Mars in just 12 weeks. Pioneer 10 was the first spacecraft to successfully navigate the asteroid belt and reach Jupiter's environment, flying past the planet after a twenty-one month cruise on December 3, 1973 passing about 130,000 kilometers above the cloud tops. Pioneer 10 made the first up-close studies of the planet and made crude images of all four Galilean moons. Pioneer 10 also conducted studies of the planet's interior, magnetic field and Jovian environment and was nearly destroyed by Jupiter's powerful magnetic field. The data provided by Pioneer 10 was crucial in designing the probes to come behind it enabling them to survive and flourish in the Jovian magnetosphere. Pioneer 11 followed its older sister to Jupiter launching on an Atlas–Centaur on April 5, 1973. Pioneer 11 arrived at Jupiter almost exactly one year after Pioneer 10 on December 2, 1974. Pioneer 11 passed only 66,000 kilometers above Jupiter's cloud tops. Although the probes did not get sharp pictures of the moons, they did learn an important fact about Io. The innermost Galilean moon orbited Jupiter in an orbital tube of ionized sulfur. Where the sulfur was coming from was a mystery intriguing enough that it would lead NASA to make an important decision about the missions to follow.

In an exquisite celestial billiard shot, Pioneer 11 used Jupiter's gravity to accelerate it to over 171,000 kph across almost 3 billion kilometers of space looping high above the ecliptic to a second flyby of Saturn where it became the first probe to ever investigate that planet on September 1, 1979.

Even as Pioneer 10 and 11 were cruising towards their dates with Jupiter, NASA was already planning more audacious missions as a critical moment arrived in the mid-1970s. Once each two centuries the four outer planets align in such a way that one spacecraft could be employed to visit all four of the outer gas giants. NASA sold Congress on a multibillion-dollar two-spacecraft mission to fly what became known as the "Grand Tour." Congress approved the Mariner Mark II program, targeting launch of the two probes in 1977. NASA later renamed the project "Voyager." Like with Viking, the mission would be launched using the high-powered, high risk Titan III with a stretched Centaur upper stage. In addition to the Centaur, each Voyager was also equipped with a Star-48 solid fuel upper stage. The principal reason again for two probes was simply that NASA was afraid of the likelihood that one of the launches would end in failure, since each rocket would require five stages to perform properly. But again, the two Titan Centaurs delivered perfection sending the Voyagers on their way to Jupiter on August 20, 1977 and September 5, 1977. Voyager 2 was actually launched first on a path to Jupiter that would create an optimal trajectory to tour all four planets but one that would arrive at Jupiter four months later than Voyager 1. Voyager 1 was then launched on a trajectory that would allow an up close study of Jupiter's innermost Galilean moon Io when encountering Jupiter, then would allow a close flyby of Saturn's enigmatic moon Titan. But obtaining the close flybys of the two moons came with a price because the resultant trajectory past Saturn would preclude Voyager 1 from going on to Uranus and Neptune. Voyager 2 did not have as favorable a flyby at either Jupiter or Saturn in order to allow it to fly on to Uranus and Neptune. In early 1979, Voyager 1 neared Jupiter after a cruise of just over eighteen months.

The images returned from Voyager were outstanding, revealing the planet in detail that Pioneer could not begin to approach, but Jupiter's moons wound up stealing the show. It was thought that the moons would be homogenous rocky boring bodies, but each moon it turns out has its own personality. Io looked more like a giant pan pizza according to stunned members of the imaging team who first saw the pictures. The second moon, Europa, was as smooth as a cue ball. Both Io and Europa were just about completely crater free, as though some force were continuously reshaping their surfaces. Europa seemed to be more like an enormous drop of water rather than a rocky object. Scientists were soon speculating that Europa harbored subsurface oceans that were warm enough to perhaps even allow aquatic life to develop. Io was more of a mystery until members of Voyager's navigation team looked back at Io. An engineer named Linda Morabito was actually looking at was a star in the background of a picture of Io obtained post flyby to be used as a navigation fix. But as Morabito enhanced the picture to brighten the star, something else appeared. A crescent shaped light extending above the moon like a plume. Voyager had caught a volcanic eruption in progress, the first active volcano known to exist on any world other than Earth. By the time the imaging team members had finished going back through the thousands of images of Io, more than a dozen volcanoes had turned up. As a final bonus, several members of the imaging team, motivated by the recent discovery of rings around Uranus demanded a quick look back to see if any light could be found backscattering off any thin Jovian rings. After much arguing, they got their pictures and got rewarded. Jupiter did indeed have a faint, dusty ring surrounding it. The rings are very dark and appear to contain no ice, unlike those of Saturn.

The Voyager results were so fascinating that NASA again began planning a multi-billion-dollar follow-up that would enter a long-term orbit of Jupiter and spend about five years studying the planet and five of its moons. The project was named in honor of the man who gave so much of himself to spread knowledge. The Galileo mission would launch in May 1986 and fly to Jupiter reaching the planet in early

1989 flying on a Centaur upper stage after deployment from the Space Shuttle *Challenger*. But the loss of *Challenger* delayed the launch of Galileo by more than three years and forced a much longer trajectory to Jupiter after NASA abandoned use of the Centaur with the Shuttle. The launch from Space Shuttle *Atlantis* was a success, but several months after launch, Galileo ran into a crippling problem when its high-gain antenna failed to unfurl. Three of the sixteen pins that secured the antenna for launch failed to release. Galileo would instead be forced to record most of its data then transmit that data back to Earth through a low gain antenna a few bits at a time. The dedication of the mission managers and engineers enabled Galileo to score a magnificent success including a front row seat for the July 1994 impact of Comet Shoemaker–Levy 9 at Jupiter. Galileo spent more than six years studying Jupiter in unparalled depth before plunging into the atmosphere of Jupiter. Galileo also dropped a probe through the atmosphere of the planet to study the atmosphere in unprecedented detail.

In addition to the five dedicated missions to Jupiter, the planet has also been studied by the Ulysses spacecraft while it was using Jupiter's gravity to deflect it into a polar orbit over the Sun. The Cassini spacecraft also used Jupiter for a gravity assist to reach Saturn, flying past on December 30, 2000. Cassini and Galileo made coordinated observations of the planet for the first time in history. In 2007, the Pluto New Horizons probe will fly-by Jupiter for a gravity assist.

Currently no major space agency has any further plans to return to the solar system's mightiest planet. As of late 2005, the proposed Jupiter Icy Moons Orbiter has been cancelled and right now there are no other missions on the drawing board for Jupiter. While the mission may some day be revived, it seems for now that there is little hope for a mission to Jupiter taking place in the next ten years.

Observing Jupiter

There is almost nothing in the universe that is easier to observe than Jupiter. Even when near solar conjunction, Jupiter is *always* bright, shining at a minimum as bright as Sirius. And Jupiter is always large in a telescope, never subtending an angular diameter of less than 30 arc seconds. Because the planet orbits so much farther outside the orbit of Earth, the planet spends a lot of time in that favorable area more than 90 degrees from the Sun, about five to six months out of each thirteen month synodic period. And when the planet is at its best, it is awesome. Even when Jupiter is near aphelion, its disk can reach 45 arc seconds and the planet shines at magnitude −2.4. When near perihelion, the planet reaches nearly magnitude −3.0 and 50 arc seconds. There is hardly ever a bad time to observe Jupiter.

Jupiter's cloud tops are organized into alternating dark bands and lighter colored zones. Remember that even though Jupiter is very bright, you should not expect to see the planet in the same kind of vivid color that photographs show. Your eyes cannot collect enough light at this distance to trigger the cones of your eyes even at magnitudes brighter than −2 so the Jupiter you will see will be in alternating browns and tans to your eye. The details in the planet emerge with careful study and waiting for that precise moment that the air picks to be steady. Then those cloud bands erupt in detail. The clouds bands are filled with swirling eddies and currents that momentarily pop into view when the planet is high and the air is still. Be patient, relax and wait. Jupiter *will* reward you.

Swirling high in the Jovian atmosphere in the mid-southern latitudes is one of the most persistent atmospheric phenomena observed anywhere in the solar system. Jupiter's mighty Great Red Spot was first noticed as long as 300 years ago in telescopes and has persisted at varying levels of intensity ever since. One of the most common misconceptions about the Spot is that it is a storm, much like a hurricane on Earth. But terrestrial hurricanes are deep low-pressure systems that spin counterclockwise in the north and clockwise in the south. The Great Red Spot rotates counterclockwise in the southern hemisphere of Jupiter therefore the Spot is not a cyclonic system, but an *anticyclone* or an area of high pressure. Close up images of the area by Galileo have shown that the Great Red Spot is covered by a high-altitude cloud cover that stands a substantial distance above the underlying gas layers. What source feeds the Spot and keeps the local pressure high is something that is still unknown after hundreds of years.

The four moons that made Galileo famous and got him in so much trouble dance nightly around the planet. Io, the innermost moon circles Jupiter once every two days. Europa, each four days and Ganymede each eight days. Jupiter's inner three big moons thus are in a 4:2:1 resonance so that all three moons return to approximately the same position relative to each other every seven days. The outermost moon, Callisto, need fourteen days to orbit the planet but is slowly drifting outward so that in a few tens of millions of years, it too will be in resonance with the three inner Galilean moons. Jupiter's powerful gravity is one of nature's great organizing tools, bringing order to an entire system of moons, each of which is large enough to be a planet in its own right. Io is larger than our Moon, Europa slightly smaller. Ganymede is the solar system's largest moon at over 5,600 kilometers in diameter and Callisto is slightly smaller, but both are larger than Mercury. The moons are all smaller than one arc second, but in an 8-inch telescope in stable conditions they will show disks at high magnification. Still the moons are far too tiny to show any detail in medium-sized scopes. Large scopes might show the subtle differences in color between the moons.

As the moons circle the planet, they will at times pass alternately behind, then in front of Jupiter. Jupiter's axis is barely tilted with respect to the plane of its orbit, only about three degrees. When the moons pass in front of Jupiter, they will cast their shadows on the planet. Even when Jupiter's poles are pointed at their maximum towards Earth, only most distant Callisto's shadow will miss the planet. Publications such as *Sky & Telescope* or the *Observers Handbook* will issue predictions for events involving Jupiter's moons. If Jupiter is nearing quadrature, the shadows will lead or trail the moon by a substantial distance. The shadow may in fact appear on the face of Jupiter while the moon itself is still several arc seconds clear of Jupiter's disk. As the moons travel around the back of Jupiter, the planet's own shadow extends further out towards the east or west of the disk. As the moon approaches Jupiter's disk it begins to disappear into the shadow several hours before it would pass behind Jupiter or may remain hidden for a few hours after emerging from behind the planet. In some cases, a moon may emerge from occultation behind Jupiter from our point of view and reappear for a few hours only to disappear again into Jupiter's shadow. Also for a period of many months near each node of Jupiter's orbit, the satellites will take turns casting their shadows on each other creating eclipses. These so-called *mutual events* will involve all four moons eclipsing and being eclipsed by each other. Each event is unique in personality depending upon whether the eclipse or occultation is full or partial. The moons, normally fifth or sixth magnitude, can actually fade to tenth or below when totally eclipsed and can completely vanish for several minutes before the eclipse ends and the moon reappears. The moons can also occult each other. It can be fascinating to watch the two pinpoints come closer and closer together until they appear to merge, then split apart again. The next round of mutual events of Jupiter's satellites will begin as the planet nears the node of its orbit in 2008–2009.

For observers with larger telescopes and some advance planning and persistence, Jupiter may sport even more moons. Orbiting inside Io is a tiny potato-shaped moon called Almalthea. Io is exquisitely difficult to see because it never strays much more than one planetary diameter from Jupiter and can just barely reach fourteenth magnitude. That would put it just barely within the reach of a 10-inch telescope. Almalthea is probably better saved as a challenge to telescopes of 16 inches or larger. If you're an owner of one of those big Dobsonians, use an astronomy program to find the locations of some of Jupiter's outer moons as well. Most are below magnitude 15 and are exquisite challenges for the largest scopes. Unlike Almalthea, which scoots around Jupiter in barely a day, these moons take many months to as long as two years to orbit the planet, so their motion must be carefully observed over time. If you can find the outer moons, note that they orbit Jupiter backwards or *retrograde*. This behavior is indicative of captured asteroids. These outer moons probably were once main belt members that drifted too close to Jupiter and were gravitationally boosted into higher orbits until Jupiter captured them into these retrograde paths.

Father of Jupiter, a World with Ears

Slowly traveling around the Sun in an orbit 1,429,400,000 kilometers out, twice as far out as is that of Jupiter, Saturn in antiquity was the faintest of the five wanderers. The ancients Romans named Saturn for the god of agriculture and father of Jupiter. During its 29-year journey around the zodiac, Saturn attracted a minimal amount of attention for it never became particularly bright at any time compared to the other four planets, all of which can outshine Sirius at times. Saturn can become as bright as magnitude −0.5 if conditions are favorable and if not, might not exceed magnitude zero. Saturn however rarely fades even close to magnitude +1.0. The slow moving, steady light made its stately way around the sky in relative obscurity until Galileo turned his first telescope on it in 1609 and was astounded by what he saw. With the rings tilted near their maximum, Galileo saw a triple orbed object through his optically imperfect telescope. "Saturn has ears," he wrote in his observing log.

Time and better telescopes revealed Saturn's ears for what they truly are. In 1659 the astronomer and optician Christiaan Huygens correctly inferred the rings' true structure. Quickly Saturn became the great showpiece of the solar system. Like

Figure 10.2. Saturn 18″ diameter, rings open 22 degrees on May 17, 2005. Image by author using the Celestron Super C8 Plus and Meade DSI.

Jupiter, Saturn is what astronomers call a "gas giant." Its chemical composition is similar to that of Jupiter, mostly hydrogen and helium. Saturn is even more oblate than Jupiter is, with an equatorial diameter of 120,536 kilometers and a polar diameter of 108,728 kilometers. The planet's rotation rate, while not as fast as that of Jupiter, is still speedy at 10 hours, 50 minutes. Saturn's density is substantially lower than even Jupiter, equaling only 0.79 grams per cubic centimeter. The combination of low density and high rotation period causes Saturn to bulge as much as it does. Saturn's density is so low that if you could find a sufficiently large body of water, Saturn would float in it! The planet has an atmosphere that is as dynamic as Jupiter, however a hydrocarbon haze makes it nearly impossible to observe from Earth. The haze is methane that condenses out of the atmosphere into visible form because it is substantially colder at Saturn than it is at Jupiter.

Saturn, like Jupiter radiates more energy than it receives from the Sun. Like with Jupiter the process that causes this is the Kelvin–Helmholtz mechanism, though this model cannot explain all of the heat emanating from Saturn. Astronomers believe one possible source of the extra energy is latent heat released by the condensation of helium raining out in Saturn's interior. Temperatures at the core of the planet reach over 12,000 Kelvin. Saturn's interior is much like Jupiter's, consisting of a small rocky core, a deep layer of liquid metallic hydrogen and then the hydrogen rich atmosphere overlying it.

Saturn's signature feature is its magnificent ring system. As telescopes became larger and of better quality over the centuries since Galileo saw "ears," Saturn's rings revealed more and more of their secrets. Astronomers began to see organized structure in the ring system and by the twentieth century astronomers had clearly identified three different ring structures. The outermost ring was dark in color and relatively narrow. Astronomers designated this the "A" ring. A broad white structure lies inside the "A" ring and this is logically called the "B" ring. This is Saturn's widest and brightest ring. Inside the "B" ring is a less well-defined ring called the "C" ring. Occasionally this ring is called the "crepe ring" because it can be very difficult to see and is so thin that it seems to taper into transparency as it nears the planet. To observers using medium size telescopes, there appears to be a gap between the A and B rings, which astronomers have named for Giovanni Domenico Cassini, the Italian born first director of the Royal Observatory in Paris. The Cassini division is an actual gap, relatively free of ring material. Gravitational interaction by the moon Mimas likely accelerates material into the A ring in much the same way that resonances with Jupiter creates the Kirkwood Gaps in the asteroid belt. An object at the distance of the Cassini division would have an orbital period that would have a two-to-one resonance with Mimas. The Voyagers showed us the rings in dramatic detail showing that the three rings were not really just three, but thousands of individual ringlets. Voyager 1 and 2 also showed that the rings were adorned with dark colored "spikes" which rotated together around the planet, even though the laws of orbital mechanics dictate that they should shear apart as they move around the planet. The spikes have oddly enough not been imaged by Cassini. The rings appear to be mostly ice with a small amount of rock mixed in, which would explain their remarkable brightness. The rings are also remarkably narrow. Though the system spans over 220,000 kilometers, the rings are barely one kilometer thick. Voyager also found a "D", "E" and "F" ring. The F ring proved particularly fascinating. The F ring appeared to be two separate strands of material intertwined with each other, looking oddly like strands of DNA strung around Saturn. The orbital dynamics that allow for this to happen have baffled astronomers since. Much about the forces that shape Saturn's rings are still poorly understood.

Saturn has 47 known moons as of October, 2005. Of these, nine are major bodies. Eight of those all orbit in low-inclination prograde orbits. The ninth, Phoebe, orbits the planet in a highly inclined retrograde orbit. Many of these other moons are in close to the ring system and play an active role in gravitationally shepherding the ring particles and keeping the ring system in order. Many others orbit well away from the planet in highly inclined retrograde orbits and are probably captured asteroids. The first moon to be discovered was Titan. Huygens found Titan in 1655. It is named for a family of giants who were the children of Uranus (which would not be found for another century). Astronomers have known for centuries that Titan was something special. For many years, it was thought to be the solar system's largest moon (later more precise measurements now show Ganymede is larger). Spectroscopic observations showed during the twentieth century that Titan has an atmosphere and a substantial one. We now know that surface pressures on Titan are about 50% higher than they are on Earth. The atmosphere was for the longest time thought to be composed primarily of methane, but the Voyager 1 flyby shocked astronomers by indicating the atmosphere almost entirely nitrogen, over 95%! While methane is present, it is primarily in the form of smog that resides high in the atmosphere that renders the atmosphere almost opaque in visible light. Despite several close flybys by the Cassini spacecraft and the landing by the Huygens probe, Titan remains a world that is frustratingly mysterious.

After the discovery of Titan, the search for more moons of Saturn was on. Cassini would find four more moons beginning with Iapetus in 1671, Rhea in 1672, then Tethys and Dione in 1684. Saturn's moon count stayed at five until William Herschel found Mimas and Enceladus in 1789. Saturn's eighth moon, Hyperion was found in 1848 by the American astronomer William Bond. The American astronomer William Pickering found Phoebe, the last of Saturn's prominent moons, in 1898. We knew little more about these moons other than their existence and somewhat accurate measures of their sizes until space probes examined them up close beginning in the late 1970s. Saturn's eight largest moons orbit the planet in a very orderly manner caused by gravitational resonances of the moons' orbital periods. Mimas and Tethys orbit in a 1:2 resonance as do Enceladus and Dione. Titan and Hyperion orbit in a 3:4 resonance.

The first mission to reach Saturn was something of a bonus. Pioneer 11, like Pioneer 10 the year before had been very fortunate to survive nearly being fried in Jupiter's powerful magnetosphere. NASA took advantage of a crude alignment between Jupiter and Saturn to send the still functioning Pioneer 11 spacecraft to investigate the planet in a maneuver the proved the viability of the Grand Tour concept for the Voyager probes that had been launched behind it already when the probe reached Saturn in September 1979. Pioneer's rudimentary results from Saturn were heavily overshadowed by the encounters of the Voyagers at Jupiter earlier that year. The Voyagers themselves arrived in the Saturn system in November 1980 and August 1981 respectively. Voyager 1 gave up the rest of the Grand Tour in order to get a close flyby of Titan. Voyager 1 flew within 4,000 kilometers of Titan and in just a few hours taught us more about the planet than we had learned in the previous 300 years. Voyager also returned stunning images of most of the other moons. Little Mimas was found to sport an enormous impact crater astronomers named Herschel (for the moon's discoverer). The impact that formed the crater was so massive that it nearly destroyed Mimas and left it looking oddly like George Lucas' "Death Star." Enceladus is a world much like Europa, and is the smoothest body in the solar system. Enceladus is almost entirely water ice and is even more reflective than Venus. Its surface is being constantly being refreshed by a process that is as yet unknown. Tethys, like Mimas sports an enormous impact crater that spans 2/5ths of the moon's diameter and the moon's remarkably low density suggests that it is almost entirely water ice. Dione and Rhea are very similar. Iapetus is the great celestial oddball. One side of the moon is bright and reflective while the other side is as dark as tar. We have known this for years but Voyager 1 gave us our first good look at it. Voyager 2 flew past the planet nine months after Voyager 1. Though Voyager 2 returned solid data from Saturn, the path of the probe past Saturn was principally designed to enable further voyages to Uranus and Neptune.

Saturn remained unexplored for nearly a quarter of a century after Voyager 2 flew past. In the early 1990s, NASA gained approval from the first Bush administration under administrator Richard Truly for a multibillion dollar mission that would utilize a Titan IV with a Centaur upper stage to launch. The spacecraft would be built mostly using off the shelf leftover parts from the Galileo project and a fixed antenna from the Magellan program (thus avoiding the problem that afflicted Galileo). The ambitious mission became a prompt target of the budget axe when first Dan Goldin succeeded Truly and then Bill Clinton came to office the next year. Goldin in particular had a disdain for big budget missions (which was amplified by the 1993 loss of the billion dollar Mars Observer) and worked to ensure that there would be no more missions of this class ever flown again. Cassini was massive, the largest interplanetary spacecraft ever built. Cassini and its piggyback Huygens probe weighed more than 5,000 pounds. Goldin derisively referred to

Cassini as "Battlestar *Galactica*."[30] Cassini was launched in 1997 on a seven-year looping trajectory that included a flyby of Venus, two of Earth and a flyby of Jupiter before reaching Saturn in July 2004. Cassini and Huygens have already rewritten most of what we know about Titan and promises to do so about Saturn as well.

Observing Saturn

Viewing the ringed planet with a telescope is relatively easy for most of the same reasons that observing Jupiter is easy. The planet spends a large amount of time in a favorable portion of the sky for viewing, about six months out of each 12.5-month apparition. Saturn's rings span a total area about as large as the disk of Jupiter, though the disk of the planet itself rarely exceeds 20 arc seconds. In one area the planet is not as easy to view as Jupiter. Since Saturn is about twice as far from the Sun as Jupiter is, the planet receives far less light than Jupiter does and so its surface brightness is much lower than is Jupiter. This in turn can make the planet a bit of a challenge for the amateur astrophotographer to make crisp images. Saturn's maximum possible brightness is only about magnitude −0.5 with the rings fully open and when the rings are closed, it may not reach zero magnitude even at opposition.

Your attention will be quickly drawn to the ring system. Most of the light you will see comes from the B ring area, which has an albedo of nearly 60%. Mid-size telescopes should easily be able to pick up the A ring and the Cassini division in between the two. The A ring is considerably darker with an albedo value of about 30%. The C ring is more difficult to see inside the B ring and makes a good challenge. The C ring is much darker with an albedo value of some 20%. None of Saturn's other rings are visible through amateur telescopes. How spectacular the rings are will be dependent upon Saturn's position in its orbit. When the planet is near one of its solstices, the rings are at their maximum visibility and in fact the back side of the rings are tipped above the planet' summer time pole. When Saturn is near an equinox, then the rings appear edge on and may in fact disappear completely. For a time during the edge on period (which will occur next in 2010) the rings will actually present their unilluminated side towards Earth allowing them to be glimpsed in silhouette in front of the planet. The rings will also cast shadows on the planet, which gives Saturn a remarkable three-dimensional appearance not evident on any other planet. When Saturn is near quadrature, watch as the planet casts its shadows on the rings, further enhancing the three-dimensional aspect of our view.

Saturn's globe is itself somewhat bland in nature. A hydrocarbon haze shrouds the planet's dynamic atmosphere from our view, leaving Saturn a plain butterscotch color. When Saturn's poles are tipped towards Earth, you may be able to notice a discoloration in the haze at the poles. Here the hydrocarbon smog is

---

[30] To many in the media, this is a reference to Cassini's immense size. The Battlestar *Galactica* is a fictional space-going aircraft carrier nearly a mile long. But there was a more insidious meaning to the insult. The Battlestar *Galactica* was also the last ship of her kind.

thinner in nature, broken down by continuous exposure to sunlight over the nearly fifteen uninterrupted years of sunlight. The process is similar to what causes the naturally occurring "ozone holes" in Earth's atmosphere.

Many of Saturn's moons are visible in modest telescopes. Titan shines at magnitude +8 and is visible to even the smallest of telescopes. Several other moons are within reach of telescopes of six inches or larger. Titan takes about two weeks to make a leisurely swing around Saturn. Unlike Jupiter's moons, which move back and forth in a straight line, Titan makes big loops around the planet unless Saturn's rings are edge on. So most of the time you will see Titan pass well north or south of the planet's disk. Inward of Titan, the moons Rhea, Tethys and Dione are all approximately magnitude +10 and should be visible with a 4-inch telescope under dark skies and easily with a 6-inch. Enceladus at magnitude +12 starts to push the limits of 8-inch scopes under a dark sky and Mimas is a challenge even for the largest of scopes because its feeble magnitude +13 glow is buried very close to Saturn itself. All of these moons orbit in the ring plane as Titan does so they follow the same types of path around the planet that Titan does. They swing behind the planet over one pole, move out to a greatest elongation then swing back in to pass in front of the planet passing the opposite pole. There are also moons to see outward of Titan. Iapetus is one of the great odd balls of the solar system. When Iapetus is west of Saturn, it points its bright trailing hemisphere towards Earth and the moon brightens to magnitude +10. But as it travels around to the east side of the planet, it points more and more of its tar-black leading side towards Earth and the moon fades to magnitude +14. Hyperion is magnitude +14 at all times and requires a telescope of 10 inches to be just barely visible. The last of Saturn's telescopically discovered moons is Phoebe. This moon, likely a captured asteroid orbiting Saturn in the reverse direction of all the other moons is fainter than magnitude +16 and more likely a target for professionals.

As Jupiter and Saturn ride high into the evening sky at opposition time, it's time to mount our scopes for an observing journey to the Kings of Worlds and explore their wonders. Each is a mini solar system in its own right, ruling its gravitational domain with an iron fist. Let's go view the wonders of Jupiter and Saturn.

Observing Projects X – Viewing Jupiter and Saturn, the Mighty Gas Giants

Observing Project 10A – The Great Red Spot and the Rotation Period of Jupiter

Here's a bit of a rehash of a project we did back at Mars, timing the planet's rotation period. This is one of the simplest things that astronomers did back as far as the 1600s to learn fundamental facts about Jupiter once the true nature of what the planet was became known after the invention of the telescope. Unlike Mars, where you can pick any feature and follow it across the face of its globe, Jupiter is not quite as consistent as is Mars in terms of position of "surface" features. But there is one thing you can find that really sticks out at you.

Base your timing on Jupiter's Great Red Spot. Ephermides for the Spot's meridian crossing time are published in many leading astronomy journals and publications. Time how long it takes for the Spot to rotate from the meridian until it passes out of sight around the far limb. Multiply your result by four to get Jupiter's rotation period. This is not as easy as it sound because the Spot will become more difficult to track as it nears the edge of the planet. At times the Spot is tough to track anyway simply because it is not always so red as its name suggests. Sometimes the Spot is a very deep red and prominent and at other times it fades to such a pale shade of brown or beige that it can be hard to separate from the planet's cloud bands. The Spot will disappear around the corner in just under two and a half hours. If Jupiter is near opposition and the Great Red Spot passed local noon early in your Earthly evening, then Jupiter will be around long enough for you to see the Great Red Spot return to noon 9 hours and 50 minutes after you observed it earlier.

The Great Red Spot is one of the most persistent atmospheric phenomena in the solar system having been continuously observed since the time that telescopes were of high enough quality to see it. Always make it a point to keep an eye on the spot. Watch it brighten and darken over the months that Jupiter is in view and over the years of your observing life as Jupiter returns year after year. Use the highest power that your telescope and the seeing conditions will tolerate. Jupiter offers plenty of light to work with. There are many other swirls and storms on Jupiter, but only the Great Red Spot is of sufficient size, contrast and persistence that it is easily visible to amateur astronomers equipped with small telescopes.

Observing Project 10B – The Ballet of the Galilean Satellites

The discovery of Jupiter's four bright satellites was one of the most important in the history of science because it proved to be the deathblow to the idea that Earth was the center of the universe and that everything else circled it. The finding of four bodies circling Jupiter proved the Copernican theory that the Sun was as the center of the solar system. For daring to oppose the Church in his beliefs, Galileo was humiliated, disgraced, imprisoned and broken.

Though the moons are bright and will stand up to high power, there really is in fact very little to be viewed in amateur telescopes. The surface features of the moons are far too tiny to be viewed with the human eye from Earth at any magnification. Use a low to medium power to study the moons, perhaps a little more magnification to split apart two moons that are very close together.

The innermost moon is Io and it shines at about magnitude +6. Io is slightly larger than Earth's moon at 3,630 kilometers in diameter and zips around Jupiter in just under two days at a distance of 420,000 kilometers. Europa is slightly smaller than Earth's moon at 3,138 kilometers and circles Jupiter each four days at a distance of 670,900 kilometers. Europa also shines at magnitude +6. The two outer moons are larger and slightly brighter. Ganymede is the third moon and circles

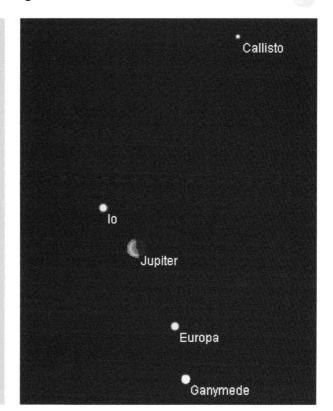

Figure 10.3. Jupiter's moons on Sept. 25, 2005. Graphic created by author with *Redshift 4.*

Jupiter each eight days at a distance of 1,070,000 kilometers. Ganymede shines at magnitude +5 and at 5,262 kilometers is the largest satellite body in the solar system. Callisto is the fourth moon and circles Jupiter every fourteen days at a distance of 1,883,000 kilometers from Jupiter. Callisto also shines at magnitude +5 and measures slightly smaller than Ganymede at 4,800 kilometers.

As the moons zip around the planet, you might note that the satellites do not move in random patterns from day to day. Watch the movement of Io and Europa relative to each other over one revolution of Europa. Where is Io relative to Europa when Europa is at greatest elongation? Now watch Europa make a full revolution around the planet. At the instant Europa returns to greatest elongation, where is Io? You will find Io in about the same place it was four days ago, except that Io has made two revolutions around the planet while Europa made one. This is not a coincidence. Europa and Io orbit in resonance with each other. The organizing effects of the gravity of both Jupiter and the satellites themselves cause the resonance. Then try including Ganymede in your experiment over one revolution of Ganymede. Where are Europa and Io compared to where they were when you first marked Ganymede at greatest elongation from Jupiter? They are both in the same place as they were eight days prior, except that Europa has gone around the planet twice and Io four times while Ganymede has made one revolution. All three moons

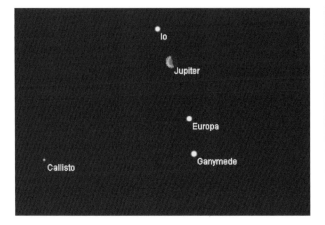

Figure 10.4. Jupiter's moons on Oct. 2, 2005. Graphic created by author with *Redshift 4.*

are part of this resonance. Europa's orbital period is almost exactly twice as long as that of Io and Ganymede's is twice that of Europa. The gravitational and tidal interaction you are watching has important effects on the moons as well. Io's interior is heated as a result of tidal interactions with Europa and Ganymede as well as Jupiter. These tides stretch Io by as much as 100 meters during an orbit. Can you imagine tides 330 feet high? The flexing of the moon is responsible for the volcanoes that wrack Io's surface. Europa may have an enormous subsurface ocean of liquid water, kept from freezing by tidal stresses. Ganymede shows signs of tectonic activity, folded mountains and dynamic surface activity.

Now let's try Callisto. Mark the positions of the three inner moons when Callisto is at greatest elongation and see what happens when Callisto returns to greatest elongation fourteen days later. The moons are in different positions. Callisto is not part of this resonance among the Galilean moons and Callisto is a very different moon. No tectonics, no warmth, no volcanoes. Callisto's surface is the oldest in the solar system, virtually unchanged by any force except for the occasional impact for some four billion years. Callisto does not even exhibit any evidence of internal structure at all with no clearly definable mantle or core. But Jupiter's gravitational organizing power is relentless and eventually Callisto will drift out away from Jupiter far enough that it will fall into resonance with the other four moons.

The eternal dance of Jupiter's moons gives us joy to watch their beautiful changing patterns from night to night. But beauty is not always random accident. The resonance of the inner three Galilean satellites is a large part of what makes Io, Europa and Ganymede as dynamic as they are while Callisto is old, dull and monolithic.

Observing Project 10C – Just me and my Shadows

As the Galilean satellites zip around Jupiter, at each orbit they will pass in front of Jupiter and cast their shadows on the planet's cloud tops, creating eclipses of the

Sun for the areas in the shadow track. The only time this does not occur is when Jupiter is near one of its solstices. At that point the three-degree inclination of Jupiter's axis is sufficient to cause the shadow of Callisto, the most distant moon, to drift off the planet's face entirely. Check with sources that publish ephermides to see when shadow crossings will occur. *Sky & Telescope* publishes monthly highlights of events involving Jupiter's moons including graphic depictions of notable moon configurations. Sometimes more than one moon will cast its shadow on the planet at the same time. On one rare occasion in the spring of 2004, three of Jupiter's Galilean satellites cast their shadows on the planet for more than a half hour. Although it was almost certainly too small to see with any instrument, my astronomy software also indicated that little Almalthea also dragged its shadow across the clouds during that event as well. Imagine four total solar eclipses going on at the same time in four different places on the same planet?

When the moons race around the back side of the planet, they will encounter Jupiter's shadow during each orbit. Again the only exception is that Callisto will miss Jupiter's shadow during the time period around the solstices. As the moons near Jupiter's limb, the planet's shadow extends out slightly to the east or west, with the effect being most prominent with Jupiter near quadrature. With Jupiter in the morning sky watch the fifth or sixth magnitude dot near the edge of the planet and then before getting there, it will fade out over a period of several minutes and disappear. Near quadrature, it is possible that the shadow extends far enough west of Jupiter that Callisto or Ganymede might actually complete the eclipse and reappear before it passes behind the planet into occultation. Io and Europa usually will not reappear before occultation takes place. When Jupiter is in the evening sky watch for the reverse to occur with the outer moons. Callisto may appear from behind Jupiter, remain in view for a few hours, and then disappear again for a time into Jupiter's shadow.

Once each six years, as Jupiter is near one of the equinox points of its orbit, the orbits of the Galilean satellites lie close to the plane of Jupiter's orbit. When that happens, the satellites cast their shadows directly on Jupiter's equator and also will begin to cast their shadows on each other. These are called *mutual events* of Jupiter's satellites. Each moon is able to cast its shadow on the other three as well as on Jupiter. Moons may also occult each other. You will be able to watch two moons draw slowly together and appear to merge, then drift apart again over the space of a few hours. A deep partial or total eclipse can be even more dramatic. A moon can lose four or more magnitudes during these events becoming completely invisible to owners of smaller telescopes. As a series of events gets underway, major astronomy publications will provide a monthly or yearly list of eclipse and occultation times.

Observing Project 10D – The Majesty of Saturn's Rings

The rings of Saturn are the signature image of astronomy, one of the first things that any new amateur is introduced to when he looks through a telescope for the first time. The behavior of the rings over the course of a fifteen-year period is the subject of this project.

In 1995, Saturn passed the autumnal equinox for the northern hemisphere and spring for the south. At that time the planets rings were presented edge-on to Earth and virtually disappeared leaving a naked Saturn. But the only time that all evidence of the rings is completely gone is in the instant that the inclination of the rings is zero. During any given year, the angle at which we view the rings changes slightly because Earth and Saturn do not orbit in exactly the same plane, so even as the rings are about edge on to the Sun, Earth will alternately travel two degrees towards the illuminated plane and then travel about two degrees to the dark side of the rings. When this happens the rings are not quite invisible but can be viewed in silhouette against the planet's disk. The rings may also paint a hairline shadow on the planet's face. By the time Saturn emerged from conjunction the next year, the rings had slowly begun to open as the planet's south pole began to point more towards Earth, we began to get a better and better view of the rings. The rings reappeared in 1996 as a bright slash through the planet and by the year after, they were open wide enough for details to begin to appear. With the rings about 8–10 degrees open, it becomes easier to differentiate the A ring from the B ring and on nights of good seeing, the Cassini division begins to appear.

By 2000, the rings were open by nearly twenty degrees and this began a five-year period where the rings are just at their most magnificent. The rings reached their maximum opening of about 28 degrees in 2003. Now is when we can see the most detail. The A and B rings are clearly visible as distinctly different structures and the Cassini division is obvious even in marginal seeing. On nights of good seeing the inner C ring begins to show itself as a distinctly different structure from the B ring outside of it. At maximum opening, the rings are so wide open that the segment of the rings that is behind the planet tips clear of the planet's disk allowing us to trace the outer circumference of the rings all the way around the planet.

Saturn's rings are bright. The middle B ring reflects more than 60% of the light that falls on them and the A ring about 30–40%. The rings are in fact brighter than the planet itself. Where there is a lot of light available there is the opportunity to use a lot of telescope power so if the air is steady, go ahead and use high power to study the rings.

As of the time this book is published, Saturn's rings are now in the early part of a seven-year long journey towards closure as southern summer progresses. By 2010, the rings will once again reach edge-on and disappear from our sights for a few months. But within a few short months we will begin to see them again as the northern face of the rings, in darkness for fifteen years sees the light of the Sun.

Observing Projects 10E – Saturn's Inner Moons

Saturn has nine major moons that are brighter than magnitude +16, which is the photographic limit of an 8-inch telescope under a dark sky. Realistically under any level of light pollution a telescope in the 8–10 inch class will be able to pick out as many as five of Saturn's moons. One of these moons, Titan is within easy reach of any telescope. Three moons inside of Titan; Rhea, Dione and Tethys are about magnitude +10 and can be viewed with some effort in moderate light pollution with a medium aperture telescope. A fifth moon also can be magnitude +10 and orbits outside Titan, which we will deal with in another project.

Finding tenth magnitude specks of light circling Saturn sounds daunting if you have tried our observing project at Mars searching for Phobos. Finding Rhea, Dione and Tethys is much easier even though they are about the same brightness as Phobos. First, all four of these moons circle much farther from Saturn than Phobos does. Secondly, Saturn does not have nearly the same surface brightness as Mars does. These two facts make locating Saturn's three inner moons much easier than the more elusive moons of Mars. One thing you should avoid is going after these faint prizes when a bright moon is near by. As the moon makes its monthly passage of Saturn, its light can make viewing the moons very difficult. Also remember your eyes and take proper care to dark-adapt prior to taking on difficult targets.

When Saturn is favorably placed in the evening sky many astronomy publications will list out the times of greatest elongation for several of Saturn's moons. *Sky & Telescope* publishes a monthly chart of Saturn moon locations for Titan and the four brightest satellites that orbit inside Titan. This would also include Enceladus, which is marginally visible at magnitude +12 in an 8-inch telescope under the darkest of conditions. The best time to try for any of these moons is when they are either near their greatest elongation from the planet and thus easiest to see, or when they are near enough to Titan so the brighter moon can serve as a guidepost.

Saturn's gravity, like that of Jupiter, has a profound organizing effect on its flock of moons. Unfortunately for us, the resonances are not so easy to notice because they usually involve the much fainter moons. Tethys resonates on a 1:2 ratio with Mimas (magnitude +16) and Titan on a 3:4 ratio with Hyperion (also magnitude +16). The one pair of resonating moons that you might be able to track is Dione and Enceladus (magnitude +12). Enceladus will be well beyond the capabilities of an 8-inch telescope unless the skies are very dark.

As Saturn's moons circle the planet, you will notice one big difference from the moons of Jupiter. Instead of moving back and forth on a straight line, the satellites instead are tracing ovals around Saturn. This is because Saturn's axis is inclined 28 degrees to its orbit while the axis of Jupiter is inclined only three degrees. An important ramification of this is that the moons of Saturn do not engage in regular eclipses as those of Jupiter do. The shadows of the moons are usually aimed well off into space and only when Saturn's rings approach edge on to our line of sight do the moons begin to direct their shadows at the planet. Like the edge-on ring period, this only occurs about once each fifteen years.

The moons of Saturn present a moderately difficult challenge that will test your ability to plan, see and execute. The reward of seeing them and beginning to push your eyes and equipment to new levels is worth the work and will inspire you to push yourself further out into the universe.

Observing Project 10F – Split-Personality Moon – Iapetus

In addition to eighth-magnitude Titan and the three tenth-magnitude moons that orbit inside of it, one other moon in the Saturn system can reach tenth magnitude. Iapetus circles Saturn well outside the orbit of Titan at a distance of about 3,560,000 kilometers from Saturn and is about 1,400 kilometers in diameter. Iapetus requires eighty-one days to make one orbit of the planet. When Cassini discovered Iapetus in 1671, he noticed something very odd about it. When Iapetus was west of Saturn, showing us its trailing edge[31], Cassini could see it easily in his telescope but when Iapetus moved around to the east side of Saturn over the next six weeks it would fade gradually until he lost sight of it. As the moon came back around in front of Saturn again he would regain sight of it as it brightened again.

In modern times, astronomers have measured the albedo of Iapetus' trailing hemisphere as being about 0.60 or about as reflective as Europa but the leading side has an albedo of only 0.04, which is as dark as tar. One of the greatest mysteries of Saturn is learning how Iapetus got this split personality. Some have theorized that the leading hemisphere of Iapetus is collecting dust that is being spun off from Phoebe. But there are two reasons why this is unlikely. First Iapetus and Phoebe are in differing orbital planes and in fact orbit the planet in different directions. Secondly, the surface of Phoebe and the dark hemisphere of Iapetus are not quite the same color, suggesting that they are made of differing materials.

Use your astronomy software to track when Iapetus comes to greatest western elongation and begin keeping track of it night by night as it circles behind Saturn and moves off to the east. The time between western and eastern elongation is just under six weeks. Keep track of the moon's changing brightness and how long you can see it for.

---

[31] Iapetus, like all of Saturn's moons, is in a captive rotation with the same side always facing Saturn. Therefore one side always faces the satellite's direction of travel while the other side always trails.

CHAPTER ELEVEN

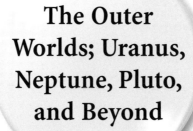

The Outer Worlds; Uranus, Neptune, Pluto, and Beyond

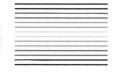

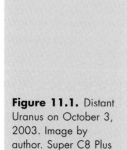

Figure 11.1. Distant Uranus on October 3, 2003. Image by author. Super C8 Plus and Meade 216 XT.

For five millennia of human history, the known solar system consisted of the Sun, the Moon and the five wanderers, which the ancients named "planets." Though the occasional comet would wander in from the depths of the firmament to strike dread into the hearts of primitive man, the limits of the known solar system remained unchanged for thousands of years. The telescope opened up the unknown outer solar system where worlds unimagined by the ancient Greeks and Romans circled. Science combined with luck brought the outer worlds of Uranus, Neptune and Pluto into the realm of our knowledge. Uranus and Neptune circle far beyond the previously known boundaries of the solar system and represent entirely different types of worlds from even Jupiter and Saturn, what scientists have come to call the "ice giants." Tiny frozen Pluto circles even farther out in the distant void in an orbit unlike any of the other planets. Each time we learn more about Pluto, it shrinks a bit more in our esteem now to the point where some no longer even consider it a planet. But Pluto is a fascinating enigma, perhaps one of the largest frozen remnants of the time of the creation known to our science. The outer worlds strain our eyes and our equipment to their limits. For these planets, the challenge is not in seeing the tiny details of their clouds or surfaces as it is for Mars and Jupiter. Here the challenge is to see them at all. Now that we've mastered the subtle nuances of planetary behavior and trained our eyes and brains on brighter targets, from here on out, things get tougher.

The God of the Heavens, a Triumph of Persistence

The British astronomer William Herschel was systematically scanning the skies of Aries when on March 13, 1781 he happened on a star he did not recall as being there. Since he was searching for comets, his first assumption was that he had found one. But after a few weeks of observing his new comet he came to realize that it neither looked like a comet, nor did it move like one. The object seemed to display a planetary disk, albeit a tiny one and it appeared to be moving in a circular orbit well beyond that of Saturn. It gradually dawned on Herschel that he had discovered a new planet, the first new planet since antiquity. Herschel actually was not the first to see it. Nearly a century before, John Flamsteed, who gained fame cataloguing stars listed Uranus as "34 Tauri" as far back as 1690. The new world was found to circle the Sun at a distance of 2,891,000 kilometers and required a full 84 years to complete one circle of the Sun. Over time the new planet's diameter was measured to be 51,118 kilometers. Suggestions abounded as to what to call the new world. Many simply named it "Herschel" for its discoverer. Herschel himself favored calling it the *Georgium Sidnus* or "Georgian Planet." The moniker honored Britain's King George III[32]. The name that eventually stuck was "Uranus." The German astronomer Johann Bode suggested the name and though it did not really become commonly accepted until about 1850, it did finally become the official name. This kept the tra-dition of naming the planets for Roman deities. Uranus was the Roman god of the heavens, father of Saturn and grandfather of Jupiter. For his historic find, George III named Herschel the Astronomer Royal. Herschel continued to study his new planet and in 1787 he found that Uranus had companions. Two faint moons circled the planet. The moons were given the names Titania and Oberon. Unlike other bodies in the solar system, Herschel's new finds were named for characters in

---

[32] The same King George III that the American colonies were rebelling against as Herschel made his discovery. One of Britain's longest reigning monarchs (and certifiably insane), he ruled from 1760 to 1820.

Shakespearean dramas. That tradition continued as more moons were found. The British astronomer William Lassell found Ariel and Umbriel in 1851. Uranus' fifth moon, Miranda, was found in 1948 by the Dutch-born American astronomer Gerard Kuiper. All of the moons are relatively small. Miranda orbits closest to the planet at a distance of 130,000 kilometers and measures 572 km across. Ariel is at 191,000 km distance and measures 1,158 km in diameter. Umbriel circles Uranus at a distance of 266,000 km and is 1,170 km in diameter. Titania is the fourth moon out and measures 1,578 km in diameter. Titania orbits at a distance of 436,000 km. The fifth moon out is Oberon at a distance of 583,000 km and is slightly smaller than Titania with a diameter of 1,522 kilometers.

The moons helped give away an amazing aspect of Uranus' behavior. As is normal, all five of the planet's large moons orbit near the plane of Uranus' equator. When Herschel discovered Titania and Oberon in 1787, he realized the moons were traveling in almost perfect circles around the planet rather than shuttling back and forth in lines like Jupiter's moons, almost as though Uranus' pole was pointed straight at Earth. As the planet circled the Sun over the next twenty years, the path of the moons became more typically straight as the equator pointed at Earth. Uranus revolves around the Sun lying on its side. In fact, one of the more heated semantic arguments that astronomers have is over exactly how to define the planet's axial inclination. If it is assumed that the planet rotates in the normal (prograde) direction, then the planet's axis is tilted beyond sideways at 98 degrees. If the planet is assumed to rotate retrograde, then the inclination is 82 degrees. Either way, Uranus' axial inclination makes it one of the true oddities in the solar system and no one clearly understands why or how this happened.

In 1978, astronomers were preparing to observe Uranus as it passed in front of a star, a rare occultation for the tiny four arc-second planet. The occultation would provide an opportunity to make precision observations of Uranus' diameter and motion. But no one was prepared for what they saw as Uranus neared the star. The brightness of the star suddenly dipped and just as quickly recovered. Then it dipped again. And again. For a total of nine times, the star suddenly faded and brightened. The star then disappeared behind Uranus and reappeared on the other side. Astronomers then noticed that once again, the star dipped in brightness nine times in exactly the reverse order that it did on the way in. The conclusion was that Uranus had a ring system. Further observations using numerous wavelengths revealed that Uranus did indeed have rings, a total of nine of them. The rings of Uranus were unlike the rings of Saturn in two important ways. First they were very narrow, not more than a few hundred kilometers wide. Secondly they were as black as coal.

Eight years later, Uranus got its one and only visitor from planet Earth. On January 28, 1986 within a span of five hours, Voyager 2 flew past Uranus and observed it rings and moons. Because the planet's south pole was pointed straight at Earth, the probe flew across the plane of the moons perpendicularly rather than parallel as it did at Jupiter and Saturn. The encounter was therefore over very quickly. Tragically the Voyager 2 flyby of Uranus was almost completely ignored by the media because just a few hours before encounter, the Space Shuttle *Challenger* erupted into flame shortly after launch from the Kennedy Space Center, plunging the world into shock and mourning.

In the five-hour encounter period, Voyager 2 took thousands of images of the planet and its moons that are still yielding discoveries to this day. Voyager 2 added ten new moons to Uranus' ledger. Five more were later discovered telescopically. The twenty-one moons of Uranus can be clearly divided into three families. The innermost eleven moons orbit in amongst the rings and play critical roles in keeping the rings organized. The moons are nearly as black in color as the rings themselves so the moons may not just be shepherding the rings, but supplying the source material for the rings as well. The outermost five moons circle well beyond Oberon in far-flung orbits. These moons range in size from the 160-kilometer diameter Sycorax to the ten-kilometer diameter Trinculo.

While Voyager 2 logged plenty of new moons during the flyby, it found very little of visual interest in Uranus itself. At visible wavelengths, the planet was a bland uniform blue and green with no sign of the vibrant cloud bands of Jupiter or even the more subtle ones of Saturn. Some years later, the Hubble Space Telescope would image Uranus at wavelengths that were not available to Voyager and image processing tools that could only have been dreamed of in 1986. Uranus grudgingly began to reveal signs of circulation in its atmosphere. Also in 1986, Uranus' south pole was pointing almost directly at the Sun. In the early part of this decade, the Sun was moving much more towards the equator in the Uranian sky. This probably causes more diurnal variations in weather in the atmosphere and thus Uranus appears to be a much more vibrant place than it was when Voyager flew by it 20 years ago. Voyager discovered that Uranus is a very different kind of world than are Jupiter and Saturn. Uranus is only 15% hydrogen, while the inner gas giants are mostly hydrogen. The planet lacks the massive liquid metallic hydrogen envelope that surrounds the cores of Jupiter and Saturn. Uranus seems to be mostly rock and ices with most of the atmosphere being dominated by the planet's relatively sparse amount of hydrogen. Most of the planet's other material seems to be evenly distributed throughout Uranus with no sign of an obvious rocky core. The icy nature of the planet and lack of hydrogen has led astronomers to call Uranus an "ice giant."

The God of the Sea, a Triumph of Mathematics

As astronomers observed Uranus in its travels around the Sun, they became troubled by their inability to precisely predict the planet's motion. Newton's laws dictated the manner in which the planet should move and yet it did not. Uranus behaved almost as though something were pulling at it from out beyond its orbit. Some predicted that another planet must be out beyond Uranus and some set about determining where it might be. In the early 1840s two astronomers, the British John Adams (who did not publish his prediction) and the French Urbain Le Verrier, independently calculated the position of the new planet and telescopes were turned to that area. In 1846, Johann Galle and Heinrich d'Arrest found the planet exactly where both men suggested it should be. Ironically while both Adams and Le Verrier correctly calculated the planet's position, but incorrectly postulated its orbit. Had the search taken place a couple of years later, the planet's position would have drifted further and further from the predicted position. The discovery created an international incident as the French and British governments (not that they ever needed an excuse to argue) over who had the right to name the new planet. Neptune, like Uranus, was actually viewed many times prior to its actual discovery. Galileo's early telescopes were not powerful enough to resolve Neptune's 2.3 arc-second disk so when Jupiter drifted by Neptune in 1613, Galileo mistook it for a background star.

Its sea-green color led the astronomical community to settle on the name "Neptune" who was the Roman god of the seas. Neptune circles the Sun once each 163 Earth years at a distance of 4,504,000,000 kilometers from the Sun. That's nearly two billion kilometers further out than is Uranus. Neptune's diameter is 49,532 kilometers, slightly smaller than is Uranus though Neptune is more massive than Uranus. Neptune's distance and tiny apparent size kept astronomers from learning a great deal about it. Very soon after Neptune was found though, the British astronomer William Lasselle found that Neptune had a moon. It was quickly dubbed "Triton" for the son of the god of the sea. Triton was once thought to be as large as Titan and Ganymede but we now know that it is actually somewhat smaller than our own moon at 2,700 kilometers in diameter. Distance precluded our really learning much of anything about it until Voyager 2 flew past in 1989. One thing that immediately stood out about Triton is that it is the only large moon in the solar system with a retrograde orbit. Because of this, Triton (presently slightly closer to Neptune than our moon is to Earth) is slowly losing orbital energy and is destined to be broken up by Neptune's gravity. Neptune's second moon, Nereid, was found in 1949 by Gerard Kuiper. Nereid circles Neptune in a moderately inclined prograde orbit about 5.5 million kilometers from the planet and has a diameter of 710 kilometers. Likely both Triton and Nereid were captured bodies that originated in orbits outside Neptune's sphere of influence.

Virtually everything we know about Neptune we learned during the flyby of Voyager 2 on August 25, 1989. Astronomers expected that Neptune would be a boring place like Uranus was but were quickly proven wrong. Neptune's axis is inclined only about 30 degrees to its orbit while Uranus was pointing it south pole at the Sun at the time of the Voyager flyby. Neptune therefore was experiencing at the time much greater day–night (diurnal) variations in its weather. Voyager found the planet to be wracked with storms. The probe found one massive Earth-sized storm in the southern hemisphere called the Great Dark Spot and many smaller storms. The Dark Spot proved not to be as enduring as its Jovian counterpart. By 1994, the newly repaired Hubble Space Telescope turned its attention to Neptune and found the Dark Spot had disappeared. Later that year, Hubble turned its attention to Neptune again and found that a new spot had emerged in the northern hemisphere. Neptune's winds are by far the fastest in the solar system, whipping around the planet at speed of up to 2,000 kilometers per hour. Neptune, like Jupiter and Saturn radiates more energy than it receives from the Sun, twice as much. Why Neptune does this is a bit puzzling since Uranus does not. Neptune is a bit more massive than Uranus, but not so much so to explain the discrepancy. Certainly though Neptune's internal heat source is a big part of what drives the planet's wild winds.

Like the other three giants, Neptune has rings. They were first observed by ground-based telescopes and imaged only as partial arcing segments traveling around the planet. Voyager 2 later imaged them as whole rings. Like with the rings of Uranus, those of Neptune are very dark. There are four major ring structures. Three of them are named for Adams, Le Verrier and Galle for the men credited with the planet's discovery. The fourth is unnamed. Voyager 2 also found six shepherding moons orbiting within the rings. In addition to these moons, there are five others, which were discovered in 2002 and 2003, which have not yet been named.

Neptune is the final major body in the solar system and stands silent sentinel at the boundaries of explored space. Voyager 2 and her predecessors continue out into deep space.[33] Yet there is more out there in the deep and distant reaches of the solar system.

The God of the Underworld, a Triumph of Luck

As the century turned, astronomers again were having difficulty accounting for discrepancies in not only the orbit of Uranus, but now that of Neptune as well. Again speculation began to run that yet another undiscovered planet was lying out beyond Neptune and the search was again on. Astronomers attempted again to mathematically calculate the position of the missing planet. Many observatories began to look for it and in 1930, the American astronomer Clyde Tombaugh hit pay dirt while rapidly flashing back and forth between two slides he had taken of an identical star field, one distant object began to jump back and forth at him across the frame. The ninth planet had been found, but purely by luck.

It turned out the calculations upon which Tombaugh based his search were in error. It just happened to be one of the greatest strokes of coincidence that the little new planet just happened to be lying there right at that moment in time. It also quickly turned out that Tombaugh's new planet was far too tiny to have any impact whatsoever on the motions of Uranus and Neptune and much too far away as well. Because of the extreme cold and relative darkness in which the new planet must have existed, astronomers named the new world "Pluto" for the ancient Roman god of the underworld. The first two letters are also the initials of Percival Lowell, the astronomer who had spearheaded the search for the ninth planet. Tombaugh was one of his successors working at the observatory that bore Lowell's name.

For decades Pluto was held in the same esteem as the other eight planets. If you would like to see a good fight, walk into any astronomy forum and say, "Pluto is not a planet." You will start a brawl faster than you would if you had walked into a Boston bar in a Yankees jersey. We know next to nothing about Pluto, except for the fact that the more we learned about it, the smaller it became. Initial estimates of Pluto's size placed it as being as big as Earth. By the early 1970s our best estimates of Pluto's mass and orbital motion suggested a size slightly larger than Mercury.

---

[33] Pioneer 10 and 11 are both now dead, their nuclear power generators having died some years ago. Voyager 1 continues in excellent health and recently overtook Pioneer 10 as the most distant object launched by man. Voyager 2 is less healthy and running low on both electrical power and maneuvering fuel, having made two extra-planetary encounters.

Pluto displays another un-planet like trait. Unlike all the other planets, which never stray far from the ecliptic plane, Pluto's orbit is inclined nearly 17 degrees causing it to arc away into many constellations that are not part of the traditional zodiac, such as Cetus and Bootes. Pluto's orbit is also very eccentric. At the perihelion point of its 248-year orbit Pluto actually comes inside the orbit of Neptune (though the two orbits do not intersect, the crossing occurs with Pluto well north of the ecliptic). At its farthest, Pluto orbits out as far as 7.5 billion kilometers.

Then came our most shocking discovery about Pluto to date. Pluto has a moon! While tremendously exciting, it also meant the mass present in that tiny dot of light was spread out over two bodies, not one. By now our best estimates, Pluto's size is about 2,274 kilometers but that value can still be in error by as much as 2%. Pluto is now known to be less than half the size of Mercury, and smaller than seven other moons in the solar system. The tiny moon is named Charon and was discovered in 1978 by astronomer James Christy who partially resolved the moon in photographic plates. Charon was the mythical figure that ferried the dead across the River Acheron into Hades. But it is also a tribute from Christy to his wife Charlene. Charon is believed to have a diameter of 1,172 kilometers and orbits only 19,600 kilometers from Pluto. Like Earth and its Moon, Pluto and Charon revolve around a barycenter which unlike Earth and its Moon, lies out in space, about one-third of the way from Pluto to Charon.

It was of great fortune that Charon was discovered when it was because a series of transits of the two worlds would begin shortly after the time of discovery. This gave astronomers a chance to measure carefully the changes in brightness as the two bodies repeatedly occulted each other over the next several years. This information was used to make an albedo map[34] of the surfaces of both worlds. This revealed some interesting things about Pluto as well. It has the highest contrast in surface features of any body in the solar system except for Iapetus. These observations were later verified by using the power of the Hubble Space Telescope. Hubble also showed us that Charon also has a higher reflectivity than Pluto and a density comparable to the icy moons of Jupiter or Saturn. Charon also appears to be covered in ice, while Pluto is not. This creates the potentially provocative idea that Pluto and Charon did not form together. Hubble has also discovered two additional tiny moons.

Pluto and Charon revolve around their common center of gravity each six days. Both bodies are tidally locked to each other with each showing the other the same face. This means that since an orbit of Charon takes as long as a rotation of Pluto, the balance of energy between the two bodies is neutral with neither body stealing energy from the other. This system is therefore stable and will remain in this balance in perpetuity, unless something comes along to disturb it.

Pluto is the only planet in the solar system never to have been visited by spacecraft. This will change in the next decade with the arrival of New Horizons, launched by an Atlas V in January 2006. The probe will require about nine years to reach Pluto. Planners then hope to plan the flyby in such a way that it will allow the probe to be redirected to a Kuiper belt object after the flyby. Perhaps then, many of the major questions about Pluto will have been resolved.

---

[34] A map of bright and dark areas on the planet.

Worlds Beyond Pluto, a Triumph of the Titans

Most comets appear to our eyes from deep in space. Edmund Halley was the first to discover the orbital nature of the comet that bears his name but as to exactly where comets originate very little is known beyond conjecture. In 1950, the Dutch astronomer Jan Oort, after studying the orbits of many comets inferred three things. First, no comet observed had an orbit that indicated that it came from interstellar space, secondly there is a strong bias towards comets with aphelia in the range of 50,000 AU and thirdly there is no preferential direction from which comets come. Oort used these three facts to infer the existence of a massive halo of perhaps a trillion comets lying out beyond the orbit of Pluto. This halo, called the *Oort Cloud*, marks the outermost influence of the Sun's gravity but its existence can only be inferred. It cannot be directly observed.

Until the 1970s it was thought that nearly all the asteroids in the solar system were confined within the orbit of Jupiter in the main asteroid belts or various Trojan groups affiliated with Jupiter. On August 10, 1977, the American astronomer Charles Kowal first caught sight of an asteroid-like body orbiting just outside Saturn ranging between 8 and 18 AU from the Sun. It was first noted as an asteroid, though later it was began to display cometary traits and so it is also listed as a comet. As comets go, it was enormous at a diameter of 170 kilometers across. Chiron proved to be the harbinger of many more discoveries out beyond the asteroid belt. Today astronomers know of eight other such objects circling between the orbits of Jupiter and Neptune. These objects now belong to a new class of asteroids called *Centaurs*. Centaurs are asteroids that are believed to have originated in the Kuiper Belt, a large area of asteroids thought to exist beyond the orbit of Pluto. As the next two decades wore on after the discovery of Chiron, some of these remarkably distant objects began to turn up in the monstrous new telescopes that were being built. Titans such as Subaru and the Keck twins were able to reel in tiny objects that were never before possible to see or image. These objects are called *Kuiper Belt Objects* or "trans-Neptunian" objects. The Kuiper Belt is a disk shaped region that extends beyond the orbit of Pluto and is home to many thousands of small icy bodies. Many of these bodies are over 100 kilometers in size. Some in the astronomical community believe the Kuiper Belt is the home of many short-period comets. Here they circle the Sun in the icy cold beyond Pluto until gravitational interaction with the outer planets or with each other causes some to fall into the inner solar system. The Centaurs are believed to be Kuiper Belt refugees.

In 2004 astronomers announced the discovery of what was at the time the most distant known object circling the Sun; designated 2003 VB 12 and provisionally named "Sedna." At a mean distance of over 90 AU Sedna is three times Pluto's distance from the Sun and moves in an eccentric Pluto like orbit. Sedna is also big. At a diameter of 1,800 kilometers, it is not much smaller than Pluto itself. It is becoming therefore more apparent that if Pluto is not a planet, then it is among the largest members of an entirely new class of objects. If this were true, then it was only a matter of time before some object larger than Pluto was found and in

July 2005, it was announced that object had been found. This object, designated 2003 UB 13 is located 97 AU from the Sun, presently near the aphelion of an orbit that will bring it to within 39 AU of the Sun. 2003 UB 13 is believed to be no larger than 2,600 kilometers and takes 557 years to complete one orbit of the Sun in an orbit inclined 44 degrees to the ecliptic. The object remained unknown for such a long time because most Kuiper Belt objects orbit in fairly low-inclination orbits. Though the lower limit of its diameter has yet to be pinned down it is no doubt substantially larger than Pluto. Caltech astronomer Michael Brown at the discovery press conference gave a highly scientific description of 2003 UB 13's size when he said, "It's a big sucker!" Brown and his co-discoverers Chad Trujillo of Gemini Observatory and David Rabinowitz of Yale University first imaged 2003 UB 13 on October 21, 2003 using a 48-inch telescope at Palomar Observatory and the 8-meter Gemini North telescope on Mauna Kea in Hawaii. It took until January 8, 2005 to note its motion when astronomers imaged that same area of the sky again. Trujillo took a near-infrared image of the object and found that its surface is covered in ice and it is surprisingly bright in visible light. In fact astronomers using a 14-inch telescope in Chile successfully imaged the nineteenth magnitude speck of light within six hours of the announcement of its discovery. Later in 2005, astronomers got yet another surprise. Images of 2003 UB 13, which astronomers had since nicknamed "Xena" revealed that it had a companion. The companion is nicknamed "Gabrielle" for the fictional Xena's loyal sidekick. The discovery of 2003 UB 13 is the pinnacle (thus far) of the search for Kuiper Belt objects. Some 800 of them are now known and it is believed the Kuiper Belt contains far more 100-kilometer class objects than the asteroid belt does and we will find more and more of them as the years go on.

Observing Projects XI – Specters in the Deep, Cold Darkness

Observing Project 11A – Uranus and Neptune at the Eyepiece

At first glance it would seem that the triumph is in simply finding the two icy giants at the edge of the solar system. But with sufficiently large instruments and a little patience and skill, you can coax some of the secrets of the ice giants out of those pale blue-green dots.

Uranus shines at a feeble magnitude +5.7 and through a telescope its disk spans not quite four arc seconds. Even if Uranus had some details to see, you could not see them through amateur telescopes anyway. What may surprise some is that the planet's moons are visible through large amateur telescopes. The two outer moons, Oberon and Titania, are both magnitude +14 when the planet is at opposition. This puts them at the extreme visual range of a 10-inch telescope, which with the advent of the Dobsonian many astronomers can now afford that aperture. Ariel is slightly fainter at magnitude +14.2. Umbriel is magnitude +15.1 and requires at least 14 inches and Miranda is the faintest at magnitude +16. It takes some work, patience and well cared for equipment but if you are willing to do the work, you can coax out all five of Uranus' moons if you have sufficient aperture available.

Many of the same tricks that you used at Mars to coax out Deimos and Phobos will work here too, although the glare of Uranus is not much of a factor here. What matters here is unlike at Phobos and Deimos, you are working at the most extreme limits of your equipment. Therefore it is absolutely essential that you avoid any extraneous light at all. If you have a dew cap, put it over the telescope and do whatever it takes to avoid any exposure to non-essential light.

In 1978, astronomers took advantage of a fortuitous event when Uranus passed directly in front of a star. As the planet neared the star, it faded and recovered to normal brightness nine times. After being occulted by the planet, the cycle reversed itself after the star reappeared. Check your ephemeris for occultations of stars by Uranus. This will present you with an opportunity to recreate one of the most important discoveries in the history of astronomy. The finding that Uranus had rings showed that Saturn's ring system was not unique in the universe, rather ring systems were commonplace around large planets. You can do this in a telescope of any size. Between 2005 and 2015, Uranus' rings will be involved in seven occultations, including three events alone in 2006. All three of the 2006 events involve stars brighter than magnitude +10.

About a billion and a half kilometers further out Neptune stands watch over the outermost fringe of the domain of the planets. Neptune's gravity shepherds the objects of the Kuiper Belt beyond its orbit maintaining many of the ice chunks beyond its orbit in a 3:2 resonance. Neptune's light is even more feeble than that of Uranus, shining at only magnitude +7.8 and subtending barely 2.3 arc seconds. Neptune's one large moon, Triton, shines brighter than any of Uranus'. At magnitude +13.4, Triton is at the extreme range of an 8-inch telescope. Again, make sure the telescope objective is shaded and avoid any light whatsoever.

Remember that Neptune also has rings, so if any prediction of an occultation of a star by Neptune appears, make sure you watch. As Neptune approaches the star, what will happen as the rings pass by? Neptune will occult eight stars between 2005 and 2015, including two events in 2006.

The subtle details of Uranus and Neptune can challenge you and your equipment to their limits. Watch the excess light, shade that scope and find a dark sky site. If you keep your scope well collimated, you might surprise yourself as to what you can really see. But there are some objects in the heavens that are just too faint for the eye to see. Yet that does not necessarily mean that you can't *image* these sights.

Observing Projects 11B – Uranus and Neptune in the Camera

A well maintained 8-inch telescope with modern coatings and operating under a dark sky can visually reveal stellar objects as faint as magnitude +14.0. Only one of Uranus' moons is actually brighter than that value, but any given telescope can actually photograph objects two full magnitudes fainter than its visual limit, so that 8-inch telescope can actually photograph down to magnitude +16.0.

If you are using film, consider two things. First use the fastest film you possibly can get. Second, store your film in the freezer until you are ready to use it. Freezing the film will push its photographic speed to double what it could do ordinarily. This is important in long-exposure astrophotography because you don't want the exposure to be any longer than it absolutely has to be. There are two reasons for this. First is that every second the shutter is open there is the opportunity for you to make a mistake. Secondly is that reciprocity failure will begin to set in if you leave the shutter open for the kinds of time that it might take to bring out the faint moons of Uranus and Neptune.

The optimized astronomical CCD is far better for this application because of its sensitivity and its speed. Make sure that chip is properly chilled and take progressively longer test images. Experiment with the settings to make sure you can process the noise out of your images and eventually you will begin to reach the level of the faint moons. Also always remember never to give up on a bad image because as your image processing skills at the computer get better, you will be able to pull good signal out of even the noisiest images. And when you're imaging down to magnitude +14, you're going to get a lot of noise.

Observing Project 11C – Distant Pluto and Places Beyond

The feeble glow of Pluto and Charon is barely brighter than magnitude +14, about as bright as Uranus' moon Titania. That means that under dark skies, observers with good technique and clean, well-collimated optics can view Pluto with a telescope as small as 8-inches. You would search for Pluto using the same techniques that you would use to find asteroids but with a couple of exceptions.

You must remember that you are looking for an object far fainter than anything you have searched for before and there is no bright object around to guide you to it. The faint moons of Uranus and Neptune at least have Uranus and Neptune to help you find them. Pluto stands alone. The other thing that makes finding Pluto difficult is that its motion is exquisitely slow. An asteroid or faint moon will move sufficiently to betray its presence in a few days, or a few hours, but Pluto is different. Event though it is now near the perihelion point of its 248 year journey around the Sun, its motion may take a couple of weeks to betray its presence. Remember to avoid exposure to any extraneous light whatsoever and make good use of your averted vision to see Pluto.

Capturing Pluto on film is an exquisite challenge that will require extremely long exposure times. Even with professional sized instruments, Clyde Tombaugh needed exposure times measured in hours to capture Pluto on his crude photographic plates. Film is better today than it was in 1930 but still imaging Pluto is a task that modern technology has put above and beyond the call of duty for film.

A modern CCD camera can easily capture Pluto in a matter of about a half hour. Better than that, the most recent versions of cameras available today allow for automatic alignment and stacking of many images. Rather than keeping the shutter open for a single thirty-minute exposure, you can instead take sixty thirty-second images and then stack them one on top of the other. This creates the equivalent of a much longer exposure with a greatly reduced risk that the exposure will be ruined by streaking because no one image will be exposed long enough to allow any stars to trail. You must make certain that your telescope is properly collimated and that the camera is properly configured, thermally cooled and stable. The CCD chip, just like film, is much more sensitive to faint light while it is cold, rather than if it is hot.

Telescopes of 14 inches and larger are capable of reaching down near twentieth magnitude and are capable of imaging objects as faint as the now celebrated Xena and Gabrielle. Deep sky photography down near the photographic limits of your telescope can prove most intriguing. You will be able to at that value image not only Pluto, but also Chiron the Centaur asteroids as well. The only problem with CCD cameras in searching large areas of sky is that their field of view is measured in arc minutes and not a lot of them at that. If you elect to undertake such a search remember that 2003 UB 13 has taught us that these large Kuiper Belt objects do not have to be located near the ecliptic. 2003 UB 13 has an orbit inclined 44 degrees to the ecliptic so the whole sky is fair game. But with enough aperture and a bit of luck, you might get to discover a Kuiper Belt object of your own.

CHAPTER TWELVE

Twinkle, Twinkle, Little Star (Now Knock It Off!)

Figure 12.1.
Thousands of hot young stars in the summer Milky Way including brilliant Deneb. Image by author.

The verse that we all grew up singing as children would of course later come to be the bane of our existence as amateur astronomers. Twinkling stars mean bad seeing and turbulent air. But what about the stars? The sky is filled with those wonderful little points of light, more than six thousand of them that you can see with your unaided eye in a dark sky. Each one of those little points of light has a story to tell. Some have been sitting there in our sky burning steady and unchanged since not long after the Big Bang. Others have been formed within the last few tens of millions of years and will end their lives in a short spectacular burst of light within the next few tens of millions years. They come in myriad different personalities. Some are bright and hot, many are faint and cool. Some travel alone and others travel in pairs, or trios, or pairs of pairs. Some stars burn steady and others just don't. Many amateur astronomers spend their time exploring the mountains and valleys and craters of the Moon, the wonders of the planets or seeking FFTs (faint fuzzy things). Yet we are remiss by not considering the simple beauty of the stars, the celestial nuclear furnaces from which all things are made. Our Milky Way galaxy is conservatively estimated to contain more than 100 billion stars. Let's go take a look at some of them.

Stars of the Main Sequence

As we learned in Chapter 6, the Sun and most stars belong to the *main sequence.* A main-sequence star is a star that is in the hydrogen-burning phase of its life as our Sun is. Let's take a look around at our Sun's immediate neighbors for other examples of main sequence stars. Lying in Cetus, some 11.9 light-years away is Tau Ceti. This is one example of a main sequence star similar to our Sun. It is spectral type G8 and therefore burns a little cooler and is somewhat deeper yellow in color than is our Sun. It has about 81% of the Sun's mass and is about 77% of the Sun's diameter. It is quite a bit older than the Sun at some ten billion years of age and is probably nearing the end of its main-sequence lifetime. One surprising fact about Tau Ceti is that for a star of its age, it is surprisingly poor in metals[35] compared to the Sun, which is much younger. Shining at magnitude +3.5 in the southern part of Cetus, it is easily visible to the unaided eye almost anywhere in the world. Not far away to the east is Epsilon Eridani, a spectral type K2 orange colored star. It too is slightly smaller than the Sun and slightly less massive but has only about 25% of the Sun's luminosity. Unlike Tau Ceti, Epsilon Eridani appears extremely young, less than 500 million years. Alpha Centauri is a star that burns slightly brighter than the Sun and is slightly larger but not very much so, it having the same spectral class (G2) as our Sun, Alpha Centauri shines in our sky at about magnitude −0.2 and is about two or three billion years older than the Sun.

But not all main-sequence stars resemble our Sun. Some main-sequence stars are very small and burn hydrogen very slowly. Such a nearby star is the red dwarf Wolf 359. Wolf 359 is the faintest of the stars within ten light years of the Sun. It shines with only 1/50,000th the luminosity of the Sun. Wolf 359 has about 9–13% of the Sun's mass and about somewhere between 16 and 19% of the Sun's diameter. Although it is only 7.8 light years away, it shines in our sky at only magnitude +13.5. If Wolf 359 replaced our Sun, it would shine with only about ten times the brightness of the full moon and an observer would need a telescope to clearly discern its disk. Wolf 359 is estimated to be some ten billion years old, twice the age of our Sun and will likely continue to burn along the main sequence for many billions if not tens of billions of years after our Sun has burned itself out.

At the opposite end of the main sequence is Sirius. It is the brightest star in our sky shining at magnitude −1.5 in Canis Major. It is a spectral type A0 (zero) slightly bluish white star located about 8.8 light years from Earth. Sirius has about 2.14 times our Sun's mass and is about 1.7 times larger in diameter than the Sun. Sirius

---

[35] To an astronomer studying the stars, a metal is anything that is not hydrogen.

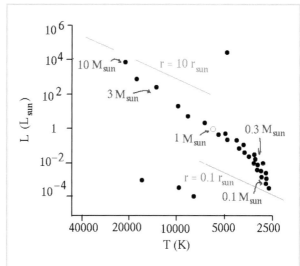

Figure 12.2.
Hertzsprung–Russell diagram illustrating the main sequence. Power Point illustration by author.

shines with about 21 times the luminosity of the Sun but will not do so for very long despite the fact that at 225 million years of age, Sirius is a celestial toddler. As stars grow larger they must burn through their fuel at an exponentially greater rate to maintain themselves in gravitational balance and not collapse in on themselves. Sirius will burn through its supply of hydrogen and evolve into a red giant within a billion years.

Stars of the main sequence are hydrogen-burning stars in the best and steadiest years of their lives. A star's lifespan is directly tied to its size and luminosity. The bigger and brighter the stars are, the faster they will run through their reserves of fuel and evolve off the main sequence into their old age.

Non-Main-Sequence Stars

Many of the brightest stars visible in our sky are not main sequence however. Marking the shoulder of Orion is the red supergiant Betelgeuse. This is one of the largest stars in all the heavens. If it were put at the center of the solar system in place of the Sun, it would swallow everything out to Mars and continue well into the asteroid belt. It shines in our sky at apparent magnitude +0.5 from a distance of 430 light years. For all its size though, Betelgeuse has only about 12 to 17 times the mass of the Sun. The star is also surrounded by a massive shell of dust of its own making caused by a powerful stellar wind that is blowing out the stars outer layers. Betelgeuse is a highly evolved star that is dying, likely fusing helium in its core into carbon and oxygen. A spectacular fate awaits it in the not too distant celestial future. Stay tuned.

Stars that are big are not always red. At the other end of Orion from Betelgeuse is Rigel. This is a massive blue supergiant that makes a beautiful contrast with red

Betelgeuse and nearby white Sirius. Rigel is 775 light years away and shines at a visual magnitude of +0.2. It is spectral type B8, which gives it a surface temperature of some 11,000 K, more than twice as hot as the Sun and among the hotter stars in the heavens. Rigel shines with 40,000 times the luminosity of the Sun. Rigel burns hotter than Betelgeuse does because it is a bit smaller and more massive. More mass requires more fusion to prevent the mass from collapsing itself. Rigel is even more evolved than is Betelgeuse and is dying. Its supply of hydrogen is exhausted and it is now fusing helium into carbon, beryllium and oxygen. Rigel will eventually share the same spectacular fate as awaits Betelgeuse.

Both Rigel and Betelgeuse are very massive stars that quickly exhausted their supplies of hydrogen, leaving a helium ash core behind that was spurned into fusion when gravitational energy spurred by core collapse caused helium fusion to begin, while residual hydrogen is fused in an outer shell driving the stars outer layers into space. These stars are massive enough that eventually they will progress to fusing carbon into oxygen and eventually oxygen into iron.

At the bottom end of the Hertzsprung–Russell diagram are the white dwarfs. Sirius has a tiny faint companion that shines faintly at magnitude +8.0. Sirius B is spectral type A2 on the H-R diagram, just slightly to the right of where Sirius A is. Despite the fact that it is smaller in size than Earth (11,200 kilometers) it appears to have about the same mass as the Sun. This means that one cubic centimeter of matter from Sirius B would weigh approximately 15 tons! Sirius B is a foreshadowing of what awaits our Sun at the end of its life, crushed by its own weight into a degenerate mass of heavy neutrons that emit heat of over 20,000 K. Sirius B is dead from the point of view that it no longer supports any nuclear fusion. The heat from this "star" is a result of energy given off through gravitational compression, a runaway of the same phenomenon that cause Jupiter and Saturn to emit more heat than they receive from the Sun. It's light is created by the crushing of the atoms under the amazing pressure created by stellar collapse.

Spectacular Stellar Deaths

Stars of the Sun's size lack the mass to create temperatures that will allow oxygen to be fused into heavier elements like carbon, silicon and sulfur. But stars the size of Betelgeuse can continue the process and as they die, they will fuse residual hydrogen, helium, beryllium, carbon, oxygen, silicon and sulfur in a series of shells and finally in the deep core, silicon and sulfur are fused into iron. Iron cannot be fused into anything else in an energy-creating reaction so after a time, a sufficient mass of iron is produced that any ongoing fusion cannot create enough outward-pushing energy to counter the power of gravity trying to collapse the star. So gravity finally wins out in the end, as it always does. When this happens, the iron atoms are compressed with such force that they are broken down into their sub-atomic particles (electrons, protons, neutrons). The entire core collapses or *implodes* to a volume only a few kilometers across. This collapse creates a shock wave (equal and opposite reaction) that rips through the surrounding fusion layers and completely destroys the star in an awesome explosion called a *supernova*. The star will increase in brightness by many thousands of times and the heat of the blast is so incredibly hot that the fusion layers of the star in their death throes will create all the elements of the periodic table up to uranium, scattering it throughout the universe. From such a spectacular death though can also come new life. If the shockwave from the supernova blast encounters an interstellar cloud of hydrogen and dust, it can cause that cloud to begin to contract and begin a new cycle of stellar birth. Such an explosion many billions of years ago started such a cloud contracting and created the Sun.

As the gases of the explosion are driven off into space they leave two markers behind. First is an ever-expanding cloud of gas illuminated by radiation created within the cloud by high temperatures and a strong magnetic field. This *nebula* is the grave of the deceased star marking where the titanic explosion took place. In the year 1054, skywatchers noted a brilliant new star in Taurus. The star grew so bright that it was visible for months in broad daylight before fading away to never be viewed again. At least that is, not until the telescope was invented. Astronomers searching that area found the remnant of the supernova and named the cloud the Crab Nebula for its crustacean-like appearance. The second marker is the core remnant itself. The core is called a *neutron star* because all that is left at the core is neutrons. The collapse of the core causes the protons, electrons and neutrons in each atom to become fused together into one enormous neutron. Because the center of the supernova explosion is not at the exact center of that core, the force of the blast imparts enormous rotational energy to the core and it begins to spin very rapidly. Each time it does so a blast of radio energy sweeps across our field of view and radio telescopes on Earth are able to detect it. This type of spinning neutron star is called a *pulsar*. The rotation rates of pulsars are extremely precise and if they are single pulsars (a pulsar that formed in a multiple star system) they will gradually slow down over time.

If the star is of sufficient mass, the combination of mass and high density produces a gravitational field so powerful that nothing, not even light can escape it. The stellar remnant then simply disappears. Material drawn into the star can be viewed spiraling down into the star and disappearing when the gravity becomes so strong that light can no longer escape. The star is now referred to with the descriptive name *black hole*. The point where light can no longer escape the black hole is called the *event horizon*.

This is the fate that awaits massive stars like Rigel and Betelgeuse. When they inevitably blow, they will do so in spectacular fashion. At 430 light-years distance, Betelgeuse will supernova with light equivalent to a bright crescent moon. It will cast shadows on Earth's surface and will be easily visible in the daytime. It will not likely happen for tens of millions of years. But it could happen tomorrow too.

Double Trouble

Something that may come as a surprise to many amateur astronomers is that a majority of those pinpoints of light you see at night are not really a point of light but two or more points. A large number of the stars in our sky are actually double or multiple star systems. Among the brightest stars in our sky, Rigel, Antares, Sirius and Castor all have companions. The pinpoint of light we call Castor, or Alpha

Geminorum[36] is actually six stars that orbit so close to each other that the only way to split them all is by spectroscope. Many double stars are easily split by modest telescopes and are sights of remarkable beauty while others will push you and your telescope to the limits of their performance.

Back in Chapter 1, we introduced you to a star called Mizar, which is at the bend in the handle of the Big Dipper. Mizar is magnitude +2.2 while it companion Alcor is +4.0. Alcor can be tough to see with the unaided eye if you live in a light-polluted neighborhood. Alcor you may remember from Observing Project 1A actually lies about three light years in the background behind Mizar. But if you turn a telescope on it you will discover that Mizar does actually have a partner of its own. The partner is actually slightly brighter than Alcor but sits very close to Mizar so a telescope is required to see Mizar's real partner.

Although everything is subject to a debate, arguably the sky's most beautiful double is at the foot of the Northern Cross (an asterism in Cygnus). Albireo (the Arabic meaning of the name is unclear due to historical mistranslation) shines there at just about magnitude +3.0 but its true beauty becomes evident immediately in a telescope. With a field of Milky Way background stars scattered like diamond dust across your field of view, the two stars of Albireo (Beta Cygni) are breathtaking. A magnitude +3.0 topaz shines beautifully next to a beautiful +4.6 magnitude sapphire. The two stars are separated by about 34 arc seconds so any telescope at any power can split this pair with no effort Albireo is a treasure sought far and wide by amateurs each summer when Cygnus is high overhead for northern hemisphere observers. I can just lose myself staring at this one all night long. The topaz component of Albireo actually has another close companion that is too close to it to be viewed in a telescope. Astronomers are not certain that the two components of Albireo are actually orbiting each other. If they were, they would have a very long period, something on the order of 7,300 years for one orbit. The blue component is believed to be single. It is however a rapidly rotating star and spinning material off itself into an orbiting disk.

In the sickle of Leo, you will find the bright star Algieba (in Arabic the name means "the forehead", we also know the star as Gamma Leonis). This star shines with somewhat more total light than does Albireo at magnitude +2.0. Algieba also is a multicolored double, with a brighter yellowish star shining at +3.0 and a slightly fainter orange star that shines at +4.0. Algieba is a bit tougher though than is Albireo because its components are so much closer together, split by only 6 arc seconds. At a distance from Earth of 170 light years, this means that the pair is separated by four times the distance from the Sun to Pluto and the pair requires over 500 years to complete an orbit of each other. For many observers though, Algieba gives Albireo a run for the night sky's most beautiful double star. Each component of Algieba is believed to be an older star, just evolving off the main sequence following the beginnings of helium fusion in their cores.

Near the brilliant star Vega, forming part of the upper triangle in Lyra is the star Epsilon Lyrae. The star shines with a collective light of just brighter than magnitude +3.9. If you can get yourself in a comfortable position and watch it as it comes

---

[36] The brightest star in a constellation is designated "Alpha" the second brightest "Beta" and so on. Gemini is odd because Alpha Geminorum (Castor) shines at magnitude +1.6 while Beta Geminorum (Pollux) shines at magnitude +1.1.

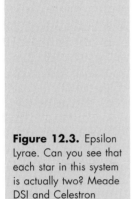

Figure 12.3. Epsilon Lyrae. Can you see that each star in this system is actually two? Meade DSI and Celestron Super C8 Plus at f/10.

overhead, you might even be able to discern with the unaided eye that Epsilon Lyrae is actually two stars with an east–west orientation. The star to the west is called Epsilon Lyrae-1 and the eastern star is called Epsilon Lyrae-2 (Eps-1 and Eps-2 for short). Now if you turn a telescope on them and use even low to medium magnification it becomes obvious that the each member of the pair is also a pair. Epsilon Lyrae is a "double-double" star system. The Eps-1 pair is separated by 2.8 arc seconds while the Eps-2 pair by 2.2 arc seconds. Each pair orbits a barycenter each 1,000 years at a distance from each other of about 170 AU, while each of the pairs orbits each other with a period of some 500,000 years. This would place the pairs some 10,000 AU apart. Sadly this wondrous quadruple star is not destined to stay together. Astronomers are at a loss to explain how this system formed in this way as the laws of motion would not allow these four stars to form in close proximity to each other and organize in this way. Likely the motions of the stars relative to each other would accelerate two of the stars out of the system leaving only a double behind. Eventually tidal forces raised by the Galaxy as a whole will break up Epsilon Lyrae into two separate pairs that will go their own respective ways. But for now, they are together and unlike other multiple star systems you can see them for what they are. Each of the stars are similar bright spectral type "A" stars.

Though not visible to a large majority of the world's populations, Rigil Kentarus (Alpha Centauri) is easily one of the most famous stars in the heavens. The Arabic name means "foot of the Centaur." It ranks as the third brightest star in the entire sky behind only Sirius and Canopus, shining at magnitude −0.1. Its remarkable brightness has little to do with any extraordinary physical characteristics but with the fact that it is the closest star to our solar system. At only 4.36 light-years distant, Alpha Centauri shines brilliantly even though it appears to be just about identical to our Sun in most ways. Alpha Centauri is by far the brightest pair of stars in our sky. The primary star is a spectral type G2 (slightly cooler and more orange than our Sun), shines with about 50% more light than the Sun and its spectra indicates that the star is much richer in metals than is our Sun suggesting that it is around seven to eight billion years old and probably within two billion years of the end of its main-sequence lifetime. It measures approximately 11–28% larger than our Sun in diameter and about 9% more massive. Alpha Centauri A shines at magnitude 0.0 by itself, so it would still be the third brightest star in the sky even without its companion. The companion is a spectral type K0 star that follows a highly elliptical orbit ranging from 11.4 to 36.0 AU from the primary. In our solar system, this would take it about to within 2 AU of the orbit of Saturn to just outside the orbit of Neptune. Alpha Centauri B is about 86.5% the size of the Sun, 90% of the Sun's mass and is about 50% as luminous. Its visual magnitude is +1.4. Both of these stars are sufficiently Sun-like to make science fiction writers wonder if either or both stars could support Earth-like planets. Either star could hold an Earth-like planet within its respective habitable zone[37] without being gravitationally disrupted by the other star. In addition the light of the other star, while significantly brighter than that of a full moon, would not be disruptive to life. If you travel to southern Florida or to Texas or points further south during the summer months, Alpha Centauri will poke itself a few degrees above your horizon. It will be low in the murk but be sure to catch this most reclusive of famous stars.

Alpha Centauri has a third member, which may or may not be gravitationally bound to the system. Called "Proxima Centauri," this red-dwarf, magnitude +10 main-sequence star is the closest star physically to our Sun. Alpha Centauri C as it is also known, is following a curved path around the main pair, but some recent observations indicate that the orbital path is hyperbolic. This is an open ended orbit indicating that Proxima Centauri is only making a flyby of the A and B stars and is not actually a true member of the system. Proxima Centauri lies about one-fifth of a light year from the A and B stars and about a tenth of a light-year closer to Earth than A and B. Proxima Centauri is a spectral class M5 dwarf with about 14% of the Sun's diameter, 12% of its mass and a luminosity of about 0.0053% of the Sun. Proxima Centauri is what is called a *flare star*, a star prone to violent sudden outbursts that can more than double its normal brightness. Even as flare stars go, Proxima is known to be unusually violent. And Proxima may not be alone. Two separate measurements of Proxima's considerable proper motion (motion against the celestial sphere, more than 10 arc seconds per year) suggest with 75% probability that it has a companion of 80% Jupiter's mass.

Not all double stars are so easy to observe. Two of the brightest stars in the sky have companions that will challenge even the best of observers. From its vantage point less than 9 light years away, Sirius rules without challenge as the night sky's brightest star at magnitude −1.5. Sirius also has a tiny companion circling it in a 50-year elliptical orbit that carries it from about 8 AU out to 31 AU. When the "Pup" as some call it, is farthest away from Sirius A, it is not too difficult to see, shining at magnitude +8.4. But when it is close to Sirius, it can be nearly impossible to observe. Sirius A only has about twice the mass of its tiny companion so the two stars actually revolve around a barycenter located about 3.5 AU from Sirius A. Sirius A follows a roughly circular path around the barycenter while Sirius B follows a highly elongated path. When Sirius A and B are at their maximum separation, they are as far apart as are the Sun and Neptune. At their closest, the

---

[37] Defined as a distance from the star where ambient temperatures would allow water to remain liquid under one atmosphere of pressure.

distance is equal to about the distance from the Sun to halfway between Jupiter and Saturn. In our telescopes at maximum separation, they are about 25 arc seconds apart, but at minimum they are barely 6 seconds apart and picking Sirius A and B apart is almost impossible. The two stars were at minimum separation in 1995 and are now slowly moving apart but right now viewing the Pup with amateur telescopes is exquisitely difficult and will be for some years to come yet.

In Scorpio, the heart of the scorpion is marked by the reddish orange star Antares (Alpha Scorpii) the name of which means, "rival of Mars." Literally translated the word "ant-Ares" means "like Mars," Ares being the Greek name for the god of war. Antares (spectral type M5) is a brilliant star, 10,000 times more luminous than our Sun at a distance of 700 light-years. Antares' luminosity and distance yield a diameter of 6 AU, so large that if it were placed at the location of the Sun, it would swallow everything in the inner solar system and most of the asteroid belt. Antares' companion resides only about 3 arc seconds away shining at magnitude +5.5. The spectral class B2 star resides about 550 AU from Antares A. The star's true color is blue-white but when contrasted with red Antares, many observers report seeing the companion as greenish. Because Antares lies almost exactly on the ecliptic, the Moon often occults it. This is how the companion was discovered when it reappeared from behind the Moon's dark limb a few seconds before Antares itself did.

The sky is filled with thousands of double or multiple star systems of varying colors and exquisite beauty. Some are easy to see and beautiful to admire while others will push you to the limits of your equipment and patience. But if you are persistent and wait for that right moment when the sky is perfectly clear, that single point of brilliant light will eventually give way and show you a tiny faint partner.

Variable Stars; Now You See Me, Now You Don't

Our lives on Earth are dominated by the presence of a star that shines with a light that never changes. The Sun is today what it was yesterday and will be precisely the same tomorrow. Not all stars are so consistent though. Many will vary considerably in luminosity from year to year or month to month or even day to day. Some will do so on a very precise schedule and others will do so on no schedule at all. And they will do so for a wide variety of reasons. Variables can be categorized into two types, *intrinsic* and *extrinsic*. Extrinsic variables brighten and dim because of a factor external to the star such as the presence of a fainter companion that regularly eclipses its brighter counterpart. Intrinsic variables change brightness due to an internal factor. Intrinsic variables either actually physically pulse in their cores causing their luminosity to vary on a regular scale or for some reason erupt violently, sometimes on a recurring basis and sometimes they are single events. Some intrinsic variables not only vary in luminosity, but in temperature and color as well. Unlike double stars, which require a telescope to see in most cases, many variables can be easily observed, enjoyed and tracked with no optical aid whatsoever.

In the constellation Perseus, the star Algol (Beta Persei) is the prototype for a type of extrinsic variable star called *eclipsing binaries*. The name Algol is derived from Arabic for "the demon," and though its nature was not scientifically documented until 1667, observers clearly knew about its regular fading for many hundreds of years before then. Algol is located 93 light years from Earth and has two components separated by less than five million miles. The system consists of a bright spectral type B8 bluish white star that shines at a temperature of 12,500 K and with 100 times the luminosity of the Sun. The star is a hydrogen-burning main-sequence dwarf with about 3.8 times the mass of the Sun. The star itself is as regular as our Sun is but exactly every 2.867 days Algol does something amazing. Its light dims from magnitude +2.1 to magnitude +3.4 or only one-third its normal luminance. The reason why is because the primary star's companion passes in front of the star at that time creating a deep partial eclipse. The secondary star is yellow-orange K type star, though the uncertainty of its exact spectral class is considerable varying from G5 to K2. It has only 2.5% of the luminosity of the primary so when it blocks the primary star's light, there is a dramatic fading in the combined luminosity of the system. The eclipse lasts for about five hours from first fading to full recovery. Algol-class eclipsing binaries are very important to astronomers because they provide a powerful tool to measuring stellar masses and sized. Thousands of these stars have been catalogued over the years.

Algol is equally famous to astronomers for another reason, which astronomers call the *Algol Paradox*. Spectroscopic observations tell us that the two stars that make up the Algol system are of about the same age. The primary "B" star is a main sequence dwarf with 3.8 solar masses while the secondary "K" star has only 0.81 solar masses and yet it is the K star that has evolved into a dying giant while the more massive blue star is still on the main sequence. The reverse should be true as more massive stars burn through their hydrogen fuel at an exponentially greater rate. The reason why is that the primary star is stripping material from the secondary at an amazing rate almost exposing its core. In looking at Algol you are observing a pair of stars in the dramatic act of one star literally stealing the life of the other in an interstellar tragedy of Shakespearean proportions.

In Cetus the Whale, each eleven months a new star appears amid the faint patterns of the mighty celestial cetacean. Dubbed "Mira" (derived from the root word for "miracle") and designated Omicron Ceti. Mira is the best known of these intrinsic long-period pulsating variables and the first to be discovered. A long-period pulsating variable or "Mira-type" variable changes magnitude by more than 2.5 and has a clearly defined period of somewhere between 100 and 1,000 days. David Fabricius, a disciple of Tycho Brahe, discovered Mira in 1596. Mira resides 420 light years from Earth so at its peak it is some 1,500 times more luminous than the Sun in visible light reaching an apparent magnitude of about +3.0. Normally though Mira is invisible being near magnitude +10 and about as luminous as the Sun. Mira is a spectral type M5 red giant that is deep into the dying stages of its life. Both hydrogen fusion and helium fusion have long since ceased within the core. Carbon and oxygen are being fused into silicon and sulfur in the core. Mira has become hugely distended in its death throes and is extremely unstable. Observations with the Hubble telescope show that the star is so unstable that it is visibly no longer round anymore. Pulsations in the slowly collapsing core cause the brightness outbursts at regular intervals. This in turn creates powerful stellar winds that are driving Mira's outer layers out into space. These in turn will eventually create a beautiful planetary nebula very close to Earth over the next few million years.

Each year as Mira builds to its peak brightness, it gradually brightens to magnitude +3.0 before fading back although at times it has become brighter. Twice during the past century, Mira peaked brighter than magnitude +2.0. William Herschel observed Mira to be "almost equal to Aldebaran" (+0.85) in November, 1779. There have also been apparitions where Mira failed to brighten beyond magnitude +5. Mira's period also varies slightly, by about a day or so from the accepted value of 332 days. The Hubble Space Telescope also discovered that Mira has a companion that orbits it once each 400 years. This discovery will now allow for precise mass determinations to be made of this most wondrous of stars.

In the polar north, in the constellation Cepheus the King is another very famous variable. Delta Cephei is at first glance a nondescript star in a nondescript constellation. This star is the prototype for perhaps the most important of intrinsic variable stars, the *Cepheid* variable. Delta Cephei is a yellow-white spectral type F5 supergiant. "Supergiant" is something of an understatement for this star, which has 40 times the diameter of the Sun and pours out the energy of 2,000 Suns. Delta Cephei is a high-mass star with about five solar masses. But pinning down Delta Cephei's spectral type is a difficult thing because unlike with the Mira-types, Cepheids change color and temperature as they oscillate. Delta Cephei varies between F and cooler G with its light variations. The star's surface temperature varies between 6,800 K and 5,500 K. Delta Cephei and stars of its type are helium-burners that pulsate at a high frequency. In the case of Delta Cephei, that period is always exactly 5 days 8 hours 47 minutes 32 seconds during which time the star varies between magnitude +3.5 to +4.3.

Cepheid-type variables have several traits in common. They are very massive, vary in spectral type as they oscillate, are powerfully luminous and are very regular. Delta Cephei and stars of its class are dying stars whose helium-burning cores cannot find the equilibrium between gravitational collapse and the countering force of energy released through fusion that most stars have. So gravity tries to collapse the core, creating more gravitational energy, which in turn causes more fusion causing the core to grow again, reducing gravitational energy and resultant fusion energy thus causing the core to shrink again. The star can never find balance. This pulsation of the core is as precise as the ticking of the finest watch. What makes Cepheids so special is that the rate of pulsation (which ranges from anywhere between one and fifty days) is directly related to their luminosity. Once we know how luminous the star is with great precision by measuring the period, we then can determine from apparent magnitude the exact distance to the star. Because Cepheids are so luminous, they can even be detected in distant galaxies. It was the discovery of a Cepheid variable within the Andromeda galaxy by Edwin Hubble that led to the first reliable measurement of the distance to Andromeda and later other galaxies. The finding of Cepheid variables in distant galaxies is one of the key projects of the mighty orbiting telescope that now bears Edwin Hubble's name.

Delta Cephei also has some naked-eye cousins including Mekbuda (Zeta Geminorum) and Eta Aquilae, both of which are actually slightly brighter than is Delta Cephei. The pulsations of Delta Cephei are easy to monitor because Delta Cephei is also a double star. The companion is a B8 located 41 arc seconds away and shines at sixth magnitude.

There are many different types of these pulsating variables that can be categorized based upon frequency of the peaks, amplitude of magnitude change and spectral type. Each type is named for the first variable of that type to be discovered. *RR Lyrae* types are very short period (less than a day usually), less than two magnitudes change in brightness and are always spectral type A. *RV Tauri* types have longer periods, between 30 and 150 days, vary by a maximum of three magnitudes and are K or G spectral type yellow supergiants. *Semiregulars* have periods of anywhere between 30 and 1,000 days and exhibit both regular periods and show periods of irregular behavior. *Small amplitude pulsating red giants* (SAPRG) are M type giants that vary between 5 and 100 days with an amplitude of one magnitude or less. These stars tend to vary because of the natural instability inherent in helium-burning giants. The effect is caused by variations in convection within the star.

Eruptive variables also come in several types. These variables are very irregular and in fact can often be one-time-only events. *R Coronae Borealis* types are very low-luminosity stars that are spectral type M or below. They maintain their normal brightness for long periods of time, then suddenly fade by several magnitudes and need months to recover. They do so on no regular schedule. The stars are hydrogen-poor, carbon-rich red dwarfs. Since they are red dwarfs and so poor in hydrogen, these stars must be immensely old. *UV Ceti* types are also known as *flare stars*. These stars will erupt on an irregular and unpredictable basis from localized spots on the surface. The star can brighten by more than two magnitudes in just a few seconds and then fade back over the next twenty minutes. Two local examples of UV Ceti types are Wolf 359 and Proxima Centauri. *Novae* are stars that show a massive eruptive increase in brightness and always occur in binary systems. Usually one star is a Sun-type star and the other is a white dwarf orbiting in close proximity. The gravity of the white dwarf draws material away from the main sequence star and begins to accumulate in a disk around the white dwarf. Eventually the material builds to sufficient mass to initiate hydrogen fusion and the mass explodes creating a massive increase in brightness that may last several months before the star fades back to normal magnitude. Often times these are single events, but in *recurrent novae* they occur repeatedly. The most amazing type of variable is the *supernova*, the catastrophic destruction of a high-mass star when its iron core collapses. The star will exhibit a sudden, dramatic (and final) brightening of as much as twenty magnitudes. They become so bright that they can be viewed from clear across the universe and become more luminous than the entire galaxy in which they reside. When they occur locally, a supernovae can become as bright in the sky as a full moon and be visible in broad daylight for weeks. There has not been a bright supernova found in the Milky Way for several hundred years, still the Milky Way is so dusty that no matter how bright a supernova might become, galactic dust clouds could block our view of it. The search of brighter galaxies for supernovae is carried out by amateur astronomers as vigorously as is the search for comets or asteroids. Some dedicated observers have found supernovae by the dozens and one has even found a hundred of them. They can occur at any time and in any galaxy. Many will peak at values bright enough to become easily visible in telescopes as small as 6 inches, while the host galaxy may in fact not even be visible. Finding supernovae may in fact be the simplest way to get your name in the astronomy magazines.

Observing Projects XII – The Beauty of the Stars

Observing Project 12A – The Showcase Doubles

After years of careful study, astronomers have come to realize that most of the stars in the local area do not circle the galaxy in isolation but have at least one partner. In our first observing project, we'll look up close at some of the most beautiful and unique in our sky. Let's take a look around the sky beginning with the summer sky and going around the sky season by season.

Let's start in the northern hemisphere summer. Earlier in this chapter we highlighted Albireo, which is at the foot of the Northern Cross. This pair is almost universally recognized as the most beautiful in the sky. The two components of Albireo are topaz orange and sapphire blue in color creating a stunning contrast. In addition, the background Milky Way stars have been described as a field of diamond dust creating a stunning setting for our multicolored jewel. The two components are both reasonably bright at magnitude +3.0 and +4.6 respectively. The star is also remarkably easy to view because its components are separated widely enough to make them easy to view in the smallest of telescopes with the lowest of magnification. The two components are spaced 34 arc seconds apart. Further, the since the eye is only capable of discerning so many different colors, the result of having two stars with such different colors is such close proximity is that the eye exaggerates the contrast between the two stars.

Another very easy double to split is at its best about at the same time as is Albireo, in late summer. The alpha star of Libra, Zubenelgenubi, is most prominent about the same time as Albireo is. This star is famous first and foremost for its tongue-twister name. Zubenelgenubi along with nearby Zubeneschamali were in antiquity the northern and southern claws of Scorpio. Libra was once a part of Scorpio (and the only constellation of the zodiac that is not emblematic of a living thing) many thousands of years ago. Zubenelgenubi is also a widely spaced double consisting of a magnitude +3.0 white spectral type "A" star and a fainter magnitude +4.8 spectral class "F" star. The two components are separated by almost three arc minutes so they are extremely easy to split and can be split in fact with the unaided eye. Unlike Albireo, the stars of Zubenelgenubi are similar in color. Can you tell the difference in color between the two?

As summer turns to fall, the string of beautiful stars rises high overhead in the constellation Andromeda. The last of stars in the beautiful line of stars is Almach (Gamma Andromedae). The name is Arabic like most star names and refers to a type of wildcat indigenous to the Middle East region. Almach is another multicolored extraordinary beauty. The two components are separated by ten arc seconds making them very easy to see. Gamma-1 is a magnitude +2 golden spectral type K giant in the process of evolving off the main sequence into a giant. Its partner, Gamma-2 is blue-green in color and shines at fifth magnitude. The companion is in itself double, though far tougher to see. Gamma-2 is composed of two fifth-magnitude spectral type "A" stars, white-blue in color circling each other every sixty years. Though the two stars are currently near maximum separation, they are still less than half an arc second apart. A 12-inch telescope at minimum is needed to split Gamma-2. But there is still more. The brighter of the Gamma-2 pair is also double, requiring a spectroscope to split. All three stars are about the same in color, so the point of light we call Almach is actually a quad.

Rising to zenith as autumn fades to winter is the brilliant Castor. Because of its affiliation with Pollux, Castor is commonly placed among the ranks of first-magnitude stars when in fact it is the brightest of the second-magnitude stars. Castor is also one of the heavens' most famous multiple stars. A telescope of 3 inches or larger splits Castor in two white components split by about two arc seconds. One component, Castor A is magnitude +1.9 and spectral-type A1; while the other, Castor B is magnitude +2.9 and spectral-type A8. Castor A and Castor B travel around each other in an elliptical orbit requiring about 400 years to complete. Currently Castor A and B are near their minimum separation so high power is required to split them. Lying about a minute to the south is a third component, a magnitude +9.0 M-class star called Castor C. Castor C's orbital period around the bright pair is unknown because we have not been able to observe it long enough to get a good look at its motion. Castor C is more than 1,000 AU from the barycenter of the bright pair. All three of the stars are visible in any telescope with adequately dark skies. But when we look even closer with a spectrograph we find that each of the three components is also double. Castor A is two identical A1 stars that orbit each other each 9.2 days at a distance less than one-tenth the distance from the Sun to Mercury. Castor B's twin stars circle even faster completing a joint circuit each 2.9 days. Castor C is comprised of a pair of nearly identical spectral-type M1 red dwarfs. The two stars each have about 0.6 solar masses and are separated by less than two million miles. These two stars race around each other in only twenty hours and either one or both of these stars are UV Ceti type variables (flare stars). Splitting the A and B pairs is relatively easy, but still a bit of a challenge. Use all the power you can and wait for a moment of still air. The C pair is far enough away to be easily found, but are your observing skills good enough to identify it? Study the system indoors first so when you go outside, you will know what it is you are looking for. A few minutes of preparation can make a tough task easy but a lack of any preparation makes easy tasks very hard.

During the later half of winter and early spring, Leo the Lion rides prominent overhead. In the Sickle of Leo, north of Regulus is the bright star Algeiba. This beautiful double, illustrated earlier, consists of a pair of stars, one yellow magnitude +3.0 and one orange magnitude +4.0. The stars are close, not so close as Castor though not nearly so easy as Albireo and Almach. Like with Zubenelgenubi, the stars are similar in color, but not identical. Zubenelgenubi is a bit easier because of the distance between the components, while those of Algeiba are split by only six arc seconds.

Observing Project 12B – The Toughest Doubles

While amateur astronomers seek many doubles because they are easy and beautiful to view, many others are sought because they are hard to split apart. Let's go around the sky once again as we did in the previous project and highlight some of the more difficult doubles.

Starting in the northern hemisphere summer, Antares rides low in the southern sky glowing reddish-orange in the warm summer nights. Antares is a double that is extremely tough to split because of the difference in the brightness in the two components. The B2 companion is magnitude +5.5 and is only three arc seconds from Antares. At that distance the companion hides in Antares +0.96 magnitude brilliance. While the pairs of Epsilon Lyrae and Castor A and B are at closer distances from each other, they are easier to split because the components are about the same brightness so one does not overwhelm the other. Often times, to split a star like Antares requires getting rid of the light of the primary. This can occur naturally when the Moon occults Antares, which it does so on a regular basis in long series of events that can last many months. But when a lunar occultation is not readily handy, you can make your own.

Back in Chapter 8, we discussed how to create an occulting bar to hide the light of a bright primary object like Mars to allow fainter objects to appear like Phobos or Deimos. This same technique will work for splitting close doubles. Positioning of the occulting bar is critical because if you are not aware of the orientations of the companion star, you may end up occulting the very star you are seeking. Set up a high-powered eyepiece with an occulting bar as we did for the Martian moons and try and find the companion. At spectral type B2, it should appear blue to the eye. Once you have found it, try removing the occulting bar and see if you can still pick out the companion. If you can still see the companion, what color does it appear to be now? Because of the contrast and the overwhelming light of Antares itself, many people who have viewed both stars together believe the companion is green. This is an illusion that is caused by the saturation of the retina with the red light of Antares, diminishing the impact of the fainter blue star on your retina and thus shifting its apparent light away from the blue end of the spectrum towards the green.

As the bright summer triangle and Scorpio and Sagittarius rotate towards the west, the fainter patterns of the fall sky come into view. The long constellation Cetus strings out along the celestial equator coming to opposition during late September and October. The fourth brightest star in Cetus is called Kaffaljidhma (Gamma Ceti). The star's full Arabic name was Al Kaff al Jidhma, which means "part of a hand." This star consists of two closely spaced components. The A-star is a main-sequence white, spectral-type A3 star. The A-star shines at magnitude +3.4. The B-star resides very close by at only 2.8 arc-seconds distant. The companion is magnitude +6.25, making it tough to pick out the spectral-type F3 companion from the relative brilliance of the A-star. Through the telescope, the stars appear to be blue and yellow but this is another example of visual contrast effect. The A and B stars are similarly yellow in color. Kaffaljidhma also has a third component, located about 15 arc minutes away is a magnitude +10.1 K5 star that shares the proper motion of the two. It appears to be a gravitational member of the system orbiting some 21,000 AU from the main pair and completing an orbit each 1.5 million years. Because the A-star is not as bright as Antares is, you should be able to split this tough pair using a small telescope at high power, but you need clear and stable air. Otherwise the twinkling of the stars will cause the two stars to blend together.

As Earth moves on around the Sun, the faint stars of autumn give way to the brilliant stars of winter. None of these are any more brilliant than is Sirius. At magnitude −1.5, Sirius is by far the brightest of all stars in our evening sky. It is also one of the toughest of all doubles in the sky to split, especially now with the two stars near their minimal separation. Sirius A is a white main-sequence star while Sirius B is a white dwarf, a dying ember that shines at magnitude +8.4. With the two stars less than five arc seconds apart, splitting them with Sirius A in view is nearly impossible given the closeness of the two stars and the ten-magnitude difference in brightness. Viewing Sirius A and B is further hampered by another complication and that is winter weather. A shivering body and tearing eyes certainly make life difficult for observing anything, never mind two tightly paired stars. The cold also makes trouble for your telescope too. If you have a telescope with a sealed tube, you will have to wait a while to try a split these stars while you wait for the interior of your telescope tube to cool down. It would probably be a good idea to leave the eyepiece out of the telescope for a time to allow the tube interior to chill down to ambient temperature. You want for this to happen as fast as possible because on a cold January night, you will not have a great deal of endurance at the eyepiece. And by the way, don't forget the occulting bar either.

Eventually the bitter cold and crystal clear skies of winter give way to the progressively more mild nights of spring. Leo is riding high into the east and Virgo is poking above the eastern horizon. In Virgo is the double star Porrima (Gamma Virginis). Porrima is an unusual star name in that it is not Arabic, but rather is Latin and honors the Roman goddess of prophesy. Porrima is a double that is exquisitely difficult to split as it is just now reaching minimum separation at barely one-half arc second. Both stars are nearly identical spectral-type "A" white stars that burn somewhat hotter than the Sun does. Porrima gives amateur astronomers a rare chance to actually catch a double star in motion. The orbit of the two components of Porrima is extremely eccentric. At maximum separation the two components are 70 AU apart, but at their closest are less than 3 AU apart. The maximum spacing of the two stars is about six arc seconds at the apoastron point of their 168-year orbit. This last occurred in 1921. At periastron in 2005, the two stars are less than one-half arc second apart and a telescope of at least ten inches will be needed to split them, and even then only in very good seeing. For the amateur with small telescopes, the challenge will be to see when you can first split Porrima as the two components begin to split again. The motion of the stars is so rapid in the second half of the 2000's that not only will an amateur be able to watch the two stars begin to move apart, but also watch the rapid change in position angle between the two. During the time of periastron, the rate of change of position angle will be some 70 degrees per year. The two stars will begin to move apart at a rate of nearly 0.2 arc seconds per year after periastron then year by year the rate will slowly begin to diminish as the stars move apart from each other.

There is scientific value to careful studies of Porrima because the orbit of the two stars is not precisely established. Even orbital elements established as recently as 1990 have been shown to be slightly in error, particularly in separation by as much as .05 arc seconds. This is a substantial error that may put the exact time of periastron in doubt by as much as two years. The 1990 elements suggest that periastron would occur in 2005, but Porrima may be two years late getting there.

As the two stars begin to drift apart over the next few years, the challenge will be to try to split them as early as you can. Astronomers can use accurate measurements of both position angle and separation over the next several years when the rate of motion of the two stars is at its greatest. By sometime around 2008, the stars should be wide enough to allow an 8-inch to split them and a 6-inch by 2010. Most of the stars in the heavens seem permanently fixed against the heavens but you can see the stars of the Porrima system rapidly move during this once in 168-year event.

Observing Project 12C – Minima of Algol and Eclipsing Binaries

Algol is the classic example of an eclipsing binary. Its secondary component orbits the primary each 2.867 days. Each time the secondary passes through inferior conjunction with its primary, it creates a deep partial eclipse causing the entire system to fade in total magnitude by about 1.4 magnitudes, or nearly 70%. This is an event that can be observed easily with the unaided eye. Each month the minima times for Algol, or time of maximum eclipse are published by *Sky & Telescope* and other publications such as the *Observer's Handbook*. When at its normal brightness, Algol is as bright as is magnitude +2.1 Almach (Gamma Andromedae) to its west, but at its faintest, it fades dimmer than nearby Epsilon Persei (+2.9) on its east.

In addition to the more famous cycle, when the companion is on the far side of Algol, it is partially eclipsed in its own right. Though the primary is much brighter, the eclipse of the secondary also produces a small light drop of about 0.2 magnitudes. You can document the rise and fall of Algol with a 35-mm camera mounted on top of your clock-driven telescope. The times of minimum brightness are easy enough to find. Make a five to ten minute exposure of the area around Algol at minimum and then make an identical exposure the next night. The difference in brightness between the two exposures will startle you when you develop the film. You can also detect the secondary eclipse event on film about 35 hours after the primary eclipse event.

There are other fine examples of eclipsing binaries in the sky that you can observe as well. Beta Lyrae is another such example of two stars that eclipse each other on a regular basis, but have a different type of light curve. Algol's primary star is reasonably spherical, but the two stars of Beta Lyrae are so close together that gravity distorts both components into ellipsoids. As a result of the shape and distance between the two stars, there is not a constant peak magnitude with Beta Lyrae, but rather a continually changing brightness throughout the 12.94-day period. Photograph it the same way as you did with Algol and be sure to follow it every day you can to document its cycle. During the fortnight of Beta Lyrae's cycle, note that there are two distinctly separate minima spaced about 6.48 days apart. Beta Lyrae cycles from magnitude +3.3 down to +4.4, back up to +3.3 then down to about +3.8 before recovering back to +3.3 to end the cycle.

ObservFing Project 12D – The Pulsations of Cepheid Variables

Because its luminosity is directly tied to its period of variability, the Cepheid variable is the most valuable measuring stick that astronomers have available to them. Their awesome brightness allows astronomers to find them in distant galaxies and thus determine the distance to those galaxies by comparing apparent magnitude against an easily and precisely determined absolute magnitude or luminosity. Here are three that are visible at differing times of the year.

Delta Cephei is the prototype for this type of star. It falls from a peak brightness of +3.5 to +4.4 over a period of exactly 5.37 days. Unlike with the eclipsing binaries, there is no secondary peak with the Cepheids. What is something of a surprise is that the light curve is not symmetrical. It takes Delta Cephei much longer to fade than it does to recover. The fall of the star's brightness takes almost four days. It recovers in less than a day and a half before starting the cycle all over again.

Zeta Geminorum is another example of a Cepheid variable, ranging in magnitude between +3.6 and +4.2. The cycle of the star's brightness takes 10.15 days to complete. Unlike with Delta Cephei, the light curve of Zeta Geminorum is symmetrical, taking about five days to fall and five days to rise back to full brightness.

Eta Aquilae is a late summer and fall variable located near Altair. Its light curve closely resembles that of Delta Cephei in that the star fades from a peak of +3.5 down to +4.4. The period is somewhat longer than Delta Cephei at 7.18 days and exhibits a similar type of asymmetric decline. There is one little oddity in the light curve of the star that you might notice: a small temporary reversal during the decline before the fall off resumes again.

Note the periods of the stars carefully as they brighten and dim and make pictures of them each night of the cycle that clear weather permits. The pulsations of the Cepheid-type variables are very important to astronomers and their regularity is one of the amazing things in the universe to watch, especially considering the fact that the reason for the pulsations is the lack of equilibrium in the star's core.

Observing Project 12E – Mira and the Pulsating Giants

No object in the heavens outside of the solar system exhibits such a radical brightness change as Mira (Omicron Ceti) does. The star normally shines at tenth magnitude, but each 332 days the star builds to a peak at magnitude +3.4. Mira should always be carefully monitored when it reaches maximum because the cycle is not always perfectly precise nor it the intensity of the peak brightness level. As recently as 1997, Mira peaked at magnitude +2.5 and remained there for nearly a month. In other years, the star has failed to reach fifth magnitude. In the middle part of this decade Mira reaches peak brightness about the same time or within a few weeks of reaching solar conjunction so we will not get to monitor Mira's peak again until later in the decade.

Mira is by far the brightest of the long-period red giant variables and by far exhibits the greatest range of magnitude change. Most naked-eye variables exhibit very little range of magnitude change, which in turn may explain how Mira became known as the "Wonder Star." It also has a reasonably short and stable period, which not all of these stars do. Betelgeuse is a massive example of this same type of star that can shine as bright as neighboring Rigel (Beta Orionis) and fade as dim as Bellatrix (Gamma Orionis). That would range between magnitude +0.3 and +0.9. Betelgeuse's cycle is poorly defined but photometric observations over a period of more than a decade suggest a cycle of about six years. Because Betelgeuse is a star that may or may not be fusing silicon and sulfur into iron in its core, its end is very near. When that end comes, it will be one of the most spectacular events viewed in our sky in many hundreds of years. Because layers of gas being blown off from its interior surround Betelgeuse, it defies attempts to measure its exact size and internal makeup. It may not erupt as a supernova for another fifty million years. But it might do so tomorrow, therefore any change in the behavior of this star would be of great interest to the astronomical community. Can you measure the changing brightness of Betelgeuse? Since good comparison stars are in close proximity in Orion, all you need to do is make regular wide field photographs of Orion and see how that red star at the top left compares to the blue one at the top right and bottom right.

Observing Project 12F – Irregular and Eruptive Variables

Not every variable star behaves on a precise or even semiregular schedule. Sometimes certain stars, giant stars most likely, may suddenly erupt and gain considerable brightness for a time before fading back. Sometimes after such an eruption, the star may actually fade considerably below its normal brightness before slowly recovering.

Gamma Cassiopeiae is an example of such a star. Throughout the recorded history of modern astronomy, Gamma Cassiopeiae maintained a constant brightness of magnitude +2.25. But suddenly in 1937, the star brightened to magnitude +1.6 as it ejected a shell of gas. By 1940, Gamma Cassiopeiae had faded all the way to magnitude +3.0 and did not recover its original luminosity until 1966. What causes a seemingly stable star to suddenly erupt in this way is not known but any large star in the sky can behave in this way.

In July 2000, Delta Scorpii (the middle star in the head of the scorpion) flared from its normal magnitude +2.3 to +1.8 and some estimates of its brightness ranged as high as +1.6 in 2002. Delta Scorpii normally shines with 3,000 times the luminosity of the Sun, but the eruption swelled that value to 5,000 Suns, a 67% increase in luminosity. Delta Scorpii has since been slowly fading back towards normal. How much it will fade is not known and the star will have to be monitored carefully for many years. Both Delta Scorpii and Gamma Cassiopeiae are spectral type B0 stars that have begun to evolve off the main sequence into the helium-fusing stage of their lives.

Just because stars are cool and faint does not mean they cannot be eruptive in nature. The tiny red dwarf Wolf 359, located in Leo the Lion, is an example of a UV Ceti type variable. Such stars are prone to sudden and violent outbursts of activity. Unlike with eruptive giants, which display activity on a long-lasting and global scale, UV Ceti types display a sudden rapid brightening that occurs over seconds and then return to normal brightness over less than thirty minutes. Wolf 359 is not particularly violent like UV Ceti or Proxima Centauri can be, but it is much more easily observed than the other two. Wolf 359 can suddenly brighten from its normal magnitude +11.0 to magnitude +9 and fade back within a few minutes. Another interesting, though unobservable aspect of flare star behavior is that while a star may brighten one or two magnitudes visually, flare stars are at their brightest in the X-ray region of the spectrum. A flare on Wolf 359 may produce X-rays 10,000 times more intense than the most powerful solar flare on our Sun. Most of the red dwarves in the Sun's vicinity are known to be UV Ceti type variables including Proxima Centauri, DX Cancri, Lalande 21185, Ross 154, Ross 248 and Barnard's Star. Despite the fact that all are less than twelve light years from Earth, all require telescopes to see, but they can be worth monitoring.

Observing Project 12G – Fast-Moving Stars

In addition to their variable nature, there is another reason to keep a close watch on these close red dwarves. They are very close to Earth and are moving in a direction that is tangential to our line of sight, so stars like Wolf 359 and Barnard's Star display a very rapid *proper motion*. Both of these stars move across the sky at ten arc seconds per year so after a time, the movement of these two stars against the background stars becomes very readily apparent.

Not all the stars in the firmament then sit still. Keep a close watch on these two stars over a period of several years and watch their positions change. Barnard's Star in particular is interesting because it is rapidly moving towards Earth, at a rate of 140 kilometers (87 miles) every second. It will approach as close as 3.7 light years to Earth in the year 11,800. It also has the fastest rate of proper motion of any star in the entire sky, moving at 10.3 arc seconds per year.

The stars are often the most ignored objects in the heavens, seeming to sit stationary and forever unmoving in the firmament of the heavens. But careful observation shows the stars are not so static as they brighten and fade, circle in pairs or even pairs of pairs and even move over time. So when the sky is too hazy or light polluted to find the faint, fuzzy things, spend some time with the stars. You will find your work will be well rewarded.

CHAPTER THIRTEEN

Faint, Fuzzy Things. Part I: Phenomena Galactica

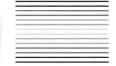

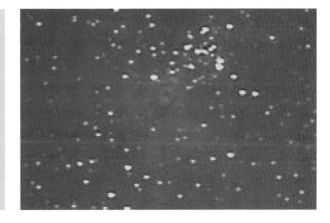

Figure 13.1. M16/ Eagle Nebula. Image by author. 10-min exposure using Meade DSI and Celestron SC8 Plus. Note the slight streaking caused by a slight misalignment of the polar axis.

As astronomers scanned across the skies with their new telescopes during the seventeenth and eighteenth century, they began to uncover in droves faint wispy patches of light that were obviously not stellar in nature. At first many professional astronomers considered them to be more problematic than interesting because they were potentially easy to mistake for comets. But as our telescopes grew more capable and our observing and photographic techniques grew more refined and sophisticated, we came to realize that these "false comets" as Charles Messier called them were each special types of objects that later on would become crucial to our understanding of the nature of the universe. Some of these patches of light represent stellar tombstones where a star met its end, sometimes in astonishing violence and glory and sometimes with but a puff of gas in an interstellar whimper. Other puffs of light represent stellar nurseries, where newborn stars are emerging from the womb beginning their lives. Some patches of light are loose communities of stars traveling together through space while others are tightly bunched gatherings of hundreds of thousands of stars that travel around our galaxy in a giant halo. Beyond the halo of clusters are the enormous "island universes," the galaxies that are host to uncountable hundreds of billions of stars and other faint fuzzy things of their own. In this chapter, we'll take a look at those objects that are indigenous to our galaxy and in the next; we'll take a look at the distant galaxies. But first, let's take a look at the man who first focused attention on the deep sky and his historic contribution to our science and hobby.

Messier and his List

The name of Charles Messier is most commonly associated with the deep sky because of his famous list of deep sky objects, but in fact Messier had very little interest in deep sky objects. Charles Messier started out in astronomy at the age of 21 when he moved to Paris in 1751 and was hired as a draftsman and observing assistant by Joseph Delisle. Within a few short years Messier had become an avid and accomplished observer in his own right. Delisle and Messier along with the rest of the astronomical community were eagerly anticipating the predicted return of what would soon become known as "Halley's Comet." Delisle had made a projection map illustrating what he thought was the path that the comet would follow as it made its return in 1758. Delisle thought that his map would give Messier an advantage in the search for the comet. Unfortunately for Messier, Delisle calculations were wrong and he was looking in the wrong place. On Christmas night in 1758, the German farmer and amateur astronomer Johann Georg Palitzsch recovered the comet. Messier did independently find the comet about a month later but was forced by Delisle to keep his discovery secret. It was not long before the news of Palitzsch's discovery reached France and Messier was deprived of his triumph. This fueled an obsession in Messier to find more comets. Over the next fifteen years, Messier would find twelve comets in all, just about every comet discovery made during that period. Messier's obsession with comets was so overwhelming that it was once said that while Messier sat at his wife's deathbed, a rival astronomer discovered a comet. When a friend came to console him, he lamented "Alas, I have found a dozen of them; Montagne had to take away the thirteenth!" before he realized his friend was talking about his late wife.

As 1P/Halley made its way across Taurus in 1759 and passed between the horns of the bull, Messier noticed a whitish patch of light near the star Zeta Tauri. It was obviously not a comet and Messier noted the position on his charts of the comet. He had found the Crab Nebula. By 1764, Messier accumulated more than twenty of these "false comets" and began to make a list of them arranged in the order that he had observed them and went to work finding more. Within seven months Messier had expanded his list to forty items. In addition to the Crab Nebula (M1), he also added the Omega (M17) and Trifid (M20) Nebulas, the Hercules Cluster (M13) and the Andromeda Galaxy (M31). Messier also included objects previously observed by Edmund Halley, William Derham and Lacaille. By 1765 he had determined precise positions for the Orion Nebula (M42 and M43), the Pleadies (M45), the open cluster near Sirius (M41) and the Beehive Cluster (M44). The catalogue was published later that same year for the first time and quickly became obsolete because by 1766, Messier had made five more entries. In 1779, Messier first began to tap the vast reservoir of galaxies in Virgo when the great comet of that year passed through the area we now call the great Virgo supercluster. By 1781, Messier had a new rival, Pierre Mechain of the Naval Map Archives in Paris. Mechain found two new comets that year and in the course of his searches also found thirty-two new nebulous objects. He communicated all of his discoveries to Messier who would confirm them and add them to the list in the order that *Messier* observed them. By late 1781, the Messier list stood at 100 when Messier was badly injured in a fall off an icehouse. Messier broke an arm, a leg and two ribs. Messier was out of commission for over a year and by the time his list was republished again in 1784, the list had grown to 107 with several items added by Mechain. The final items in the list were areas of nebulosity found around the Owl Nebula (M107). The French revolution and the economic and political turmoil that ensued brought Messier's work to a halt in the 1790 as the collapse of the monarchy cost Messier his naval pension. When stability returned after the rise of Napoleon. Messier did find one last comet and lived long enough to be elected to the newly formed Academy of Sciences and received the Legion of Honor from Napoleon. Messier passed away in 1817 at the age of 86. His catalog, intended primarily to identify false comets so that they would not be mistaken for real ones is today the standard by which observers of the faint fuzzy things execute their right of passage into the realm of accomplished deep sky astronomers. In the pages ahead, let's take a look at some of the best objects in Messier's list that reside in or immediately around our galaxy, work on techniques for finding them and then learn how to find the details when there is seemingly none.

Galactic phenomena basically consist of two types of structures. *Star clusters* are gravitationally bound groups of stars that travel together through space as they orbit the galactic center. These come in two types. Open clusters are small and orbit within the galaxy while globular clusters orbit out in the galactic halo and may contain many millions of members. *Nebulae* are structures associated with stellar evolution, marking either stellar nurseries or stellar tombs. Basic nebula types include *diffuse nebulae, planetary nebulae, supernova remnants* and *dark nebulae*. Each one of these types of objects tells a different story about the evolution of stars, the galaxy and the universe at large. They tell stories of birth and stories of death. Some stellar deaths are long and lingering others spectacular and explosive. All their stories are contained in the nebulae that mark their locations.

Open Star Clusters

Star clusters generally speaking come in two types, *open clusters* that reside within the Milky Way and *globular clusters* that surround the galaxy in a halo. Open clusters are associations of stars that likely all condensed at the same time from a common cloud of interstellar dust and gas, thus all the member stars share approximately the same orbit around the galactic center. To astronomers, stars in an open cluster are worth studying for some of the attributes that they always share in

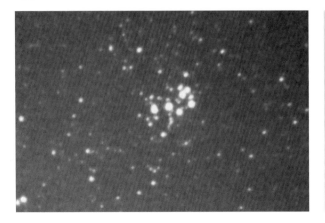

Figure 13.2. Open star cluster M45. Image by author using 35-mm SLR Camera, five-minute exposure.

common and the things that make them different. All the stars are of the same age, all of the same physical distance and all have similar chemical composition. The only significant property about the stars of an open cluster that differs is their mass. The mass of the stars in an open cluster can run from extreme to extreme. Some stars are dozens of times more massive than the Sun while others are less than a tenth of the Sun's mass. Most stars in open clusters are relatively young with the average being about 100 million years. The reason why is simple orbital mechanics. Stars in the cluster that reside closer to the galactic center, even if the difference is just a few light years, will orbit the galactic center faster and those farther away from the galactic center orbit slower. Other factors such as gravitational interactions with massive stars or other cluster members or interstellar dust clouds also act to pull cluster members apart. The result of all this is that over time, the inner part of the cluster pulls away while the outer part falls behind and the cluster becomes elongated until all of the stars become galactic *field stars* that orbit the galactic center alone in space. Very few open clusters have an age that can be measured in billions of years. Most of the stars of an open cluster are likely all born in the same stellar nursery and spend the first several hundred million years of their lives traveling together until gravitational interactions pull them apart. Open clusters are probably the first deep sky objects to be identified for their true nature. Open clusters such as the Pleiades, the Beehive and the Hyades have been known since antiquity and even the open cluster M7 has been viewed as early as AD 138 by Ptolemy. Galileo was able to resolve the open clusters into stars with his first telescopes in 1609. The telescope then allowed many more open clusters to be identified. By the time the bulk of the Messier catalog was compiled in 1782, the list contained twenty-seven open clusters and some thirty-two others were known. Today we know of more than 1,100 open clusters in our galaxy and by some estimates there are as many as 10,000 clusters.

Not all open clusters are created equal, so naturally a system is needed to type and classify open clusters. The American astronomer Harlow Shapely was the first to develop a simple system to classify clusters grading them "c" through "g" starting from loose and irregular and gradually building to very rich and concentrated.

Later, a more sophisticated system was developed by R.J. Trumpler that graded open clusters in three categories: concentration, range of brightness and richness. Concentration is graded on a scale of I through IV with grade I for a "detached" cluster that stands out strongly from the background field stars to IV, which can barely be identified from the surrounding stars. Range of brightness is graded 1 through 3. A "1" denotes relatively uniform brightness while a "3" means that the cluster has a wide range of bright and faint stars. Richness is given a grade of "p," "m" or "r" meaning poor, moderate or rich. A poor cluster has less than fifty stars typically while a moderate cluster has between fifty and a hundred and a rich cluster has more than a hundred. In addition if the cluster has an "n" after its listing, it means that there is nebulosity associated with the cluster.

Among open clusters none are more compelling or beautiful than M45. The cluster can be easily found by locating the bright red star Aldebaran and looking a few degrees to its northwest. It has been known since times of antiquity. It was mentioned in Homer's *Ilias* as far back as 750 BC and again in the *Odyssey* (720 BC). It is known by several names throughout the world. The Persians called it *Soraya* and an ancient Iranian empress was named for it. The Japanese know it as *Subaru* and the name for the popular line of automobiles is taken from the famous cluster. In fact the company logo is a representation of the famous cluster. In the western world it is called the *Pleaides*. There are several possible derivations of the name such as from the Greek words for "sail" or the word *pleios* meaning "full" or "many." Some believe it is derived from the mythological name Pleione, the mythical mother and taken as one of the names of the brighter stars. It is also ideally placed for viewing in the northern hemisphere, high in the sky during the crystal clear winter evenings of February. The cluster is a classic I, 3, r, n open cluster. The cluster shines with a total light equal to magnitude +1.6 and its single brightest star, Alcyone shines at magnitude +2.9. It is a bright and concentrated cluster with a wide variety of stellar brightness. The cluster contains more than 500 members, though the cluster is most famous for its seven brightest naked-eye stars. In addition, the cluster is also enshrouded in a bluish nebulosity. Though the cluster is young, at 100 million years of age it is average for an open cluster, the nebulosity is not remnants of the dust and gas cloud from which the cluster was formed. The gas and dust in the Pleiades is actually traveling in a completely different direction in space and the cluster and gas clouds have coincidentally wandered across each other's paths. The distance of the Pleiades has often been in dispute, but is generally accepted to be around 400 light years. Observations using the Hipparcos spacecraft yielded a distance of 380 light years, but follow-up observations using telescopes at Mount Wilson and Mount Palomar and the Hubble Space Telescope have given a new distance that accounts for the faintness of many of the cluster's stars, 440 light years, plus or minus six. In Chapter 1, we discussed how many stars might be viewed with the unaided eye; later on in this chapter we will survey how many of the cluster's members can be viewed with a telescope.

Many of the cluster's stars rotate rapidly, which is a very common trait of main-sequence A and B stars. As a result, many of the cluster's members are not spheres but rather are oblate and therefore prone to spinning off large amounts of material. Pleione in particular is known to have done this during the period between 1938 and 1952, blowing off a large shell of gas. Pleione is variable between about magnitude +4.77 and +5.50.

In the summer months another well-known open cluster comes into view. Near the tail of Scorpio is the open cluster M7. This is another cluster that has been known about since ancient times. It is situated near the tail of Scorpio. You can simply follow the body and tail of the Scorpion from bright red Antares to find M7. M7 is classified as I, 3, m. The cluster shines with a total light of magnitude +3.3 from a total of some eighty stars brighter than magnitude +10. The cluster spans about 80 arc minutes across its largest dimension. At a distance of some 1,000 light years, that would make the cluster some 25 light years across. The cluster is a bit on the old side for an open cluster at an estimated 220 million years. The brightest star in M7 is a yellow G8 shining at magnitude 5.6.

Right by M7 in the Scorpion's tail is the open cluster M6. It has a classification of II, 3, m as determined by Trumpler but the last value is often disagreed on. Some list it as "p" and some as "r." With about eighty members, the "m" designation seems most correct. The total light of the cluster is about +4.2. Original estimates had it somewhat fainter, but they were from the northern hemisphere where the cluster rides low on the horizon. M6 is approximately 1,600 light years distant. Both M6 and M7 are visible in the same field of view in binoculars or a wide field telescope. Though both are listed as type I clusters, they in fact look rather different when viewed side by side.

Another winter cluster of note is the Beehive Cluster, entered in Messier's list as M44. This is also a bright and distinct cluster, classified by Trumpler as I, 2, r. The Beehive has more than 250 confirmed members (by observing their proper motion). The cluster shines with a total magnitude of +3.7. The cluster spans some 95 arc minutes across (just over the width of three full moons). The cluster is 577 light years distant. M44 is old by cluster standards at nearly 800 million years. Curiously the age of this cluster and its proper motion is the same as the nearby Hyades cluster, suggesting that they originated from a common gas and dust cloud. The stellar content of both clusters is much the same. We'll study this later on.

Later on in this chapter in the observing projects we'll look at a wider variety of open clusters, both well defined and those that are poorly so.

Globular Star Clusters

The open clusters that we have studied thus far tend to be small, young, made up of a relatively few number similar types of stars, orbit within the disk of the galaxy and tend to be relatively short lived. Globular clusters are very different. They tend to be large, heavily populated and can be immensely old. One thing they do have in common with open clusters is that the stars tend to be of similar types and tend to plot fairly close on the Hertzsprung–Russell diagram. Charles Messier listed 29 of these objects in his catalogue but only was able to resolve one of them into stars. The first to be discovered was M22 in 1665, the discovery of which is credited to Abraham Ihle. In 1677, on a trip to the isle of St. Helena, Edmund Halley discovered Omega Centauri, the greatest of all globulars. Currently there are 151 known globulars associated with the Milky Way.

Spectroscopy tells us a great deal about the age of the stars within. Globular clusters appear to be much lower in heavy elements than is the Sun. This is significant

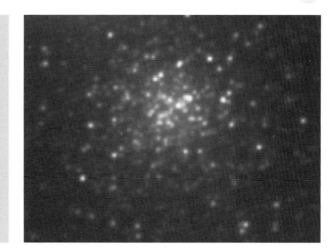

Figure 13.3.
Globular cluster M13. Image by author. sixty-second Meade DSI and Celestron Super C8 Plus at f 6.3.

because it tells us that the stars were formed before many of the heavy elements could be created in the early billions of years of the life of the universe. Plotting the position of stars within globulars shows that very few of the stars in any given cluster belong to the main sequence, which in turn also tells us that the stars are primarily very old.

Globulars also move in a very different manner than do open clusters. Open clusters orbit within the galactic disk. Globulars surround the galaxy's core in a halo and move in highly eccentric elliptical orbits that can take them many hundreds of thousands of light years away from the galactic center. This is reflected by studies of their proper motion that reveal globulars advancing and retreating from our local area at speeds of hundreds of kilometers per second. Distribution of globulars is also very different from open clusters. Open clusters are located all around the sky. Globular clusters are concentrated in a relatively small area of the sky. Of the 151 known globulars of the Milky Way, 77 of them are located in only three constellations; Sagittarius (33), Ophiuchus (25) and Scorpius (19). No other constellation has more than seven globulars and 45 of the 88 have no globulars at all. The globulars tend to be in greatest proportion around the galactic center. In fact if you divide the sky into two hemispheres, with the galactic center at the pole of one, you would find that of the 151 known globulars, 138 (91.4%) are located in the hemisphere facing the galactic center while only 13 are located on the opposite side. Of those 13, four of them (including M79) are believed to be members of the Canis Major dwarf galaxy, discovered in 2003, which is currently being gravitationally shredded and integrated into the Milky Way.

Globular clusters, like open clusters have only a limited lifetime although it is exponentially longer than it is for open clusters. Within the clusters themselves, gravitational interactions between stars eject stars from the clusters. As the clusters swing through the galactic plane near the periapsis of each orbit, tidal forces tear stars away from the outer portion of clusters. Globulars may also encounter other greater masses in their travels such as large dust clouds or other globulars that in turn may also raise tidal forces. Another force acting against globulars is that they exhaust their supplies of gas and dust in star formation, resulting in mass loss as their member stars age and die. These factors lead to globulars dying through a process called *deflation*. Over time a globular cluster's member stars and other mass is completely depleted, scattered into the galactic halo. This destructive process has reduced a population of many thousands of globular clusters around the Milky Way down to perhaps 200 or so. The process of destruction continues and probably half of the remaining clusters will be destroyed within the next ten billion years.

Oddly enough while globulars are extremely old in the Milky Way and in our large neighbor, the Andromeda Galaxy, there are globulars in other local area groups that are very young, particularly in the Large and Small Magellanic Clouds. This is not too surprising since these irregular galaxies are very young. But the spiral galaxy M33 also has a population of very young globulars. This is surprising, since a well-organized galaxy is by definition middle age since organization into a spiral form requires many billions of years.

Harlow Shapely, who developed the first simple system for cataloging open clusters developed along with Helen Battles Sawyer-Hogg, the system we use today for classifying globular clusters. A cluster is assigned a Roman numeral I through XII. The system is referred to as the Shapely–Sawyer system as it was developed before Helen Battles Sawyer was married. The higher the number, the less concentrated the cluster is. The more stars and more tightly concentrated they are, the lower the number is.

Perhaps the most beautiful globular in the sky easily accessible to northern hemisphere viewers is the Great Hercules Cluster M13. M13 shines in our sky at magnitude +5.8 and spans about 20 arc minutes in a telescope. The cluster at a distance of 25,000 light years thus spans 125 light years across. The cluster is Shapely–Sawyer class V. M13 is exquisitely old. Some estimates of its age used to be as high as 24 billion years, though more modern studies place its age at around 14 billion years. This is not much younger than the modern estimate of the age of the universe. At declination +36 degrees, it passes just about directly overhead during summer months for northern hemisphere observers making for perfect observing conditions. Observers with large aperture telescopes can resolve M13 deep into its core on nights with clear seeing. M13, like most globulars is at its best in summer months since globulars as we already know tend to congregate in the half of the sky centered on the galactic center.

Shining much lower in the sky but just about as bright at magnitude +5.6 in Scorpio is the class IX globular cluster M4. Conveniently located just 1.3 degrees from Antares just south of the line that connects it with Sigma Scorpii, this globular might well be far more spectacular than M13 were it not for the low altitude in the sky and because heavy interstellar dust clouds obscure our view of it. The dust also shifts the color slightly towards the red, just as a setting Sun is reddened by passage through a long stretch of atmosphere, so M4 is often noted to have a slight orange-brown appearance. The cluster is about 36 arc minutes across. At a distance from Earth of 7,200 light years, its diameter spans some 75 light years. This is actually slightly larger than the volume that the cluster should be capable of gravitationally controlling, so M4 is losing stars to the galactic halo at a high rate. The cluster is one of the most diffuse in the entire sky. At class IX, its compressed central core is only about 1.66 arc minutes in diameter. The cluster's half-mass radius is only eight light years, meaning that half the mass of the cluster is located within eight light years of the center.

For those taking a trip south, don't pass up the chance to see mighty Omega Centauri. This cluster is so bright that it was once mistaken for a star, hence the star-like name. But it is in fact a conglomeration of some five million times the mass of the Sun. The cluster, which is cataloged as NGC 5139 (Messier could never see it from France), is the largest cluster orbiting the Milky Way by a factor of ten. This class VIII cluster is also unusual in that it does not appear to have formed all at the same time, unlike other globulars. It is also as massive as many small galaxies. The cluster's stars are also not of a uniform age, but were formed over a span of some two billion years with several bursts of star-forming activity. These facts suggest that Omega Centauri did not begin life as a globular cluster, but rather was once the core of a modest galaxy since destroyed by gravitational interaction with the Milky Way. The cluster is 16,000 light years distant, shines at a magnitude of +3.68 and spans some 36.3 arc minutes making it bigger in the sky than a full moon. In the Observing Projects for this chapter, we'll revisit these three globulars and some others of different visual properties.

Planetary Nebulae

The term *nebula* has had several different meanings throughout the history of visual astronomy. Originally this term was generically applied to any object in the sky that was non-stellar in nature including globular and open clusters. Most of the objects identified as nebulae during the pre-telescopic age have now been shown to be open star clusters. The telescope has shown us that nebulae are very different in character from one to the next depending upon how they were formed and how they are illuminated. There are four basic types of nebulae: *planetary nebulae, diffuse nebulae, supernova remnants* and *dark nebulae.*

Charles Messier is credited with discovering the first planetary nebula when he found the Dumbbell Nebula (M27) on July 12, 1764. Antoine Darquier is credited with the discovery of the Ring Nebula (M57) in January 1779 while he and Messier were both tracking a nearby comet. Pierre Mechain later found the Little Dumbbell Nebula (M76) in 1780 and the Owl Nebula (M97)in 1781. These are the only four planetary nebulae to make their way into the Messier catalogue. William Herschel invented the name itself after he began to uncover more and more of them with his large telescopes. Herschel found the fifth planetary in 1782, which he called the Saturn Nebula. To Herschel, the little nebula strongly resembled the faint disk of the planet he had just discovered, Uranus.

A planetary nebula is formed by a dying star of Sun-like mass. As the star exhausts its hydrogen and then fuses all its helium into carbon and oxygen, fusion finally ceases except for within the outer helium shell. The star at this point becomes a long-period Mira-type variable as strong stellar winds blow away the star's outer layers. As those outer layers depart the star, the hot remnant of the collapsed core burns on as a degenerate white dwarf emitting high levels of radiation. This radiation excites the gases of the departing gas shell to glow. Since the nebula glows as a result of light created by the excited gas, it creates emission lines in its spectrum, thus these nebulae are also called *emission nebulae*. Nearly all the visible light created by planetary nebula, between 90% and 95%, is centered on 5,007 Angstroms, right in the heart of the visible green area. This is often called the *chief nebular line*. Because the human eye is most sensitive to light near the blue end of the spectrum, planetary nebula in many cases are easier to see than they are to photograph. A prototypical planetary shines about two magnitudes brighter visually than it does photographically. A planetary nebula is also very difficult to get a true color photograph of because while the eye is very sensitive to green light, most films are not. Most planetary nebula light is from the forbidden line emissions of doubly ionized oxygen or [O III][38], with strong hydrogen beta emissions and some helium as well.

As planetary nebulae age, the gases that compose them scatter into interstellar space and the nebula fades away or the white dwarf remnant cools to the point that it can no longer excite the gas and the glowing of the nebula stops. Because of this, planetary nebulae tend to have short life spans. They generally are only visible for a few thousand to perhaps as long as 100,000 years. This is perhaps the reason that while there are many tens of billions if not hundreds of billions of Sun-like stars in our galaxy, there are only about 1,500 planetary nebula that have been detected. This short life span also explains why there are so few planetary nebula found in open clusters, since an open cluster will likely dissipate long before any of its member stars can age enough to create a planetary nebula. In addition, there have only been four planetary nebulae found within globular clusters.

Planetary nebulae are categorized according to a schedule that assigns a number 1 through 6 to each of six different attributes that a planetary might have. One is for a stellar image, two is assigned to nebulae that show a smooth disk. An add-on of "A" is given to nebulae that are brighter towards the center, "B" to nebulae that show uniform brightness and "C" to nebulae that fade towards the center indicating a partial ring-like structure. Three is assigned to nebulae that show an irregular disk. An add-on designation of "A" denotes a nebula that is very irregular in brightness distribution while a "B" denotes the presence of ring-like structures. A four denotes full ring structure. A five is assigned to nebulae that show an irregular form. Six is assigned to nebulae that show an anomalous form. A nebula may be given more than one designation if for example it shows more than one of the listed traits. For example if the nebula displays a smooth appearance with a full ring structure, it will be designated 4 + 2c.

The nebula most commonly associated with planetary types is the Ring Nebula (M57) in Lyra. The nebula shines at magnitude +8.8 in the southern area of Lyra. M57 is very easy to find about one-third of the way along the line from Beta Lyrae to Gamma Lyrae at the base of the lyre. The nebula is classified as 4 + 3 and is located at a distance of 2,300 light years. We view the nebula from above the pole of the dying star. Studies have since shown that the nebula's ring shape is not really a ring but a function of our point of view. Viewed from a different angle we would see that

---

[38] This would be an oxygen atom that has had two of its electrons stripped away. The "O" represents oxygen. A Roman numeral one indicates a neutral atom with eight protons and eight electrons. Each successive numeral represents one lost electron. The brackets represent the *forbidden line*. This represents the presence of light that cannot be viewed from Earth but shows up spectrographically.

the nebula is more torus-shaped and would look more to us like the Dumbbell Nebula. Twenty-one years after the nebula's discovery, the central star was found by German astronomer Freidrich von Hahn. The star is a planet-sized white dwarf that has completed blowing off it outer layers during its Mira-type variable dying stage and now is undergoing its final cool down into a black dwarf. Determining the age of the nebula is a fairly simple matter of noting its rate of expansion compared to its size. The nebula is expanding at about one arc second per century, so given a size of 60–80 arc seconds the nebula is some 6,000 to 8,000 years old.

M27, the Dumbbell Nebula, was the first planetary nebula to be discovered by Charles Messier in 1764. This nebula is much like the Ring Nebula, but observed from a different perspective, looking at the plane of the ejecting star's equator, rather than from over its poles. The nebula is much larger than the Ring Nebula at 15 arc minutes by 6 arc minutes. It shines at magnitude +7.4, which makes it only slightly fainter than the Helix Nebula (+7.3). The Dumbbell is very close to Earth at a distance of only 1,250 light years, though there is considerable uncertainty about this. The value is derived upon CCD imaging techniques used to derive a trigonometric parallax. Observations with the Hubble Space Telescope will hopefully nail down the distance with greater accuracy, if it survives long enough to complete the project. Other existing estimates range between 490 and 3,500 light years. The nebula may or may not be even younger than is the Ring Nebula, less than 4,000 years old based on Soviet astronomers' estimate of its rate of expansion of 6.8 arc seconds per century. Other estimates suggest the expansion rate is only 1.0 arc second per century. This would yield an age much older, on the order of nearly 40,000 years. This estimate is considered more likely because the original estimates of rate of expansion were based on a distance of less than 750 light years. M27's originating star is an 85,000 K temperature blue-white dwarf that shines at magnitude +13.7, spectral type O7.

Planetary nebulae are the grave markers that show the location of dying embers of stars that have reached the end of the fusion portion of their lives and have blown off their outer shells in a final puff as the core degenerates into a white dwarf. While this end is not the final blast of glory of a massive supernova, it leaves behind a beautiful monument to itself that will endure for a comparatively short period of time but during that time are avidly sought after targets by amateur astronomers. They present us with many mysteries, such as the difficulty of estimating their distances and for years they baffled us as to the source of their brilliance. They shine brighter to the eye and more in their true color than they do on film. Many are bright enough to make easy and beautiful targets for telescopes of just about any size. Now let's go and take a look at nebulae that mark stars at the other end of their lives.

Diffuse Nebulae

Diffuse nebulae are thin conglomerations of dust and gas that if they acquire sufficient mass may begin to form stars within them. Some may form a handful of stars while others may form stars by the hundreds, creating large open clusters. Diffuse nebulae therefore are not stellar tombstones, but nurseries, where stellar lives begin. As stars begin to shine from within the dust cloud, if they are sufficiently hot and massive, their high-energy radiation will excite the gas to shine just as they do in planetary nebulae. These types of nebulae are also called *emission nebulae*. If local temperatures are not hot enough to excite the gas, the dust in the cloud will reflect the light of the stars, creating what is called a *reflection nebula*. Many emission nebulae also have an outer area associated with them that is also a reflection nebula. Diffuse nebulae are also often called H II regions because they principally consist of ionized hydrogen (hydrogen atoms that have lost their electron).

Diffuse nebulae, like planetary nebulae tend to have relatively short lifetimes because the gas and dust that it is created of will either be consumed into the newborn stars within it or be blown away by the hot stellar winds of those same new born stars. Eventually all the matter contained within the stars is blown away to space or accreted into the stars or accompanying planets (something we are discovering more and more stars have). After all the stars have moved clear of the gas and dust clouds or after the cloud has been completely dispersed, all that remains is a new open cluster of infant stars beginning their main sequence lifetimes. A typical lifetime for a diffuse nebula is about a million years, though that number can vary wildly depending upon the rate of star formation and the mass of the stars created.

The Orion Nebula (M42), visible to the unaided eye, has been known since antiquity and was observed telescopically very shortly after the invention of the telescope in the year 1610. Its nearby companion M43 was discovered by telescope in 1731, then the Omega, sometimes called "Swan" nebula (M17) in 1745. Until the 1860s all these wondrous sights were thought simply to be unresolved star clusters until the emerging technique of spectroscopy revealed their gaseous nature. Five of the six diffuse nebulae in the Messier catalog are primarily emission nebulae, though some do have reflection components. Only M78 is a pure reflection nebula. There are however many fine examples of reflection nebula that are not in Messier's catalogue. Take another look at the star field picture at the front of Chapter 12 and see the gas cloud with the outline that strongly resembles North America (and is called by coincidence the North America Nebula). This gas cloud shines primarily from light reflected from the massive nearby star Deneb, as does the Eagle Nebula slightly to the south and west (right). Another such area of nebulosity is easily imaged around Antares in Scorpius.

Just 1,600 light years away and hanging from the belt of Orion near the celestial equator, which happily allows the nebula to be viewed from anywhere on Earth, lies the brightest of the diffuse nebulae in our sky, the magnitude +4.0 Orion Nebula. Orion, located about 1,600 light-years from Earth, is a vast star-forming region that delights all who view in the smallest of binoculars to the mightiest of observatories on Earth and in space. The visible portion of the nebula extends to about 85 by 60 arc minutes and is many light years in size. It is the most easily visible part of a much larger dust and gas cloud that spans more than 10 degrees of sky, spreading over many hundreds of light years. The overall span of the complex contains several other areas famous on their own including Barnard's Loop and the Horsehead Nebula. The Orion Nebula is on its own so large, bright and rich in detail that many astronomers refer to various features by name as though they were discussing topographic features of the Moon or Mars.

M42 is a massive star-forming region and in the middle of the nebula, in an area where gases are clearly being blown away by stellar winds (called the Fish's Mouth) are four bright stars that make up a formation known as the *Trapezium*. Huygens found the first three members in 1656 and Abbe Picard found the fourth member in 1673. Over the next two hundred years, four more faint members of the cluster were found. The cluster members are designated "A" through "H." Today we know that both the "A" and "B" stars are both Algol-type eclipsing variables. The "A" star varies between magnitude +6.7 and +7.5 each 63 days. The "B" star varies between magnitude +7.9 and +8.5 each 6.5 days.

As the Orion Nebula graces the winter's skies so the Trifid Nebula (M20) graces the summer in the constellation Sagittarius. The nebula was so named by John Herschel because of its three-lobed appearance. Like with many nebulae, the distance to the Trifid is somewhat uncertain with published estimates ranging from 2,200 light years to over 9,000. The beauty of the Trifid Nebula lies in dichotomy of colors, a reddish emission nebula contrasting with a conspicuous blue reflection nebula that is particularly bright surrounding the north side of the emission nebula. Magnitude estimates of the nebula also vary widely. Brightness estimates range from magnitude +6.8 to +9.0. The brighter magnitude estimates are probably largely based upon the brightness of the triple star system at the center of M20. The three stars shine at a combined magnitude of +7.0. The brightness of the triplet makes the task of estimating the overall brightness of the nebula difficult.

Nearby M20 is an even larger nebula, M8, also known as the Lagoon Nebula. When Messier originally described this area, his description more resembled the nearby star cluster and in fact his stated position more closely defines the position of the cluster than the nebula, but the nebula is generally today considered to be M8. The open star cluster is designated NGC 6530. The Lagoon Nebula shines at an estimated magnitude +6.0 and estimates of its distance range from 4,800 to 6,500 light years. The nebula is very big, spanning some 90 by 40 arc minutes. If we use the commonly accepted distance of 5,200 light-years, this would imply at that distance that the nebula is about 140 by 60 light years in size. Near the center of the nebula is a bright region that is called the "Hourglass Nebula." The intrinsically bright young star Herschel 36, a spectral type O7, magnitude +9.5 celestial toddler, illuminates the Hourglass. The brightest star associated with the nebula is magnitude +5.9, spectral type O5 9 Sagittarii. Both of these nebulae are also active stellar nurseries, cranking out young stars to wander about the galaxy.

These three nebulae represent half of the only six diffuse nebulae that appear in the Messier catalogue. Today we have catalogued many thousands of diffuse nebulae throughout our galaxy and in many others. They are sights of awesome multicolored beauty and show us the very beginnings of stellar evolution. In the planetary nebulae, we have viewed the relatively peaceful ending of Sun-mass stars. Some nebulae represent the spectacular deaths that await the most massive of stars.

Supernovae Remnants

On the fourth of July, routinely our skies erupt to life with beautiful displays of fireworks but none quite like the one that occurred 722 years before the signing of the Declaration of Independence. On July 4, 1054, Chinese astronomers noted the appearance of a "guest star" in the constellation Taurus. The Chinese took very meticulous records of changes in the sky and authentically recorded some seventy-five such stars between 532 BC and AD 1064. The star flared four times brighter than Venus, reaching magnitude −6.0 and remained visible in daylight for over three weeks and was visible to the unaided eye for nearly two years.

Supernovae events in any given galaxy are very rare events. If we limit the accounts to only those events that we are absolutely certain about, there has not been a supernova in the Milky Way galaxy for over three hundred years and only six in the last millennium (though supernovae in many parts of our galaxy would be obscured or even completely hidden by dust clouds). When one flares, it can emit more light than its entire host galaxy. In addition to the six in the Milky Way, a supernova was observed in 1885 in Andromeda (M31) and one in 1987 in the Large Magellanic Cloud. After time, the star fades away into the darkness but it leaves something beautiful behind it. The remnant gases begin spreading out and form a nebula marking where the cataclysm took place. The British astronomer John Bevis found the remnant of the 1054 supernova in 1731 and Charles Messier found it independently in 1758 while seeking Comet 1P/Halley.

Lord Rosse's drawings of the nebula in 1844 led to its popular name, the Crab Nebula (M1) being generally accepted in the astronomical community. Late in the nineteenth century, spectrographic observations revealed the gaseous nature of the nebula and led to its being classified as a planetary nebula, a classification that was not disproved until 1933. One fascinating discovery was the nebula's amazing expansion rate. At two-tenths of an arc second per year, the nebula was judged to be very young, about 900 years old. With an apparent size of 6 by 4 arc minutes and a distance of 6,300 light years, this makes a nebula that is more than 10 light years across! Further studies of historical records tied the nebula's location to that of the supernova of 1054. The nebula is now known to consist of matter ejected in the supernova explosion and that its properties are very different from planetary nebulae. The Crab Nebula emits light across a wide area of the spectrum whereas planetary nebulae emit almost all their light from one wavelength of the spectrum. The Crab Nebula emits two different types of visible light, red light from hydrogen emissions and a bluish light caused by high-energy electrons moving rapidly through a strong magnetic field. This emission of light is called *synchrotron radiation* and is a property not present in planetary nebulae.

In 1948 astronomers also discovered that the Crab Nebula was a strong source of radio emissions. In 1963, a small rocket-borne x-ray detector showed that the Crab was also a strong source of x-ray radiation as well. Then in 1968 came the most shocking discovery. The Crab Nebula was also a source of pulsating radio energy, emitting a blast of energy on a very precise and regular schedule, flashing once each 33.085 milliseconds. Astronomers had discovered a special type of neutron star called a *pulsar*. The star is very rapidly rotating each 33.085 milliseconds and each rotation brings a hot spot on the neutron star's surface into our view. Later on it was realized that the radio source corresponded with the position of a magnitude +16 star at the center of the nebula that brightens slightly in visible light at the same period as the radio pulses. Indeed the star is also an optical pulsar and has been given the variable star designation CM Tauri. At magnitude +16, by the way from the Crab's distance of 6,300 light years, the absolute magnitude of the pulsar would be about +4.5, or slightly brighter than the Sun. In x-ray radiation is where the pulsar really shines though, producing 100 times the luminosity of the Sun. All of this luminance comes from the rapidly spinning remnant of the 1054 supernova explosion; a remnant that contains mass equal to the Sun yet measures less than thirty kilometers across.

The Crab Nebula represents one of the most magnificent tombstones in all the heavens, a beautiful testament that glows virtually across the entire electromagnetic spectrum to the amazing power that was displayed there, to the power of the creation of every naturally occurring element in the universe heavier than iron. It has spread across more than 10 light years of space in only 950 years. Can the human mind even conceive of the power required to propel matter at that speed? The Voyagers and Pioneers will require tens of thousands of years to reach the closest local area stars even traveling over 100,000 miles per hour! One of your most powerful observing tools is the power of your imagination, fueled by the knowledge of what created that little gray patch of light in your telescope.

Dark Nebulae

A dark nebula is a cloud of dust and gas as all nebulae are and are similar to reflection nebulae. The main difference is the perspective of the observer. If the illuminating star is in front of the dust cloud, then we see a bright reflection nebula, but if the cloud is in front of the star and the star's light cannot penetrate the cloud, then a dark nebula is formed.

The most famous example of a dark nebula is the Horsehead in Orion. Located north of the Great Orion Nebula, the Horsehead is a protruding dark cloud that forms the silhouette of a horse's head. The nebula sits near the gas cloud that forms the reflection nebula IC 434. The nearby star Sigma Orionis illuminates the nebula. The gas that forms both the dark nebula and the reflection nebula are probably the same dust and gas cloud. That portion which forms the Horsehead sits between Earth and Sigma Orionis while that area which forms the reflection nebula sits behind it. The geometry of this arrangement is not unlike what causes Mercury or Venus to show us more and more of their dark sides as they move from behind the Sun to a position in front of it.

Dark nebulae are among the most difficult objects in all the heavens to observe because frankly, they're dark. Your only hope of being able to find one is if the source that silhouettes it from behind close enough and bright enough to create the required contrast without being so bright that it washes out the dust cloud. Finding such nebula can be the ultimate test of your observing skills.

Now let's take what we learned about distant groupings of stars and the various nebula that represent the birthplaces and tombstones of stars and go and explore them. We'll see what makes them special and why they are important to us. Open clusters represent some of the youngest stars in the galaxy and diffuse nebulae represent the birthplace of those young stars. Planetary nebulae show where stars of modest mass puff away most of their mass as they die a relatively placid death. Supernovae remnants are where the universe's most cataclysmic deaths take place. Let's go explore some of them.

Observing Projects XIII – From Nurseries to Tombstones, Life and Death in the Galaxy

Observing Project 13A – Open Star Clusters

Let's start by taking a gaze at some of the sky's most attractive open clusters and what they offer to the stargazer as we look around the sky.

The sky's most beautiful open cluster and the brightest entry in Messier's catalogue (+1.6) is the Pleadies cluster (M45). In Chapter 1, we listed out in detail each of the seven brightest naked-eye stars. The Pleadies offer far more to see through the telescope. By some estimates, the cluster contains more than 500 stars. Using a low-power eyepiece, preferably a wide field type, see how many members of the Pleaides you can see with your telescope. Also as you look at the cluster, can you make out the bluish reflection nebula that surrounds the star cluster? The cluster's nebulosity remember is not part of the cluster's birth cloud but rather is a chance crossing between the cluster and a passing gas cloud, each of which exhibits a different proper motion. The cluster is Trumpler-type II, 3, r. The designation means detached (stands out from the background), a wide variety of stellar brightness and richly populated. Do you agree?

High in the sky as darkness falls in March is the open cluster M35, near the feet of Gemini. The cluster is just bright enough to be visible to the unaided eye under very dark skies. This cluster is approximately 2,800 light years distant and shines at magnitude +5.8. The cluster spans about 25 arc minutes. M35 is listed as Trumpler type III, 3, r. The cluster has a wide variety of star brightness and note the color of the stars. Many are yellow and orange, indicating that the cluster is of roughly average age for an open cluster, maybe about a hundred million years. It does not stand out as clearly from the background as do some of the others open clusters that we've discussed thus far. Use the widest field eyepiece you have available because M35 offers more treats if you're prepared and know what to look for and where it is. Just 15 minutes to the southwest of M35 is the small open cluster NGC 2158. This tiny cluster is only 5 arc minutes across and shines at magnitude +8.6. I estimate it as being Trumpler-type II, 1, r. It is so compact that at one point it was actually considered to be a globular. The cluster is also far distant in the background at approximately 16,000 light years distant. The open cluster of young stars makes a fascinating contrast with the much older cluster of cooler stars in the distance. And there is yet more for owners of larger telescopes. More to the west of M35, by about 50 arc minutes is the open cluster IC 2157. The cluster is about the same size and brightness as NGC 2158 but is very star-poor and loosely associated. With a field of view of about 1.5 degrees and sufficient aperture, you should be able to see all three.

Rising nearly half-way to the zenith during the months of June and July in Serpens is the Eagle Nebula and forming from it is the open cluster M16. The nebula and cluster lie about 7,000 light years away in the next arm inward in the

Milky Way. M16 has a Trumpler classification of II, 3, m, n and with an angular diameter of about 7 minutes spans a linear distance of about 15 light years. The much larger nebula spans some 70 by 55 light years. The cluster is extremely young at an estimated age of less than six million years old. Star formation is still on going in the cluster and the nebula. This is one of the youngest formations you will ever lay eyes on in the sky. In a 4-inch telescope, you should be able to make out about twenty members of the cluster in a dark sky. The members of the cluster are extremely hot and bright, most of spectral type O6. This makes the cluster shine at an absolute magnitude of −8.6; perhaps the brightest cluster in the sky. The Eagle Nebula and M16 together is one of the heaven's most spectacular examples of babies leaving the womb and beginning their travels through the universe.

Observing Projects 13B – Globular Star Clusters

While open clusters are visible year around, M13 and nearly all the globulars have a distinct viewing season. Since they are gathered around the galactic center they are most prominent when the center of the galaxy is in opposition to the Sun. This occurs in late June when the Sun is in Gemini and Sagittarius is at opposition. A majority of the globulars are located in Sagittarius and two adjoining constellations. So globular hunting is a very popular summertime activity for amateur astronomers.

M13 is not only the most beautiful of the northern hemisphere globulars but also one of the oldest structures easily visible to the amateur. It is some 14 billion years old and may contain as many as a million stars. M13 is very easy to find on the line connecting the two stars that make up the west side of the Hercules keystone. Note the color of the stars within the cluster. Nearly all are yellow and orange in color indicating that the stars within are very old main-sequence stars of what are called "Population II" stars. With globular clusters, use all the magnifying power the weather conditions will tolerate and see how close to the center of the cluster you can resolve individual stars. M13 is categorized as a category V globular, is heavily populated and very concentrated towards the center. Test the Shapely–Sawyer system out and see if you agree.

As summer moves towards its later months and the lackluster constellation Aquarius comes to the meridian at midnight, you will have your best chance to observe the tightly packed category II globular cluster M2. This cluster shines at magnitude 6.5 and spans 16 arc minutes across. At a distance of about 37,500 light years, the cluster is huge, spanning some 300 light years. The measure of how tightly packed this cluster is would be through the measure of *half-mass radius*. This measure for M2 is only 0.93 arc minutes. This means that in a globular cluster that spans 16 arc minutes (a radius of 8 minutes) half the mass of the cluster is compacted within less than one arc minute of the center. This in turn means that the cluster's tidal radius is enormous, about 233 light years. The cluster would likely capture any object passing within 233 light years of that center. You can see very easily that the structure of this cluster is very different from M13. Finding the cluster is very easy. Find Alpha Aquarii with an equatorially mounted telescope and scan west to find M2. It is 5 degrees due north of Beta Aquarii.

Nearly all the globulars in the heavens are in the half of the sky centered on the core of the Milky Way. One of the few that is not is M79. This globular in Lepus is also therefore one of the few that is best viewed in winter. The cluster is located some 40,000 light years from Earth, but 60,000 light years from the galactic center. The cluster is magnitude 7.7 and spans just a bit less than 10 arc minutes. This makes for a linear extension of 118 light years. The cluster is rated as a category V globular, the same as M13. We know that M79 is not a native to our galaxy but rather was a member of the Canis Major dwarf galaxy, a small irregular galaxy that was once a satellite of the Milky Way and is being gravitationally destroyed and consolidated by our galaxy.

Most visitors to Lyra go seeking the double-double Epsilon Lyrae or the famous Ring Nebula M57. But Lyra is also host to a faint, but interesting globular listed as M56. This little globular is graded as category X on the Shapely–Sawyer system. The cluster is barely three minutes across in most amateur telescopes and shines at magnitude +8.3. In larger instruments, the total extent of this cluster is about 9 arc minutes. The cluster is a long way from Earth at more than 34,000 light years.

These four Messier globulars show you the full range of qualities of globular clusters from large to small and from dense and compact to loose and sparse. All of these clusters are home to stars of extreme age, including many RR Lyrae type variables. Globular clusters teach us what the oldest stars in the universe are about.

Observing Project 13C – Planetary Nebulae and Supernovae Remnants, the Stellar Tombstones

In this project, we will look more closely at two prototypical nebulae that represent two very different types of stellar deaths. M57 in Lyra is the one of the best-known examples of a planetary nebula while M1 is the only supernova remnant on Messier's famous list.

M57 is a summer time object in bottom of Lyra and you will be able to locate the Ring Nebula sitting right on the line between Beta and Gamma Lyrae about a third of the way from Beta. The class 4 + 3 nebula is 2,300 light years from Earth and spans about 60 by 80 arc seconds indicating that at its currently known expansion rate the nebula is some 6,000–8,000 years old. It was formed by a Sun-like star puffing away its outer layers during the final collapse of its core when helium fusion ceased some 6,000–8,000 years ago.

Because the nebula's light is concentrated into a relatively small area, the surface brightness of M57 makes it fairly easy to see despite its relatively weak magnitude of +8.8. When you observe M57, remember that you are looking at the pole of a tubular shaped structure. It had been previously thought that the nebula was spherical in shape and we were seeing the Ring as the projection of the densest part. We now know that the nebula is actually shaped more like twin-lobed Dumbbell Nebula.

Observing the nebula gives us our first chance to use one of the more interesting tools in your observing bag, the light-pollution rejection filter. Most of these filters are designed to only pass certain wavelengths of light while completely blocking all others. These filters are designed to reject the emission lines of mercury vapor and sodium while enhancing the light of hydrogen-beta, hydrogen-alpha and most importantly in this case, doubly ionized oxygen. Remember that all planetary nebulae emit light primarily on that one wavelength at 5007 angstroms. Just as when we were viewing the Sun's chromosphere, it is important to remember that sometimes to see more, you actually need to see less. By enhancing the light we want to see while suppressing that light we don't want to see, we can see nebulae like M57 with a great deal more clarity and contrast.

Averted vision is a powerful tool for use on objects this faint. Use the periphery of your field of view so that the far more light-sensitive cells in the outer retina can be used to detect the detail in the nebula. If you look at the nebula straight on, you will see it, but it will appear uniform in light intensity across the entire span of the nebula. If you use averted vision, you may be able to view the fading at the center of the nebula and see a true "ring." It is not worth it really to use the color sensitive cones at the center of the eye because there is not enough light energy to see the color anyway with the less sensitive cones.

M1 is a classic example of a supernova remnant. It shines from some 6,300 light years away at magnitude +8.4 and spans about 6 by 4 arc minutes. Unlike the Ring Nebula, which emits visible light in primarily one wavelength, the Crab Nebula emits energy across virtually the entire spectrum from high-energy x-rays to low energy radio waves. It's one of the few objects in the heavens that you can both *look* at and *listen* to. It was formed in the wake of the catastrophic explosion of a massive star that was noted on Earth to occur on July 4, 1054.

The Crab Nebula is an extremely low surface brightness object. It shines with about the same total light as the Ring Nebula, but the light of the Crab is spread over a much wider area than that of the Ring. There is also no central concentration of brighter light like there is in globular or even concentrated open clusters. Any background light at all may render the Crab impossible to see. The light-pollution filter is useful here because it will suppress background light from mercury vapor and sodium emissions but it does not really do much to enhance the visible light of the nebula. Most LPR filters, whether broadband or narrowband, are primarily sensitive to the wavelengths of light we discussed on the previous page and these wavelengths are not prevalent in the Crab.

It is important to keep perspective when looking for the Crab when you actually see it. The bright pink and blue images are very tough to see and are the result of long-exposure images often taken with special filters through large-aperture professional instruments that can gather light in quantities millions of times greater than what the eye can. If you are successful in finding it, you will see nothing more than an oblong patch of gray light. The nebula to be easy to see at all requires at least a 5-inch telescope in crystal clear and dark skies. If the slightest amount of light pollution is present, then you will have a hard time viewing it with even an 8-inch scope. In small and medium size amateur scopes, the triumph is in finding it at all. To see any detail in the nebula, you need a Dobsonian "light-bucket" to bring it into view. A telescope of at least 16 inches in aperture will begin to show the nebula's beautiful structure and tendrils, at least barely so.

Observing Project 13D – The Detail in the Great Nebula of Orion

It sprawls over more than one full square degree of the sky, equal to four full moons. It is the crucible and nursery where new stars are born. In open clusters, we have viewed the beginning of the journey of infant stars. In the cluster M16 and the adjacent Eagle Nebula, we have viewed newborns leaving the womb. Now in Orion, we can see the star creation engine in progress.

The nebula is so huge and so relatively close to Earth that we can see an amazing wealth of detail in its structure. Remember that the nebula you see is just the brightest part of a much larger cloud of gas and dust that sprawls over hundreds of light-years in size and will be forming stars for millions of years to come. Astronomers have actually mapped the detail of the nebula and given names to various features as though they were craters on the Moon or Mars. In the picture on the left you can clearly see two separate gas and dust clouds, a large main cloud at the center of the picture and a small fainter cloud at the upper right. The smaller cloud is designated M43. The two nebulae are actually the same nebula separated by a dark lane of dust. The dark nebula that separates the two forms a semicircular indentation in M42 that is called the "Fish's Mouth." From the sides of the Fish's Mouth two wing shaped features spread out to the top left and middle right of the picture that are called "wings." The extension at the right (east) side of the Fish's Mouth is called the "Sword." The bright nebulosity to the lower right of the Fish's Mouth is called the "Thrust." The fainter wing on the left (west) is named the "Sail." Emerging from the Fish's Mouth is a cluster of stars called the "Trapezium." Over the years, eight of these stars have been found. A telescope of 2 inches or more will easily show the two brightest. Two more are visible to telescopes of at least 6 inches under light-polluted skies or smaller scopes under dark skies. The others are rather faint and only visible in the largest amateur telescopes. These stars are among the youngest known in the sky and may be less than 100,000 years old.

Figure 13.4. Great Orion Nebula, image by author (with some tutoring) using Meade DSI and Celestron Super C8 Plus at f 6.3. Ten-minute exposure. Processed with *Photoshop*.

Even under the most heavily light-polluted skies the Orion Nebula can awe one with its beauty. Other nebulae are not quite so bright, beautifully detailed or so well placed but with the right techniques and the right tools, other nebulae will reveal their beauty to you. LPR filters are very helpful here because they will enhance the light of hydrogen-beta, which is one of the primary wavelengths that diffuse nebulae shine in. In the summertime, don't forget to turn your scope to M8 and M20 above the Sagittarius Teapot and the Eagle Nebula, associated with M16 in Serpens. Some of these emission nebulae are too far spread out to be easily viewed in telescopes but by aiming a 35-mm camera at them mounted on a clock-driven telescope can easily show large structures like the North America Nebula, which is excited to shine by light from nearby Deneb. Just aim the camera at Deneb and lock the shutter open for ten or fifteen minutes. The nebula will be easily visible in a wide field exposure. Your star atlas is full of these sprawling star-clouds so do a little advance planning and figure out when some of these beauties are visible.

Now having looked around the galaxy at beautiful stars, clusters of astonishing youth and astonishing age and spectacularly beautiful stellar tombstones and nurseries, let's now look out beyond the Milky Way and into intergalactic space.

CHAPTER FOURTEEN

Faint, Fuzzy Things. Part II: The Island Universes

Figure 14.1. M31, the Andromeda Galaxy. Image by author using a piggybacked 35-mm SLR. Exposure time is five minutes. Look carefully and you can see M32 and M110.

Each October, it rises high overhead in the northern evening sky crossing the zenith for observers at 40 degrees north. The small nebulous patch had been known for many years. It was first noted in writing by the Persian astronomer Abd-al Rahman al Sufi who noted it in his *Book Of Fixed Stars* that was published in AD 964. Messier catalogued the nebula in 1764 and, unaware of the earlier discoveries, ascribed the discovery of the nebula to Simon Marius, who first observed it with a telescope in 1612. Marius though never claimed to have discovered it. Messier assigned the nebula the number 31 in his famous list. For many years, astronomers thought that M31 was one of the closest of the nebulae. It was William Herschel who first correctly surmised that M31 was a relatively close-by *island universe*. While Herschel's thoughts about the true nature of M31 were correct, he was way off on his estimates of its size and distance. Herschel believed that M31 was about 2,000 times the distance of Sirius (17,000 light years) and some 850 Sirius-distances in diameter and 155 Sirius-distances thick. It turns out he was off just a little bit in that regard. It was William Huggins, the pioneer of the science of spec-troscopy, who obtained the first definitive proof of what M31 was when he obtained spectra of the nebula in 1864. The nebula's spectra were continuous, rather than being dominated by emission lines like planetary nebula. M31's spectra looked more like those of the stars. We began to learn more about its nature in 1887 when the astronomer Issac Roberts photographed it for the first time and began to reveal its spiral structure. The final critical discovery that nailed down the first reason-able distance for M31 came in 1923 when Edwin Hubble discovered a Cepheid vari-able in M31. Though Hubble's estimate was incorrect by a factor of two, his discovery gave us our first glimpse of the awesome scale of intergalactic distances. Later observations beginning with the 1953 completion of the Hale Telescope at Mount Palomar provided more exact Cepheid information, pinning down the distance to M31 to about 2.9 million light years. M31 has become by far the most studied of the island universes but it was also just the beginning of our expansion into a universe far larger and more amazing than anyone could ever have envisaged.

Our Own Galaxy, Galactic Structures and Types

Our knowledge of galaxies and their structures has grown by leaps and bounds through the twentieth century including knowledge of our own galaxy. Our knowl-edge of the Milky Way continues to grow and evolve in fits and starts because in visible wavelengths we just cannot see a great deal of it. The core of the galaxy is completely hidden from our view in visible light because of thick interstellar dust clouds. We know that our galaxy is spiral in shape with several arcing structures or arms that wrap around the core. The Milky Way is some 100,000 light years in diameter and contains a mass equivalent to one trillion Suns. The best current esti-mate of the total galactic population is around four hundred billion stars. The Milky Way is now known to be spiral in layout. This was discovered by the Amer-ican astronomer Harlow Shapley, who made radio observations of hydrogen clouds in the galaxy and from that was able to postulate the spiral structure. Shapley also correctly pinned down the approximate location of the galactic center in Sagittar-ius, near the borders of Scorpius and Ophiuchus. Subsequent studies, crowned by astrometric observations using the Hipparcos spacecraft have confirmed that the solar system lies approximately 28,000 light years from the core and 22,000 light years from the edge and is almost perfectly centered within the galaxy, lying only 20 light years above the plane of the galaxy's disk, the galactic equator if you will. We know now that the Milky Way is structured like most spirals with two distinctly different components. First are the *spiral arms*, which wrap tightly around the galaxy surrounding a bulging *core*, from which the arms emanate. The nature and content of the two structures are very different. The spiral arms contain most of the galaxy's dust and gas and therefore most of the star-forming regions. The stars in the arms tend to be young and hot. We call these *Population I* stars. The core contains no or very little in the way of star-forming materials and most of the stars are very old and have evolved off the main sequence. These stars are called *Population II* type stars. The environment of the galactic core very much resembles an enormous globular cluster with one important additional object. This object is a supermassive black hole around which the entire galaxy revolves. The region of the black hole is also one of the most powerful known radio sources in the sky, a point that coincides with the galactic center called *Sagittarius A*. Other such so-called supermassive dark objects have been located in the core areas of M31, M32 and six other Messier galaxies. Some galactic nuclei display tremendous levels of activity and radiate enormous amounts of energy across the entire spectrum from x-rays to radio. Remember that many galaxies contain supermassive central objects at the center of their cores. The electromagnetic activity is caused by interstellar matter falling into the central object at extremely high speeds. These galaxies are called Seyfert-type galaxies after their discoverer, Karl Seyfert. The brightest Seyfert galaxy in our sky is M77. Galaxies with even more exotic nuclei become so compact and bright that they outshine their entire host galaxy pouring out astonishing amounts of energy. Because these nuclei appear almost stellar within their galaxies, they have been given the name "quasi-stellar objects" or *quasars* for short. Quasars however are extremely remote, the closest being more than two billion light years away. No quasars appear in the Messier catalogue, the New General Catalogue or the Index (IC) Catalogue.

The Milky Way has five distinct spiral arms that have been clearly identified. Our solar systems orbits within the *Orion Arm* of the galaxy. This arm is also designated as "0" or the *Local Arm*. Arms that are outward from our position from the galactic center are numbered with a positive number while those arms inward of our position are noted with a negative number. The Orion Arm is actually a spur arm, a bridge connecting the *Sagittarius Arm* (–I) towards the center with the *Perseus Arm* (+I), which is outside our position. The innermost arm is called the *Scutum–Crux Arm* (–II), while the outermost arm is called simply the *Outer Arm* (+II). The arms are named for the constellations in the sky where the light of the arms is most heavily concentrated. There may be as many as three more arms that have not yet been positively identified. There are many other such spiral galaxies in the universe but not all of these spirals are the same in construction. Later on we'll look more closely at spiral galaxy make-up.

One thing that should have become very evident in our outward journey through the universe by now is that the universe for all its chaos is a very highly organized and well-ordered place. Organization does not stop with galactic structure. The intergalactic realm is also very highly organized. The Milky Way is not floating alone in space but belongs to an association of approximately 46 other galaxies gravitationally bound together[39]. Andromeda is the largest of these galaxies however the Milky Way may actually be more massive. Each of the two major galaxies has accumulated a large number of satellite galaxies. The third major member

---

[39] We are not certain that every single one of these satellite galaxies are actually bound to their assumed primaries so the use of the word "approximately" is appropriate until further study of proper motion of these galaxies can be verified.

of the Local Group is the Triangulum Spiral M33, though some believe it may be a large satellite of Andromeda. Of the Milky Way's satellites, the most prominent are the Large and Small Magellanic Clouds, though they are only visible from the southern hemisphere. Others include the recently discovered Canis Major Dwarf, a galaxy that is being disrupted and integrated by the Milky Way, dwarf galaxies in Ursa Minor, Carina, Sextans, Sculptor, Fornax, Draco, Sagittarius and two in Leo. The Andromeda Galaxy has two companions that appear on the Messier list (M32 and M110) both of which are conspicuous in amateur telescopes. Andromeda has as many as 10 additional satellites. M33 has at least one satellite and possibly three others, though their orbital nature is highly in doubt. Several other small galaxies travel as part of the local group though without being gravitationally bound to one of the three large galaxies. The Local Group also is involved in gravitational inter-action and member exchange with other groups. The giant elliptical galaxy Maffei 1 has split off from the Local group and has since become a group on its own with its satellite galaxies. The Sculptor or South Polar Group is dominated the galaxy NGC 253 and contains a handful of smaller galaxies and there are two other groups dominated by the galaxies M81 and M83. Maffei 1 and M83 seem to be heading away from the local group while the Local Group itself seems to be moving in the general direction of the Sculptor group. As for future interactions, the Milky Way and Andromeda galaxies are moving slowly towards each other and will likely merge into a massive elliptical galaxy, while the entire Local Group will eventually merge into the giant Virgo supercluster.

The Virgo supercluster is the closest major association of galaxies to our Local Group. Messier noticed an unusual accumulation of nebulae in the northern wing of Virgo. By the time the list was completed, Messier had listed sixteen nebulae in

Figure 14.2. The central portion of the Virgo supercluster. The image is centered on M87. NASA image, equipment unspecified.

this one area of Virgo. The cluster is so massive that many member galaxies that are on the far side of the center of mass and are being accelerated towards us in their orbits show the highest blueshifts of any galaxies in the heavens. Conversely the galaxies that are moving away from us exhibit some of the largest known redshifts in the universe. The galaxy IC3258 is approaching Earth at 517 km/sec and the lines in its spectrum are shifted towards the blue end of the spectrum. Since the entire cluster is moving away from us at 1,100 km/sec, IC3258 is traveling around the cluster's center at 1,600 km/sec. The galaxy NGC 4388 is receding from Earth at over 2,535 km/sec, giving it an orbital velocity of about 1,400 km/sec. These superclusters represent the largest structures in the universe. The Virgo supercluster is now known to have more than 2,000 members. Near the center of the cluster is the giant elliptical galaxy M87. This massive galaxy appears to be the most dominant gravitational influence in the cluster.

Galaxies come in many shapes and sizes. The most visually compelling systems are the spirals, though not all spirals are created alike. Just like with everything else in the universe there is a system by which spiral galaxies are categorized. Edwin Hubble devised the scale. Elliptical galaxies are denoted with "E" then assigned a number 0 through 7 depending on the degree of elongations with 0 being assigned to galaxies that are closest to spherical and 7 to the most "football" shaped. Irregulars are given the letter "I". With spirals, the initial designation is with an "S". Galaxies are then sub-characterized by lower case letters "a" through "d." A spiral galaxy that is graded "Sa" is one that has a large central nucleus as its most dominant feature. If the galaxy also sports conspicuous arms with a dominant core, it is called "Sb." A spiral galaxy that has an identifiable core and dominant arms is graded "Sc." If the galaxy has visible spiral arms and no visible core at all, then it is graded "Sd." A spiral may also have an intermediate grading consisting of two lower case letters in case it does not exactly fit one sub-class or another. In addition, many spirals exhibit a structure in their cores that extends the classic elliptical structure into a bar shape. In such cases, the class of the galaxy will be noted as "SB" followed by the lower case letter denoting arm–core dominance. In addition, some galaxies may show a spiral shaped disk but with no definable features. Such a galaxy is called a *lenticular galaxy*. The loss of spiral structure may be attributable to extreme age or a lack of gravitational interaction over a long period of time. Lenticular galaxies are graded as "S0" galaxies. The Shapely grade for the Milky Way is SBbc. This indicates a barred spiral with roughly equally dominant core and arms. The discovery that the Milky Way is a barred spiral was one of the more important discoveries in galactic astronomy in the last twenty years and in fact some astronomers now believe that the bar is more dominant in the core than had been originally believed. The bar is now believed to be as long as 27,000 light years in length or more than one-quarter of the galaxy's diameter. Let's now see how the Milky Way compares to other relatively easy to view spirals.

Spiral Galaxies

Having discussed the classification and structure of spiral galaxies, let's take a look around the universe at some typical examples in Messier's catalogue. There are more spiral galaxies in Messier's catalogue than any other type object. Messier listed 27 of these "nebulae without stars." Using Shapley's system let's have a look around the sky at some of the Messier spirals.

Virgo is the home of many of the sky's most beautiful spirals, in fact as we have already noticed home to many, many galaxies period! Few are more beautiful to look at as M104. The famous Sombrero Galaxy is the first member of the Messier list to be added after its final published version was released. Messier added it by hand to his personal copy in May 1781. It was officially added to the Messier list in 1921. The galaxy shines at magnitude +8.0 and spans approximately 9 by 4 arc minutes. The relatively large size means that M104's light is spread over a fairly wide area making it much more difficult to see in telescopes than its total magnitude would indicate. Another part of the reason for M104's faintness is its remoteness. When you find M104, that faint little smudge of light you are looking at is about fifty million light years away. In amateur telescopes, this galaxy's most prominent feature is its enormous central bulge. Only in professional exposures do the spiral arms show up in the outermost portion of the galaxy. Another important feature that amateur telescopes can see is the dark dust lane that runs across its galactic equator that seems to divide the galaxy in two. To the amateur, this galaxy with a virtually invisible disk overrun by the enormous core area is a classic example of a Shapely class Sa galaxy. The galaxy is also listed in the New General Catalogue as NGC 4594. M104 is also the dominant member of a small group of galaxies simply called the M104 group. The association also contains seven other members. M104 is at its best during the early part of spring when the western area of Virgo is at opposition to the Sun.

As the humid nights of summer give way to the cool and crisp nights of autumn, the Great Andromeda Galaxy rises nearly to zenith. The Andromeda Galaxy (M31) is the farthest object in the heavens that you can see with the unaided eye. But it is in the telescope that M31 displays its true grandeur. M31 shines in the sky at magnitude +3.4 and spans almost three full degrees in length by just over one degree wide. That's the equivalent of six full moons by two full moons in size. Any telescope will clearly show M31's large central core. Under dark skies, a telescope of about 8 inches will begin to show Andromeda's inner spiral arms. The nucleus is the dominant feature of the galaxy though Andromeda does have clear spiral arms; therefore Andromeda is Shapley type Sb. Andromeda also has two bright companions that are also on the Messier list. M32 is a small elliptical galaxy that appears in close to the core of M31. The other small elliptical galaxy on the opposite size of the core in M110, the last item on the Messier list. M32 is magnitude +8.1 and spans 8 by 6 arc minutes. M110 is harder to see because it is both larger and fainter than is M32. It spans 17 by 10 arc minutes and shines at magnitude +8.6. M110 can become lost very easily in any amount of light pollution. M31 and M32 are undergoing an interesting interaction. When imaged with observatory class telescopes, it is obvious that the bluish gas cloud arms (neutral hydrogen) are split out by about 4,000 light years from the stellar arms. The gas arms cannot be followed continuously in the areas closest to M32. It is obvious from these facts that M31 and M32 are undergoing a major interaction that is drawing the gas out of the arms of M31 and tearing loose many thousands of stars from M32 into the halo of the parent galaxy. Even more detailed observations by the Hubble Telescope show that M31 has a double nucleus, possibly caused by a dynamic interaction that likely happened early in the history of the Local Group. It is also possible that the double nucleus is an optical illusion caused by a lane of dark dust.

In the northern part of Canes Venatici resides the beautiful spiral galaxy M51. At declination +47 degrees, it like the Andromeda Galaxy passes almost directly overhead for mid-northern observers in the early spring. M51 shines at magnitude +8.7 from a distance of about thirty-seven million light years. It spans an angular distance of 11 by 7 arc minutes. Unlike Andromeda or the Sombrero, we view M51 just about completely face on. Messier himself saw M51 as two nebulae and recorded in his log "it is a double and each has a bright center." Because Messier saw two objects, both nebulous areas are considered by some to be part of M51. They are two different galaxies undergoing an interaction following a close pass. The brighter primary is a type Sc spiral while the companion is irregular. In the NGC catalogue, they are given separate numbers. The large spiral is NGC 5194 and the smaller interacting irregular is NGC 5195. Most astronomers prefer to keep the M51 designation reserved for the spiral itself. It was the first of the island universes in which spiral structure was discovered. Lord Rosse made detailed drawings of it in 1845. The galaxy is Shapley type Sc because its bright and dominant spiral arms stand out more than the tightly condensed core. The core itself is now known to be active and is classified as a Seyfert type galaxy.

In Ursa Major lies one of the skies most gorgeous galaxies. M81 also lies in a near face-on orientation. M81 is also one of Messier's brightest galaxies at magnitude +6.9 from a distance of twelve million light years. M81 is a Shapely type Sb spiral that has a nearby companion, M82. The galaxy has a huge central core and two beautiful sweeping spiral arms that span a total area of 21 by 10 arc minutes. The galaxy was discovered along with M82 on December 31, 1774 and many still call the galaxy "Bode's Nebula." Observers here can see the results of galactic interactions. The irregular M82 is visibly disrupted into its current shape and the close pass by M81 has caused an outburst of star-forming activity. Though M82 shines at only magnitude +8.4 in visible light, when viewed in infrared it is the brightest galaxy in the sky. The companion of M51 exhibits similar behavior. M81 is also beautiful in the ultraviolet and ultraviolet images returned from the ASTRO 1 Ultraviolet Imaging Telescope[40] (UIT) showed outbursts of star-forming activity ongoing in both spiral arms. Together M81 and M82 are the showpiece galaxies of the northern sky. Only M31 is more beautiful, but M81 and M82 are more spectacular as a pair than any other galaxy accessible to most of us. Even under light polluted skies, they can both be viewed easily.

Further south and west, near the bottom of the pot of the Big Dipper at the handle end is the galaxy M109. This is a classic barred spiral galaxy. It is rather faint shining at magnitude +9.6 over its 7 by 4 arc minute span. Because it is small though, it has a fairly high surface brightness and can be easily viewed with medium-sized telescopes if skies are reasonably dark. M109 is Shapely type SBc. The capital "B" indicates that the galaxy is a barred spiral. In amateur telescopes,

---

[40] ASTRO 1 is a package of four telescopes flown in the cargo bay of Space Shuttle *Columbia* between December 2 and 10, 1990. The mission suffered several delays due to fuel leaks. Forecast bad weather at the planned landing site cut the mission short a day with only 70% of the mission objectives completed. Still *Columbia's* flight was considered a pioneering success for space-based astronomy.

the galaxy appears somewhat pear shaped with a bright nucleus at the center. Long-exposure imagery begins to bring out faint spiral arms.

The skies of mid to late autumn bring Cetus the Whale to its opposition and here is the galaxy M77. Earlier we mentioned that there is something special about this galaxy's nucleus. M77 is what is called a Seyfert galaxy and it is the brightest galaxy of this type on the Messier list. The galaxy is Shapely type Sb indicating that it has both definable spiral arms but a more dominant core. But that is only the beginning of the story. M77's core is not only very large, but also extremely bright. This is caused by the very rapid motion of gases through the area, exciting those gases to shine brilliantly across the spectrum. Large amounts of radio energy are emitted from gases that are escaping the core while high-energy x-rays are indicative of matter falling into a black hole or other supermassive object at the center of the galaxy. ASTRO 1's UIT also found intense star-forming activity going on within the nucleus within 12 to 100 light years of the center.

Visually M77 shines at magnitude +8.9 over an area of 7 by 6 arc minutes. Because it is so bright at the center, it is very easily visible. That bright center makes the galaxy relatively easy to see. Some of these galaxies are not so easy to see. We'll revisit all of these galaxies and show you how to find them and how to observe them in the Observing Projects.

Elliptical and Lenticular Galaxies

Elliptical galaxies are very different from spirals in that they have very little in the way of internal structure. Even to the largest telescopes, they only appear to be uniform patches of light that form an elongated shape. These galaxies have very little in the way of interstellar matter such as dust and gas traveling between stars. This gives the suggestion that the elliptical galaxies are extremely old. Elliptical galaxies vary tremendously in size from the tiny companion galaxy M32 to the massive M87. These galaxies are characterized in the Shapely system with the letter "E" followed by a number 0 through 7 with the lower numbers being the closest to spherical in shape. In reality the elliptical galaxies for the most part are shaped like footballs, with different measurements on each of its three axes. The Messier catalogue lists a total of eight elliptical galaxies.

Lenticular galaxies are disk-shaped galaxies that also lack any internal structure, so they were often mistaken for elliptical galaxies. The Shapely system actually gives them an "S" designation followed by a zero. These "S0" galaxies can also be found among the Messier list in small numbers, there are three and possibly four. Galaxies take on a lenticular shape as the result in many billions of years of uninterrupted organization. In both elliptical and lenticular galaxies, the common theme is extreme age.

With the many thousands of galaxies in the western Virgo area, there have to be some choice elliptical galaxies to look at and our first is M49. The best time for viewing this galaxy is during early spring when western Virgo is in opposition to the Sun. At a distance from Earth of some sixty million light years, it shines at magnitude +8.4 and spans about 9 by 7 arc minutes. At that distance from Earth, that means the galaxy shines at absolute magnitude −22.8 and joins M60 and M87 as one of the truly giant galaxies of the Virgo Cluster. It is about 160,000 light years along its *major axis*, or the longest dimension of the ellipsoid. That number is subject to considerable uncertainty because with an elliptical galaxy it's hard to tell exactly what direction that major axis is pointed in. The galaxy is very elongated and rates an E4 on the Shapely system, which suggests that it may well be shaped exactly like a football. The galaxy has an integrated, or average spectral type of G7. This makes its overall color somewhat more yellow than typical galaxies in the Virgo cluster. This suggests that it may be a bit younger than the average age of galaxies in the cluster.

M87 is the signature galaxy for elliptical galaxy chasers. This monstrous galaxy may also be the gravitationally dominant galaxy of the entire Virgo supercluster. Unlike the more football-shaped M49, M87 is a Shapely-type E1, indicating it is nearly spherical in shape. It is also large. Spanning 7 arc minutes of sky from sixty million light years away makes it some 120,000 light years across. The galaxy shines at magnitude +8.6 from that distance also making it very intrinsically bright, around absolute magnitude −22. Though M87 is slightly smaller in size than M49, it appears to be significantly denser than M49. M87 is also noteworthy for three other reasons that are not readily observable to amateurs. First, it has the largest known system of globular clusters of any galaxy that we've been able to account for. While the Milky Way has some 200 globulars (of which we know of 151) M87 may possess as many as 15,000 of them! A second fascinating feature that required very large telescopes to observe is a jet emanating from the galaxy's core. Early Hubble images (predating the 1993 repair) show the jet clearly emerging from a supermassive central object at the core that may possess as much as three billion times the mass of the Sun. The jet was first discovered in 1918. In 1966 a second jet was discovered extending in the opposite direction, although since it points more away from our line of sight, it is not as obvious. Third, M87 is a powerful source of electromagnetic energy from low-energy radio waves (designated Virgo A) to being the center of an enormous sphere of high-energy x-rays. The center of M87 is extremely active although there is very little gas and dust in the galactic core are, it does qualify as having an active galactic nuclei although without large amounts of gas to be excited, it cannot become a Seyfert-type galaxy. Still not even many Seyferts have a feature like M87's powerful jet. M87 is one of the most amazing places to study in the entire universe as it shines in every wavelength of light in the spectrum. It is conveniently placed in Virgo allowing anyone on Earth to see it. And it may well be home to the largest known suspected black hole in the cosmos. There's quite a bit in that little smudge of gray light.

From the enormous to the tiny, the elliptical galaxy M32 is a satellite of mighty Andromeda. This galaxy also shines at magnitude +8.1, which actually makes it slightly brighter than giant M49 or M87, but this is a function of perspective since M32 is less than 1/20th the distance from Earth of the two giant ellipticals. The galaxy spans 8 by 6 arc minutes and has a mass of about three billion Suns. That's about the same mass as the supermassive central object at the center of M87. The galaxy spans a linear diameter of some 8,000 light years along its major axis. A large amount of mass in M32 is concentrated close to the core; some 5,000 solar masses per cubic parsec and these stars are in rapid motion around a supermassive central object in the core. M32 has an average spectral type of G3, which would indicate that this old galaxy is contaminated with a sprinkling of younger stars of about two to three billion years in age. Otherwise, M32's stellar population is largely intrinsically faint orange and red dwarves and non-nuclear shining degenerate white dwarves. M32's population strongly resembles that of a much larger elliptical, say like M49. It is possible and even likely that M32 was once a much larger galaxy that has since lost of its stellar mass to Andromeda.

Going back into the Virgo supercluster once again, just about 1.3 degrees to the west of M87 is the lenticular galaxy M86. This Shapely type S0 galaxy shines at magnitude +8.9 and spans about 7 by 5 arc minutes from sixty million light years away. This makes it another intrinsically bright giant like M49 and M87 spanning perhaps a diameter slightly larger than that of the Milky Way. Another interesting facet of Virgo supercluster galaxies that we talked about previously is their speeds as they move around the cluster's center. M86 has the largest blue shift of all the Messier galaxies, approaching Earth at some 419 km/sec. M86, though it is not an elliptical by shape, is probably populated much like ellipticals, consisting of intrinsically faint small main sequence dwarves and degenerate white dwarves.

While spirals are galaxies that are vibrant with life, forming new stars regularly in its arms, ellipticals and lenticulars are galaxies that have ended their star-forming lives and shine with the light of expired white dwarves and faint low-mass main-sequence stars. Let's next have a look at unusual galaxies that seem to have no organized structure at all.

Irregular Galaxies

Galaxies that are classified as irregular are those that simply fit no other organizational category for galaxies such as lenticular, spiral, or elliptical. Since the universe is an orderly place by definition, irregular galaxies become so because they are being acted on by some other force, usually a much bigger galaxy. We have already noted the irregular companion of M51, designated NGC 5195 and M82, the Cigar galaxy, which is the only irregular galaxy in the Messier catalogue. Two other irregulars of consequence that are not in the Messier catalogue (because he could not see them) are the Large and Small Magellanic Clouds deep in the southern sky. These two galaxies are gravitationally disrupting each other as well as by the Milky Way as well. The two Magellanic Clouds have populations of stars, clusters and nebulae that are remarkably similar to the Milky Way and other star forming spirals. This tells us that irregular galaxies are relatively young and that the only difference between them and young star-forming spirals is the gravitational interactions that are destroying their structures. In some cases, remnants of that structure can still be viewed. Careful study of the Small Magellanic Cloud reveals that it may at one time have been a barred spiral. Both galaxies are very bright. The Large Magellanic Cloud shines with a total light of magnitude +0.1, the Small Magellanic Cloud at +2.2. The LMC spans about 11 degrees by 9 degrees while the SMC covers about 5 by 3 degrees. The LMC is about 170,000 light years away while the SMC is about 210,000 light years distant. They are the second and third nearest galaxies to the Milky Way. Only the Sagittarius Dwarf Elliptical galaxy is closer than the LMC.

M82 is the relatively bright companion of M81. It shines at magnitude +8.4. Both galaxies can be easily viewed in the same wide field eyepiece field of view. Under a dark sky a medium-sized telescope will easily show you what has been done to M82. The galaxy is undergoing tremendous bursts of star-forming activity and several dark lanes can be viewed crossing the galaxy in varying directions. The galaxy clearly at one point had a disk shape. The enormous movement of gases within the galaxy, rushing into star-formation areas, creates large amounts of radio noise. In fact it is the strongest such source of noise in Ursa Major. Irregular galaxies therefore represent the manifestation of enormous amounts of gravitational energy at play, ripping galaxies asunder and setting off astonishing bursts of star-forming activity.

When spring comes around and the great Virgo supercluster comes into view, galaxy season kicks into peak time. Six months later, the early autumn skies offer our best look at our closest large intergalactic neighbors. Let's break out the telescopes and look beyond our own island universe into the deepest parts of the cosmos.

Observing Projects XIV – The Realm of the Galaxies

Observing Project 14A – The Great Andromeda Galaxy and Friends

Early October brings the Great Square of Pegasus high overhead. This asterism, part of the fabled winged horse, is one of the most easily identifiable patterns in the autumn sky. From here we can hop from star to star across to M31.

After locating the Great Square, begin at the star on the northeast corner (top left if you are facing south and the Square is at the zenith). This is Alpheratz (spectral type B9, magnitude +2.0) and from here we'll begin to journey to M31. Alpheratz (Alpha Andromedae) is also the westernmost member of a three-star arc that runs off towards the northeast. The center star is called Mirach (Beta Andromedae), a magnitude +2.1, type M0 red giant. From Mirach, scan north-northwest just under four degrees to magnitude +3.8 Mu Andromedae. Extending the line from Mirach to Mu Andromedae about another four degrees brings us to M31.

Galaxies, particularly spirals like M31 stand up well to high power in the telescope, but M31 is very large so that wisdom does not necessarily hold here. With M31 we want to see as much we can so use your widest field eyepiece and a telecompressor if you have one. Start with M31's nucleus and spend some time looking at this area. Andromeda, like the Milky Way, is a very active star-forming galaxy. If you look carefully and your optics and sky are both in optimal condition, you *might* see a faint dark lane strewn across the nucleus. This could be as we illustrated before a lane of dust, or it could be a dynamic double nucleus. Whatever it is, once you have found this, try following the dark lane away from the nucleus in either direction. The dust shows the directions of the spiral arms around the galaxy. If your skies are dark and your optics clean and well aligned you may be able to see the faint glow of the arm stretching away in telescopes as small as 6 inches. Any light pollution will render the arms almost impossible to see. In larger telescopes

Figure 14.3. The core of M31. Meade DSI and Celestron Super C-8 Plus at f/6.3. One-minute exposure. The stars are in the foreground. Image by author.

the arms show up very nicely. Don't forget to use averted vision and if you are in a light-polluted area a light-pollution filter. While they do not enhance the light of galaxies, they will suppress the unwanted wavelengths of light pollution.

In a wide field eyepiece, just about any telescope will show a second soft glow near the nucleus of M31. This is the dwarf elliptical galaxy M32, a companion galaxy that sits within 130,000 light years of the M31 nucleus. The galaxy is sufficiently close that the two are significantly disrupting each other. M32 has lost an enormous amount of mass to the halo of its host galaxy. Don't waste too much energy looking for the detail in M32. It is a typical elliptical consisting mostly of much older Population II stars, though as we mentioned earlier it is contaminated with a large number of younger stars. On the opposite end of M31 is another small elliptical galaxy, M110. This galaxy is also an old elliptical galaxy that is somewhat farther from the center of M31. If you compare it to M32, you will notice that it is much more elongated in shape than is M32. M32 is a Shapely type E1 galaxy while M110 is an E4. A telescopic field of view of just over one degree should show all three galactic cores in a single field of view.

M31 also makes a pretty photographic subject without using complicated imaging techniques. Try a few long-exposure images using various lenses on your 35-mm SLR camera. Remember that larger lenses will require longer images. The image on page 201 was made with a 35-mm camera and stock lens. I used ISO 400 film and a five-minute exposure. Unfortunately the image is smeared because someone with the same initials as myself accidentally kicked the tripod during the exposure (oops). For the next picture, try attaching the camera to the telescope at prime focus with the telecompressor and make images of M31's nucleus area. If your telescope has a modern drive and is precisely aligned, you should be able to leave the shutter open for several minutes and get a nice exposure of the nucleus. If the telescope has a short enough focal length, you may also be able to image the core of M32 as well. If you own a CCD camera and telescope combination capable of "auto-mosaic" (such as a Meade imager with an LX-200), use it to make a nine-image sequence of the core area. This technique allows you to create an image with a much wider field than the miniscule area available to the camera. Do this on a regular basis when M31 is visible because many of these stars in the core area are much older Population II stars that may be nearing an iron-core collapse and trigger a supernova explosion. The last (and only) supernova in Andromeda was discovered on August 20, 1885 so it may be that this galaxy is long overdue to spring one. Pay careful attention to that area as you visit and revisit it over the years. Remember that when you are imaging, patience is a virtue. You will make many mistakes, smear many frames of film and even take some exposures of empty space when you thought there was a galaxy there. Time and practice will reduce your mistakes. Don't get mad about bad pictures. The only way you will learn to take good pictures is to take many bad ones.

Looking through a spectroscope, you will notice something else about Andromeda. Its spectrum looks much like many typical stars, but the emission lines of the spectrum are shifted slightly towards the blue end of the spectrum. This indicates that Andromeda and company are moving towards us and it is in fact on a collision course with the Milky Way. When Andromeda and the Milky Way meet several billion years from now, someone way across the cosmos will have a spectacular view of the collision.

Observing Project 14B – Touring the Virgo Supercluster

The closest supercluster of galaxies to our Local Group is the Virgo supercluster in the western part of that constellation. There are a total of sixteen Messier galaxies in the cluster. Perhaps the most dominant member of this cluster is the giant elliptical galaxy M87. To find M87, start at the tail of Leo and find the magnitude +2.1 blue star Denebola (Beta Leonis). Denebola is easy to find as it marks the end of the tail of Leo. About 18 degrees to the northeast of Denebola is the magnitude +2.8 star Vindemiatrix (Epsilon Virginis). M87 lies about exactly on a line between these two stars about 8 degrees southwest of Vindemiatrix. From Vindemiatrix, as you start towards M87, you will encounter a triangle of stars, the brightest of which is 6.1 magnitude 41 Virginis, which is the northern point of the triangle. The two seventh-magnitude stars make a straight line along with Vindemiatrix those points to M87. From that last star, you are within one finder-field-width of M87. Just follow that line for another 5 degrees from the last seventh-magnitude star to find M87. You are now looking into the heart of the Virgo supercluster. There are many other potential paths to M87 but many of them depend on your being able to find other galaxies or stars along that path that may be fainter. Using this path from Vindemiatrix, you travel in a straight line all the way to the target.

M87 is an E1 type elliptical and is a giant among galaxies, spanning about 160,000 light years along its major axis. The E1 rating means that the galaxy is close to spherical in shape. M87 shows little signs of any detail within the round gray blob, though we know that M87 is in fact extremely active with a supermassive central object at its core that is as massive as all of M32. M87 is among the brightest of the galaxies. If the "seeing" conditions are such that you cannot see M87 on a given night, then the Virgo supercluster tour is best left for another night.

With your wide field eyepiece and a telecompressor installed, you should have a field of view of easily 1.3 degrees in an f/10 telescope. If you scan to the west and slightly to the south, M87 is on the edge of your field of view at about the ten o'clock position.[41] Directly opposite where M87 is, the lenticular (S0) type galaxy M86 should be entering your field of view. After a short look at M86, you will readily see why many professionals mistake these lenticulars for ellipticals. It is just looks more like an E4 type elliptical galaxy. It took some time before astronomers realized that the galaxy was not an ellipsoid, but a disk.

Just 15 arc minutes further to the south and west of M86 is the lenticular galaxy M84. This galaxy is a bit smaller than M84 at five arc-minutes and a bit fainter at magnitude +9.1. If the seeing starts getting tough, don't be afraid to go to the LPR filter. This galaxy, like M87, has a very active nucleus and a supermassive central object with a mass of some 300 million Suns. That supermassive central object also like M87 spews two jets from the galactic core, a characteristic of huge black holes. Like many lenticulars, M84 was also classified for many years as an E4 elliptical galaxy, but further study revealed its disk shape. Now that were getting down below magnitude nine with some of these objects, don't be afraid to use those rod cells with averted vision. There is no color to see in these galaxies so there is little need to use the color-sensitive cones of your retina from here on.

Let's push our equipment and eyes little harder. From M84, scan 2.1 degrees to the west across the border into Coma Berenices to find the faint spiral galaxy M99. This galaxy glows feebly at magnitude +9.9 and spans about 5.4 by 4.8 arc minutes. M99 is a Shapely class Sc spiral and is noteworthy because it has the highest redshift of any Messier object, receding from Earth at 2,324 km/sec. For the amateur astronomer working with small and medium telescopes, the triumph is in simply spotting the tiny faint glow. Professional telescopes show that this galaxy is somewhat unusual in that its nucleus is not exactly at the center of the spiral structure. This suggests that this galaxy may have been disrupted by a recent encounter with another galaxy, possibly the nearby galaxy M98. To find M98, go about 0.8 degrees to the west to find the relatively bright (compared to what we've been looking at) magnitude +5.1 blue star 6 Coma Berenices. From there, travel about 0.5 degrees southwest to the galaxy M98.

M98 is a nearly edge on spiral of Shapely type Sb. This galaxy measure about 9 by 3 minutes and shows at magnitude +10.1. This galaxy is even fainter than M99 because its nucleus is obscured from our view by enormous clouds of light-choking dust that hide large parts of it from our view. It is a galaxy that is in a turbulent state, which gives rise to the possibility that it is the galaxy that that disrupted M99. As the galaxies get fainter and fainter, it becomes more and more important to ensure that you are using good observing techniques such as averted vision and proper dark-adaptation.

The final galaxy in our tour of the Virgo Supercluster is M100. From M98, scan 1.7 degrees north to find M100. This galaxy is slightly brighter than M98 at magnitude +9.3 and spans 7 by 6 arc-minutes. It is a spiral of Shapely type Sc, with a slightly subdued nucleus compared to its spiral arms. The galaxy is oriented close to face-on relative to our line of sight. The overall spectral type of this galaxy is

---

[41] That assumes a normal north up and east left view. In a telescope's inverted view, the galaxy would be at four o'clock. If you are using a star-diagonal, the image is upright, mirror reversed. So the galaxy would be positioned at two o'clock.

shaded slightly toward the blue end of the spectrum and in large telescopes the galaxy's spiral arms have a distinct blue color. This is caused by a burst of stellar formation triggered by interaction between M100 and two other galaxies that lay just a few arc minutes away.

These are just seven galaxies of the Virgo supercluster. There are more than two thousand known members of this supercluster for you to explore. With a telescope of 6 inches or larger and dark skies, you should be able to see all sixteen of the Messier galaxies within this cluster. There are many hundreds of objects in the NGC and IC catalogues for viewing that are within the reach of large "light-bucket" owners. Study your charts, pick your targets and take a good look around.

All of these galaxies make relatively easy targets for owners of light-sensitive CCD cameras. Start with the larger brighter ones with exposures totaling ten minutes to start (one ten minute exposure or twenty thirty-second exposures or any such combination). Align and stack your images and watch your target appear. Photography of such faint targets again is so very tough because you must have everything exactly right from the setup of the camera to the alignment of your telescope. Wide field exposures on film in a dark enough sky might reveal several Virgo supercluster galaxies. Use your largest telephoto lens, or couple your camera to a smaller telescope such as a 90-mm Schmidt–Cassegrain and piggyback it on your main scope, then while tracking on M87 or M84 for about a half hour, lock the shutter open. Expose the film for as long as the local conditions will permit. Remember that if your skies are light polluted, this period of time will be less because soon you will start recording more background fog than deep sky objects.

Observing Project 14C – Showcase Spirals and Their Oddball Companions

Two showcase galaxies that pass high overhead in the evening sky for northern observers are M51 and its companion NGC 5195 in Canes Venatici and the spiral M81 and its irregular companion M82 in Ursa Major.

M81 is the easier galaxy to find because it is so much brighter. To find M81, first locate the bluish star Dubhe at the end of the pot of the Big Dipper. To star hop to M81 from here is difficult because there are few descript stars here that stand out and the galaxy is a full ten degrees from Dubhe. Scan about 5 degrees to the north and west of Dubhe to find a group of four fifth-magnitude stars, three of which form a triangle. The brightest of these stars is magnitude +5.7 32 Ursae Majoris. Move past this group in roughly the same direction, perhaps slightly more to the north. As the four fifth-magnitude stars vanish out of the edge of the field, the magnitude +4.5 magnitude star 24 Ursae Majoris appears in your finder. Move your telescope then about 2 degrees back towards the south and east to center on M81. You might have passed M81 on the way between 32 and 24 U Ma. These stars are obscure so make sure you have the map handy!

M81 has a bright central core that is easy to see in just about any telescope. The challenge is to see any trace of its spiral structures in a medium or small telescope. That dominant core earns it a Shapely class Sb. The arms are areas of active star forming perhaps triggered by interaction with the nearby companion galaxy, M82.

The two galaxies are separated by less than 0.6 degrees and are easily visible in the same telescopic field even without extra wide field setups. M82 is the only irregular galaxy in the Messier catalogue. Called the "Cigar Galaxy" for its stogie-like profile, this galaxy may have once been a spiral or lenticular galaxy before being gravitationally disrupted and deformed by M81. M82 is also relatively bright shining at magnitude +8.2. I can see them both at zenith from my light-polluted yard in Newark, DE on a night with good transparency. Together they make one of the great showcase pairs of galaxies in the northern sky.

A rival pair of galaxies that vies for that title is M51 and its irregular partner NGC 5195. To find this galaxy, start at the other end of the Big Dipper where the magnitude +1.8 star Alkaid marks the end of the handle of the Dipper. From Alkaid, go south and slightly west by 2.2 degrees where you will easily find the magnitude +4.7 star 24 Canes Venatici. There is no other star around bright enough to confuse it with. From 24 C Vn, go south and slightly east by about 2 degrees to find a nearly equilateral triangle of three magnitude-seven stars. M51 can be found just to the south of this triangle.

The Whirlpool Galaxy can be tough to see despite a relatively bright total magnitude of +8.4. It has an extremely low surface brightness that makes it at times very tough to find. M51's companion, NGC 5195, is an irregular type galaxy that appears to be just emerging from a close encounter with M51. Because of the manner in which Messier described the two structures, NGC 5195 is sometimes called M51B. The galaxy passed M51 traveling away from us along our line of sight and is now slightly behind M51. NGC 5195 appears at one point to have been a disk-shaped galaxy, maybe a lenticular or a spiral, we don't know for sure but has now become an irregular galaxy because of the gravitational disruption caused by M51. Images of the pair make it appear that NGC 5195 is touching one of the outer spiral arms. Light pollution will pretty quickly make this galaxy impossible to see, so seek M51 on nights of good transparency and under a dark sky. As with other galaxies, an LPR filter will help, but only as far as suppression of sky glow is concerned.

Observing Project 14D – High-Energy Galactic Cores

When supermassive central objects at the cores of galaxies accelerate gases to high speeds, the gases become excited and emit tremendous quantities of energy. When this occurs, even in Sc type spirals or galaxies with even smaller cores, that core can glow with more light than the entire rest of the galaxy. These types of very active galaxies are called Seyfert type galaxies. The Seyferts represent a little cousin of a type of galaxy that is the most powerful source of energy in the universe.

The most energetic such galaxy on the Messier list is M77 in Cetus. To find M77, start at the magnitude +2.5 magnitude Menkar (Alpha Ceti) and go just under 5 degrees to the southwest to find magnitude +3.4 Kaffaljidhma (Gamma Ceti). From here, go just over 3 degrees to the south to find magnitude +4.1 Delta Ceti. M77 is just about 50 minutes of arc to the east of Delta Ceti.

Once centered in your main scope, it becomes obvious why M77 is special. The core of the Shapely type Sb galaxy is disproportionately bright compared to the rest of the galaxy. This is due to the excitation of gases moving at high speed around the supermassive central object at the galaxy's core. M77 also has three bright spiral arms that can be viewed with larger telescopes or medium sized scopes under very dark skies. The galaxy is inclined to our line of sight by about 51 degrees. The arms are knotted with large clusters of young stars. Observers with larger telescopes will notice the irregularity in the brightness of the arms. It would seem that the high energy levels in the core are triggering starburst events in the arms of M77. M77 is also very large in linear size. The easily visible part of M77 extends over 120,000 light years, making it one of the largest Messier spirals.

Spectroscope users can tell that the galaxy is the site of unusual activity because the emission lines in the galaxy's spectra are unusually wide, particularly at the wavelengths of hydrogen-beta, doubly ionized nitrogen and singly and doubly ionized oxygen. If you have a spectroscope compare the galaxy's emission lines to those of M31. Andromeda's emission lines are much narrower than are those of M77. The more energy that is produced by the rapidly moving matter, the more the light of the galaxy at those particular wavelengths is enhanced creating the wider emission lines in the spectroscope.

Seyfert galaxies are fascinating because they represent a lower energy cousin of one of the most enigmatic phenomena in all the cosmos. Super-active galactic nuclei create quasi-stellar appearing points at the center of distant galaxies that are many thousands of times brighter than the entire host galaxy. These quasi-stellar nuclei or *quasars* are examples of Seyfert galaxies gone completely wild. Unfortunately the nearest quasar is two billion light years distant and far beyond the reach of amateur telescopes. M77 represents on a small and local scale what is going on in the incredible depths of the universe.

Observing Project 14E – The Messier Marathon

The Messier Marathon is considered to be the ultimate test of observational skills, in a single session testing your ability to see, track, plan and execute a very precise set of observations in a very limited period of time. Charles Messier created the list in the eighteenth century as a means to avoid false comet discoveries. Its enduring legacy is as a repository of the most beautiful objects in the sky.

Random chance would seem to dictate that the 110 Messier objects would be evenly scattered throughout the sky but interestingly enough this is not true. Many Messier objects are crowded along the plane of the Milky Way while many others are concentrated among galaxy clusters, such as the sixteen members of the Virgo Supercluster. Nearly all are north of declination −30 because of Messier's observing position in France. Because of this, there is a fortuitous gap in the distribution of the Messier objects. There is only one Messier object (the very northerly M52) in the sky between RA 21:40 and 00:40 and nothing at all between 21:40 and 23:20. This creates a roughly four-week window from the second week of March to the first week of April where all 110 Messier objects are visible in a dark sky. It is not possible to conduct the marathon during each night of the window because it is critical that the Moon be at or near new so that its light does not flood the sky and wash out many of the objects. The actual window each year therefore is only a few nights wide around the time of new Moon. Because of the angle of the ecliptic with the horizon, the window is actually skewed a few days prior to new Moon because the waning crescent rises deep in twilight for several days before new Moon, while the ensuing waxing crescent will be present in a dark sky within two days after new Moon.

The concept of the Messier Marathon was invented independently by several North American astronomers, the best known of whom is Don Macholz during the 1970s. It is generally accepted that the first amateur to successfully bag all 110 in one night is Gerry Rattley from Dugas, Arizona on the night of March 23–24, 1985. Just one hour later on the same night Rick Hull of Anza, California also completed the marathon. There are several ways to make the marathon more complex for observers who are experienced and have completed it once. The simplest way to do it is with a computerized telescope with GO TO capabilities. The marathon gets more complicated when you remove the computer and have to find the objects manually, though the observer may use star charts, setting circles and a checklist. The marathon gets more and more difficult when you sequentially take away the setting circles, then the charts and finally the checklist, forcing the observer to perform the entire marathon from memory. You can then enhance the marathon experience by adding other challenges to it, such as throwing in the solar system detour, taking time off the marathon to bag all eight planets. That is not possible every year because some of the planets may be near solar conjunction during the marathon period. It will be the year 2020 before you can get all eight planets during the marathon again. Others like to add NGC objects to their hunt, especially in that area between midnight and three in the morning where if you are efficient early, you can have a long break. Don Macholz took advantage of such an opportunity to track down a mind-numbing 599 objects in one night!

Planning is absolutely critical. You must know where to begin and how to plan your evening almost minute by minute. Getting ready in plenty of time is also important. Performing the Messier Marathon makes a Space Shuttle launch seem like a spontaneous event in many ways. You must finally have a proper horizon to view low objects, especially to the west and to the southeast where you will be looking for M74 in the early evening and M30 as morning twilight approaches respectively.

Generally speaking, the flow across the sky during the night is from south to north and from west to east. If two given objects are at the same right ascension, the southern one goes first because you have the least amount of time to view it before it sets while more northerly object stay in the sky far longer. The reasons why these things are so critical is that the marathon begins with a bang. The first target in the marathon is M74 and on March 25, 2006 (the preferred date for that year), it sets only 122 minutes after the Sun does. Add to that the fact that twilight is somewhat longer than normal at that time of year and you may have only about twenty minutes or so to find it. The pressure is on right out of the gate. Then comes M77, which sets only eleven minutes later. It is essential that you be ready to start observing as soon as the brightest stars appear and know how you are going to navigate to these two galaxies. You must bag them both within a few minutes after twilight ends or you will have lost the marathon before you have even started.

Once M74 and M77 are recorded in the observing log, then comes another long string of objects that must be found in short order. During the second hour after twilight ends, you must secure M33. It follows M77 below the horizon by just about an hour so it is next on the list. Since it is a fairly bright galaxy, just a shade above magnitude +7, you should be okay if you get to it fast. Life gets a bit easier with M31, M32 and M110 all in one field of view and fairly bright. If you want to get fainter objects first, you can pass up these three for now because since they are nearly circumpolar you will have another shot in the morning sky, but it is better to get them now. Further north are the two open clusters in Cassiopeia, now well placed for viewing; M103 and M52. Six more objects must follow in quick sequence. In order during the next hour after M31 and company you must find M53, M103, M76, M34, M45 (Pleaides) and M79 in less than an hour. Once you have found these, the pressure eases a bit and you can slow down some. M42 (Great Orion Nebula), M43 and M78 are the next logical targets in Orion, then up north to get M1 (Crab). Now with eighteen objects on record in less than two hours after the end of twilight, the pace becomes a bit more leisurely.

The next two hours leading up to about eleven PM local time is dominated by the open clusters in Gemini, Auriga, Cancer, Monoceros, Canis Major, Puppis and Hydra. In order they are: M35, M37, M36, M38, M41, M93, M47, M46, M50, M48, M44 (Beehive) and M67. This makes twelve open clusters in all bringing your total for the night to 30. Open clusters can be tough because they do not always clearly stand out from the background and are not always heavily populated like M41 and M48. You are no longer in a rush, so take your time over the next two hours and carefully list each of these twelve objects.

As the midnight hour nears, Leo and Virgo are coming towards their highest altitude of the night and the focus shifts from open clusters to galaxies. There are five Messier galaxies in Leo and they can be tough. These are M95, M96, M105 (small and faint), M65 and M66. This brings your tally to thirty-five. Go north for a bit to get the six Messier objects in western Ursa Major. You already know about M81 and M82, the paired galaxies, then M97, the Owl Nebula. M108 and M109 are tough spiral galaxies because they are very faint. Perhaps toughest of all is M40, which is not really a deep sky object at all, but a double star. After you have added these eleven objects we are now at 41 of the 110 objects.

From Ursa Major, work your way south across Canes Venatici. There are five objects here to catch to get to a total of forty-six. M106, M94, M63 and M51 are all spiral galaxies. M3 is a globular cluster. In easternmost Ursa Major you can also go back and get M101. This area puts you near the Draco border so while you are here go and pick up the lenticular galaxy NGC 5866, which is considered to be M102 though historians pretty much agree that M102 is a duplicate of M101.

The marathon then moves south into galaxy country in Coma Berenices. There are eight objects here to get to fifty-four total. These are the globular cluster M53 and then the lenticular galaxy M85. The others are the spiral galaxies M64, M98, M99 (very faint), M88, M91 and M100. Then we move south and east into Virgo to pick up more galaxies. There are eleven more galaxies here to pick up and when you are done with all of these you will have found sixty-five objects. These are M84, M86, M87, M89, M90, M58, M59, M60, M49, M61 and M104. Take great care while negotiating the cluster area because if you have a large telescope in particular there are hundreds of NGC galaxies in the area as well that are bright enough to confuse with the Virgo Messier objects.

By now the clock is probably nearing one in the morning and believe it or not you actually now have a chance to rest for a while. I'm a flight instructor and I'm used to shift work. Being up in the wee hours is nothing unusual for me and I'm familiar with how to adjust my body clock for nights like this. You may not be used to it, so now may be a good time to take a nap for about an hour and a half. If you are an experienced observer you could even make it about two hours. By two-thirty to three in the morning, it's time to get going again.

The sky is now rotating Hydra, Serpens Caput, Hercules and Lyra towards the meridian. Hydra has two targets, the globular M68 and a spiral galaxy M83. Serpens Caput has the beautiful globular M5. Hercules has two globular clusters M92 and the Great Hercules Cluster M13. In Lyra, there are two objects, the open cluster M56 and the famous Ring Nebula M57. Cygnus is a bit further to the east and it has two sparse open clusters, M29 and M39. South of Cygnus is the constellation Vulpecula with one Messier object, the pretty Dumbbell Nebula M27. Sagitta is a tiny constellation underneath Vulpecula and is home to the globular cluster M71. A lot of the objects ahead are globular clusters; remember a majority of the known clusters are located in this area. You have now found seventy-six objects as we enter the gates of globular country.

The sprawling constellation Ophiuchus has seven Messier objects, all globulars. They are in order of recommended viewing: M107, M10, M12, M14, M9, M19 and M62. Scorpius has two more globulars, M4 and M80 as well as two open clusters, M6 and M7. That brings the total to eighty-seven. The little constellation Scutum has two open clusters M11 and M26. Between Scorpius and Scutum is Serpens Cauda with the lovely open cluster M16 and the Eagle Nebula (not a Messier object, but who cares?).

Now with ninety prizes in our log, the finish line is in sight, but at three in the morning time is running short. Twilight is now about an hour and a half away. Sagittarius will keep you really busy. Fifteen of the last twenty objects are in this constellation. Start with the easy M17, the Omega Nebula. Then comes the open cluster M18 (tough sparse cluster). Next comes something really tough, the "Milky Way Cloud" as M24 is called. Then come three open clusters; M25, M23 and M21. Once you have these, you can view the two closely paired diffuse nebulae, M8 (Lagoon) and M20 (Trifid) above the top of the Teapot. Then through the rest of the Archer are seven globular clusters. In recommended order they are M28, M22, M69, M70, M54, M55 and M75. Once you have finished in Sagittarius, you are running out of time and objects. You have now found 105 objects. You must now begin working your way towards the southeast horizon and with speed for the last of the objects we must find are exactly where the sky will first begin to brighten with the onset of morning twilight. If by chance you passed up M31 and its two Messier cohorts earlier in the evening, then go and get them now. The globular cluster M15 sits on the very western edge of Pegasus, well west of the Great Square. Not far away to the south is the globular cluster M2 is Aquarius. Aquarius hosts two more clusters on the list, the globular M72 and the open cluster M73. The final target on the list is the globular M30. Going for M72 and M73 first buck conventional wisdom but because Capricornus is a lot further south than Aquarius, M72 and M73 rise higher even though they rise later. You need a bit more time for M30 to gain enough altitude to see. The race will be on to find it as twilight begins to brighten the eastern sky.

Completing the Messier Marathon is one of the great crowning jewels in an observer's career and is the ultimate test of the amateur astronomer's ability to see, to plan, to execute a program of serious science. It also shows you have mastered the technique of marrying the brain, the eye and the telescope into a single integrated observing system. If you can do the marathon, then you have completed your journey from frustrated amateur at the eyepiece of the department store telescope to accomplished amateur astronomer. The emphasis on that phrase is no longer on the "amateur" but on "astronomer." You the amateur make many of the key discoveries in our science that grab the headlines. Now you have the tools and knowledge you need to go and explore the heavens. Enjoy the view.

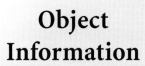

Object Information

The Solar System–Major Bodies

Name	Diameter	Moons	Density	Dist and from Sun	Period	Rotation
Mercury	4,880	0	5.4	0.38	88 days	59 days
Venus	12,103	0	5.2	0.72	225 days	−243 days
Earth	12,756	1	5.4	1.00	365 days	23h 56m
Mars	6,794	2	3.9	1.52	687 days	24h 37m
Jupiter	142,984	63	1.3	5.20	11.9 years	9h 50m
Saturn	120,536	34	0.7	9.54	29 years	10h 40m
Uranus	51,118	21	1.3	19.22	84 years	15h 30m
Neptune	49,532	13	1.6	30.06	168 years	15h 00m
Pluto	2,274	1	2.1	39.05	248 years	6.4 days

Diameter is in kilometers. Density is in grams/cc. Distance in AU.

Notable Asteroids

Name	Diameter	Dist and from Sun	Class	Orbit Period	Eccentrically	Inclination	Rotation
1 Ceres	960 × 932	2.767	C	4.6 years	0.07	10.5	9.0 hours
2 Pallas	570 × 525 × 482	2.774	U	4.6 years	0.22	34.8	7.8 hours
3 Juno	240	2.669	S	4.4 years	0.25	12.9	7.2 hours
4 Vesta	530	2.362	U	3.6 years	0.09	7.1	5.3 hours
243 Ida	58 × 23	2.861	S	4.8 years	0.04	1.4	4.6 hours
433 Eros	43 × 13 × 13	1.458	S	1.8 years	0.22	10.8	5.3 hours
1862 Apollo	1.6	1.471	S	1.8 years	0.56	6.4	3.1 hours
2060 Charon	180	13.630	B	50.7 years	0.38	6.9	5.9 hours
4179 Toutatis	4.6 × 2.4 × 1.9	2.512	S	4.0 years	0.63	0.4	130 hours

Diameter is in kilometers. Distances in AU.

Notable Comets

Name	Perihelion	Aphelion	Orbit Period	Eccentricity	Inclination
1P/Halley	0.586	34.930	75.8 years	0.967	162.3
2P/Encke	0.335	4.103	3.3 years	0.849	11.9
6P/d'Arrest	1.353	5.639	6.6 years	0.613	19.5
9P/Tempel 1	1.505	4.739	5.7 years	0.518	10.5
33P/Tempel – Tuttle	0.976	19.790	33.0 years	0.906	162.5
81P/Wild 2	1.585	5.307	6.2 years	0.540	3.2
109P/Swift – Tuttle	0.960	50.932	120.0 years	0.963	113.5
153P/Ikeya – Zhang	0.507	100.893	150.0 years	0.990	28.1

Stars–Brightest

Name	Bayer Des.	Apt. Mag	Abs. Mag	Spect. Type	Distance
Sirius	Alpha Canis Majoris	−1.5	+1.4	A0	8.8 l.y.
Canopus	Alpha Carinae	−0.6	−5.5	F0	312.5 l.y.
Rigil Kentarus	Alpha Centauri A	−0.1	+4.3	G2	4.3 l.y.
	Alpha Centauri B	+1.4	+5.7	K1	4.3 l.y.
Arcturus	Alpha Bootis	+0.0	−0.3	G3	36.6 l.y.
Vega	Alpha Lyrae	+0.0	+0.5	A0	25.2 l.y.
Capella	Alpha Aurigae	+0.1	−0.5	M1	42.2 l.y.
Rigel	Beta Orionis	+0.1	−6.6	B8	772.5 l.y.
Procyon	Alpha Canis Minoris	+0.4	+2.6	F5	11.4 l.y.
Achernar	Alpha Eridanus	+0.4	−2.8	B3	143.7 l.y.
Betelgeuse	Alpha Orionis	+0.5	−5.1	M2	427.3 l.y.

Stars–Variable

Name	Variable Type	Mag. Range	Period	Spect. Type	Distance
Algol	Eclipsing	+2.1–+3.4	2.9 days	B8	92.7 l.y.
Mira	Pulsating	+2.0–+10.1	332.0 days	M5–M9	418.5 l.y.
Delta Cephei	Cepheid	+3.5–+4.4	5.4 days	G2	981.9 l.y.
RR Lyrae	RR Lyrae	+7.0–+8.1	0.5 days	A5–F7	744.3 l.y.
Gamma Cas . . .	Eruptive	+1.5–+3.0	Irregular	B0	612.8 l.y.
R Cas . . .	Pulsating	+4.7–+13.5	430.4 days	M6	347.9 l.y.

The Messier List

M#	NGC #	Constellation	RA	Dec	Comments
M1	NGC 1952	Taurus	05:31:30	+21 59	Supernova remnant Crab Nebula
M2	NGC 7089	Aquarius	21:30:54	−01 03	Globular cluster diameter 7 secs
M3	NGC 5272	Canes Venatici	13:39:54	+28 38	Globular cluster
M4	NGC 6121	Scorpio	16:20:36	−26 24	Globular cluster
M5	NGC 5904	Serpens	15:16:00	+02 16	Globular cluster fine globular
M6	NGC 6405	Scorpio	17:36:48	−32 11	Open cluster naked-eye
M7	NGC 6475	Scorpio	17:50:42	−34 48	Open cluster
M8	NGC 6523	Sagittarius	18:01:36	−24 20	Nebula Lagoon Nebula
M9	NGC 6333	Ophiuchus	17:16:12	−18 28	Globular cluster
M10	NGC 6254	Ophiuchus	16:54:30	−04 02	Globular cluster
M11	NGC 6705	Scutum	18:48:24	−06 20	Open cluster lovely fan shape
M12	NGC 6218	Ophiuchus	16:44:36	−01 52	Globular cluster none
M13	NGC 6205	Hercules	16:39:54	+36 33	Globular cluster naked-eye
M14	NGC 6402	Ophiuchus	17:35:00	−03 13	Globular cluster
M15	NGC 7078	Pegasus	21:27:36	+11 57	Globular cluster bright and
M16	NGC 6611	Serpens	18:16:00	−13 48	Nebula + cluster easy object
M17	NGC 6618	Sagittarius	18:17:54	−16 12	Omega Nebula
M18	NGC 6613	Sagittarius	18:17:00	−17 09	Open cluster sparse
M19	NGC 6273	Ophiuchus	16:59:30	−26 11	Globular cluster
M20	NGC 6514	Sagittarius	17:58:54	−23 02	Nebula Trifid Nebula
M21	NGC 6531	Sagittarius	18:01:48	−22 30	Open cluster about 50 stars
M22	NGC 6656	Sagittarius	18:33:18	−23 58	Globular cluster
M23	NGC 6494	Sagittarius	17:54:00	−19 01	Open cluster about 120 stars
M24	NGC 6603	Sagittarius	18:15:30	−18 27	Open cluster rich. Some 50 stars
M25	NGC I 47	Sagittarius	18:28:48	−19 17	Open cluster contains u sgr
M26	NGC 6694	Scutum	18:42:30	−09 27	Open cluster
M27	NGC 6853	Vulpecula	19:57:24	+22 35	Planetary Dumb-Bell Nebula
M28	NGC 6626	Sagittarius	18:21:30	−24 54	Globular cluster
M29	NGC 6913	Cygnus	20:22:12	+38 21	Open cluster 20 stars
M30	NGC 7099	Capricornus	21:37:30	−23 25	Globular cluster
M31	NGC 224	Andromeda	00:40:00	+41 00	Galaxy Great Andromeda Spiral
M32	NGC 221	Andromeda	00:40:00	+40 36	Galaxy type E2
M33	NGC 598	Triangulum	01:31:00	+30 24	Galaxy type Sc. Local Group
M34	NGC 1039	Perseus	02:38:48	+42 34	Open cluster about 80 stars

The Messier List *(Continued)*

M#	NGC #	Constellation	RA	Dec	Comments
M35	NGC 2168	Gemini	06:05:48	+24 21	Open cluster naked eye
M36	NGC 1960	Auriga	05:32:00	+34 07	Open cluster none
M37	NGC 2099	Auriga	05:49:06	+32 32	Open cluster superior to M36
M38	NGC 1912	Auriga	05:25:18	+35 48	Open cluster cruciform
M39	NGC 7092	Cygnus	21:30:24	+48 13	Open cluster about 25 stars
M40	None	Ursa Major	12:22:24	+58 05	Double Star
M41	NGC 2287	Canis Major	06:44:54	−20 41	Open cluster naked eye
M42	NGC 1976	Orion	05:32:54	−05 25	Nebula Great nebula in Orion.
M43	NGC 1982	Orion	05:32:54	−05 25	Part of the great Orion nebulae M42
M44	NGC 2632	Cancer	08:37:24	+20 00	Open cluster Praesepe. Naked-eye
M45	None	Taurus	03:44:06	+23 58	Open cluster The Pleiades
M46	NGC 2437	Puppis	07:39:30	−14 42	Open cluster
M47	NGC 2422	Puppis	07:34:18	−14 22	Open cluster naked-eye
M48	NGC 2548	Hydra	08:11:12	−05 38	Open cluster -
M49	NGC 4472	Virgo	12:27:18	+08 16	Galaxy type E4
M50	NGC 2323	Monoceros	07:00:36	−08 16	Open cluster none
M51	NGC 5194	Canes Venatici	13:27:48	+47 27	Spiral galaxy whirlpool galaxy
M52	NGC 7654	Cassiopeia	23:22:00	+61 19	Open cluster not distinctive
M53	NGC 5024	Coma Berenices	13:10:30	+18 26	Globular cluster
M54	NGC 6715	Sagittarius	18:52:00	−30 32	Globular cluster
M55	NGC 6809	Sagittarius	19:36:54	−31 03	Globular cluster
M56	NGC 6779	Lyra	19:14:36	+30 05	Globular cluster rich area
M57	NGC 6720	Lyra	18:51:42	+32 58	Planetary Ring nebula.
M58	NGC 4579	Virgo	12:35:06	+12 05	Galaxy type Sb
M59	NGC 4621	Virgo	12:39:30	+11 55	Galaxy type E3
M60	NGC 4649	Virgo	12:41:06	+11 49	Galaxy type E1
M61	NGC 4303	Virgo	12:19:24	+04 45	Galaxy type Sc
M62	NGC 6266	Ophiuchus	16:58:06	−30 03	Globular cluster
M63	NGC 5055	Canes Venatici	13:13:30	+42 17	Spiral galaxy type Sb
M64	NGC 4826	Coma Berenices	12:54:18	+21 57	Galaxy type Sb black-eye galaxy
M65	NGC 3623	Leo	11:16:18	+13 22	Galaxy type Sa
M66	NGC 3627	Leo	11:17:36	+13 16	Galaxy type Sb
M67	NGC 2682	Cancer	08:47:48	+12 00	Open cluster famous old cluster
M68	NGC 4590	Hydra	12:36:48	−26 29	Globular cluster none
M69	NGC 6637	Sagittarius	18:28:06	−32 33	Globular cluster
M70	NGC 6681	Sagittarius	18:40:00	−32 21	Globular cluster
M71	NGC 6838	Sagitta	19:51:30	+18 39	Open cluster very distant

The Messier List *(Continued)*

M#	NGC #	Constellation	RA	Dec	Comments
M72	NGC 6981	Aquarius	20:50:42	−12 44	Cluster globular cluster
M73	NGC 6994	Aquarius	20:58:54	−12 38	Double Star
M74	NGC 628	Pisces	01:34:00	+15 32	Galaxy type Sc
M75	NGC 6864	Sagittarius	20:03:12	−22 04	Globular cluster none
M76	NGC 650-	Perseus	01:38:48	+51 19	Planetary faintest messier object
M77	NGC 1068	Cetus	02:40:06	−00 14	Galaxy Seyfert galaxy
M78	NGC 2068	Orion	05:44:12	+00 02	Nebula reflection nebula
M79	NGC 1904	Lepus	05:22:12	−24 34	Globular cluster
M80	NGC 6093	Scorpio	16:14:06	−22 52	Globular cluster
M81	NGC 3031	Ursa Major	09:51:30	+69 18	Galaxy type Sb
M82	NGC 3034	Ursa Major	09:51:54	+69 56	Galaxy irregular. Radio source
M83	NGC 5236	Hydra	13:34:18	−29 37	Galaxy type Sc
M84	NGC 4374	Virgo	12:22:36	+13 10	Galaxy type E1
M85	NGC 4382	Coma Berenices	12:22:48	+18 28	Galaxy type Ep
M86	NGC 4406	Virgo	12:23:42	+13 13	Galaxy type E3
M87	NGC 4486	Virgo	12:28:18	+12 40	Galaxy type Eo. Radio source
M88	NGC 4501	Coma Berenices	12:29:30	+14 42	Galaxy type Sb
M89	NGC 4552	Virgo	12:33:06	+12 50	Galaxy type So
M90	NGC 4569	Virgo	12:43:18	+13 26	Galaxy type Sc
M91	NGC 4571	Coma Berenices	12:34:18	+14 28	Galaxy identification uncertain
M92	NGC 6341	Hercules	17:15:36	+43 12	Globular cluster
M93	NGC 2447	Puppis	07:42:24	−23 45	Open cluster loose
M94	NGC 2436	Canes Venatici	12:50:54	+41 07	Spiral Galaxy, type Sb
M95	NGC 3351	Leo	10:41:18	+11 58	Galaxy type SBb
M96	NGC 3368	Leo	10:44:06	+12 05	Galaxy type Sa. Near M.95
M97	NGC 3587	Ursa Major	11:12:00	+55 18	Planetary Owl Nebula.
M98	NGC 4192	Coma Berenices	12:11:18	+15 11	Galaxy type Sb
M99	NGC 4254	Coma Berenices	12:16:18	+14 42	Galaxy type Sc
M100	NGC 4321	Coma Berenices	12:22:00	+15 49	Galaxy
M101	NGC 5457	Ursa Major	14:03:00	+54 21	Spiral galaxy
M102	?	?	?	?	Duplicate of M101?
M103	NGC 581	Cassiopeia	01:33:00	+60 42	Star cluster
M104	NGC 4594	Virgo	12:39:00	−11 37	Galaxy
M105	NGC 3379	Leo	10:47:00	+12 35	Galaxy
M106	NGC 4258	Canes Venatici	12:19:00	+47 18	Galaxy
M107	NGC 6171	Ophiuchus	16:32:00	−13 03	Star cluster
M108	NGC 3556	Ursa Major	11:11:00	+55 41	Galaxy
M109	NGC 3992	Ursa Major	11:57:00	+53 22	Galaxy
M110	NGC 205	Andromeda	00:40:00	+41 41	Galaxy

APPENDIX B

Scales and Measures

Terrestrial Units of Measure

2.5 centimeters = 1 inch
30 centimeters = 12 inches = 1 foot
1 meter = 100 centimeters = 39.3 inches
1 kilometer = 0.62 statute mile
1 nautical mile = 1.15 statute miles = 6,080 feet = 1 minute of Earth latitude
1.0°C = 1.8°F
Freezing point of water = 0°C = 32°F
Boiling point of water = 100°C = 212°F
Absolute zero = −273°C = −459°F

Astronomical Units of Measure

1 astronomical unit = 92,900,000 statute miles = 149,838,000 kilometers
1 light year = 6,600,000,000,000 (trillion) statue miles = 10,645,000,000,000 (trillion) kilometers
1 parsec = 3.26 light years
Speed of light = 185,871 statute miles per second = 299,792 kilometers per second
1.0°C = 1.0 K
Freezing point of water = 273 K
Boiling point of water = 373 K
Absolute zero = 0 K

Stellar Brightness

The ancient Greeks classified stars on a brightness scale of 1 (brightest) to 6 (faintest). The system has since been modified on a logarithmic scale so that a magnitude +1 object is exactly 100 times brighter than a magnitude +6 object.

Sun	−26.7
Full Moon	−11.9
Venus	−4.8
Jupiter	−2.9
Mars	−2.9
Sirius	−1.5
Canopus	−0.6
Vega	−0.0
Saturn	−0.5

Asteroid Types

C-type: Carbonaceous chondrite materials; extremely dark
S-type: Metallic iron and nickel-iron, iron, magnesium-silicates
M-type: Pure nickel-iron

Stellar Classification

O: Blue, temperature above 31,000 K
B: Blue-white, temperature 9,750–31,000 K
A: White, temperature 7,100–9,750 K
F: White-yellow, temperature 5,950–7,100 K
G: Yellow, temperature 5,250–5,950 K
K: Orange, temperature 3,950–5,250 K
M: Red, temperature 2,000–3,950 K
L: Red-infrared, temperature 1,500–2,000 K
T: Infrared, temperature 1,000–2,000 K

Open Star Cluster Classification

Concentration
I   Detached and concentrated towards center
II   Detached with weak concentration towards center
III   Detached with no concentration towards center
IV   Not well detached from surrounding star field

Range in Brightness
1 Small range in brightness
2 Moderate range in brightness
3 Large range in brightness

Richness of Stellar Population
P   poor, less than fifty members
M   moderate, fifty to one hundred stars
R   rich, more than one hundred stars

Globular Star Cluster Classification

Globular clusters are assigned a Roman numeral I through XII. The lower the number the more dense the cluster is packed. Globulars are also measured by their *half-mass radius*. This is the measure of how far out from the center you have to go to find half the cluster's total mass. Clusters with lower numbers will have a half-mass radius that is fairly small.

Classification of Galaxies

E: Elliptical
Elliptical galaxies are sub-graded 0–7. A zero is spherical and each increasing number represents an increasing ellipsoidal shape.

S: Spiral
B: Barred
Sa or SBa: Spiral or barred spiral with dominant nucleus and barely detectable arms.
Sb or SBb: Spiral or barred spiral with definable arms and nucleus, nucleus dominant.
Sc or SBc: Spiral or barred spiral with definable arms and nucleus, arms dominant.
Sd or SBd: Spiral or barred spiral with dominant arms, minimal nucleus.

S0: Lenticular galaxy

I: Irregular
Type I: No definable structure.
Type II: Traces of spiral or lenticular structure detectable.

APPENDIX C

Resources

Sky & Telescope Internet Web Site, Cambridge, MA 2005
http://skyandtelescope.com

Delmarva Stargazers Internet Web Site; Smyrna, DE 2005
http://www.delmarvastargazers.org

The Solar and Heliospheric Observatory Internet Web Site; NASA HQ Washington, D.C. 2005
http://sohowww.nascom.nasa.gov

Meade Telescopes, Microscopes, Meade Instruments Internet Web Site; Irvine, CA 2005
http://www.meade.com

Celestron Telescopes Internet Web Site; Torrence, CA 2005
http://www.celestron.com

Orion Telescopes and Binoculars; Internet Web Site; Pasadena, CA 2005
http://www.telescope.com

Mount Cuba Astronomical Observatory; Internet Web Site; Greenville, DE 2005
http://www.physics.udel.edu/MCAO/

LPI: Lunar Orbiter Photographic Atlas of the Moon; Internet Web Site; Washington, D.C. 2005
http://www.lpi.usra.edu/resources/lunar_orbiter/

The Nine Planets Solar System Tour; Bill Arnett Internet Web Site, 2005
http://www.nineplanets.org/

NASA Asteroid Fact Sheet: Internet Web Site, Pasadena, CA 2005
http://nssdc.gsfc.nasa.gov/planetary/factsheet/asteroidfact.html

Asteroid/Comet Connection; Internet Web Site, 2002
http://www.hohmanntransfer.com

Solstation; Internet Web Site, 1998–2001
http://www.solstation.com/

Stars; Jim Kaler Internet Web Site, Champagne, IL 2005
http://www.astro.uiuc.edu/~kaler/sow/sowlist.html

SEDS: Students for the Exploration and Development of Space; Internet Web Site, 2005
http://www.seds.org/

A Man on the Moon; Andrew Chaikin, 1994 Penguin Books
Cosmos; Carl Sagan, 1980 Random House Books

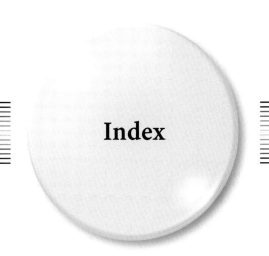

Index

Page numbers followed by f indicate figures.

A

Absolute magnitude/apparent magnitude, 104
Absorption lines/emission lines, 40, 116–117
Adiabatic lapse rate, 9
Air pressure, weight, and stability, 6, 9–10, 12–13
Airy disks, 61
Al Sufi, Abdal Rahman, 249
Albedo values
  defined, 161–162
  of Iapetus, 194
  of Saturn's rings, 186
Albireo (Beta Cygni), 212, 219
Alcor, 10–11, 212
Algeiba (Gamma Leonis), 73, 78–79, 212, 220
Algol (Beta Persei), 216, 223
Almach (Gamma Andromedae), 219–220
Alpha Centauri, 208
Alt-azimuth mountings, 25
Amour asteroids, 132, 161
Analemma, 114–115
Andromeda galaxy (M31)
  Almach (Gamma Andromedae) in, 219–220

Andromeda galaxy *(continued)*
  characteristics of, 254–255, 259–261
  companion galaxies to, 252
  discovery of, 249–250
  viewing of, 13–14
Anemic hypoxia, 7
Anorthosite, 85, 94
Antares (Alpha Scorpii), 215, 221
Aperture filters, 32–33
Aphelion/perihelion, 17
Apoastron point/periastron point, 222
Apollo 11 landing site, 96–98
Apollo asteroids, 132, 161, 272
Apollo Program, 86–88
Apparitions
  defined, 17–18
  of Mars, 146
  of Mercury, 128, 130
  of Venus, 123–124
Ariel, 197, 203
Asterisms. *See* Star clusters
Asteroids
  asteroid belt, 160–163
  camera tracking of, 135–136
  characteristics of, 159–160, 272
  classifications of, 132, 161–162, 202, 278
  impact risk from, 132–133

Asteroids *(continued)*
  near-Earth (NEAs), 161
  observing projects for, 169–171
  space missions to, 162–163
  viewing of, 163–164
Astigmatism, 7
ASTRO 1 Ultraviolet Imaging Telescope (UIT), 255
Astrometeorology, 111
  See also Weather conditions
Astronomical units of measure, 277
Astronomy
  in deep sky, 229
  misconceptions about, 1, 51
  publications on, 15, 44, 53, 67–68
Astronomy clubs
  dark sky sites of, 5, 73
  membership advantages in, 81–82
  support for "homebuilts", 47–48
Astronomy magazine, 15, 53, 67–68
Astrophotography
  of Algol and Beta Lyrae, 223
  of Big Dipper, 79–80
  cameras for. *See* Cameras
  of M31, 260–261